Solutions Manual for

Introduction to Composite Materials Design

Ever J. Barbero
October 26, 1998.

Preface

This solution manual was prepared to aid the instructor with the material in the textbook. The exercise problems should be worked out by the students, with help from the instructor, to learn-by-doing the material in the textbook. Exercise problems not only illustrate but also augment the material in the textbook. Taylor and Francis holds the copyright for this solution manual. The solution manual is for the instructor and should not be distributed in any way. However, you are welcome to copy and distribute solutions to selected problems provided said problems have been previously assigned to the students to whom the solutions are provided. Distribution must be restricted to the students enrolled in the class and it must be personal; you are not allowed to post solutions on the WEB or any other media that can be accessed outside your class. Please refer any comments or questions regarding this solution manual to Dr. Ever J. Barbero by e-mail at ebarbero@wvu.edu. A discussion users' group will be set up in the near future for the textbook and the software.

SOLUTIONS MANUAL FOR

Introduction to Composite Materials Design

by

Dr. Ever J. Barbero

Taylor & Francis

Taylor & Francis Group

New York   London

ERRATA for "Introduction to Composite Materials Design"

Ever J. Barbero, Taylor and Francis, 1999.

Contributors

Receive email when this page changes

Enter email address

· **Powered by Netmind** ·   **Click Here**

Page	Correction
Back cover	"written for the upper-level graduate student" should read "written for the senior undergraduate student"
7	Replace "Examples 5.1, 5.4 and 5.5" for "Examples 5.4 and 5.5"
16	"72.3 GPa" should read "3.5 GPa"
66	Section 4.1.5, 4th line, substitute "any pair of planes parallel to the fiber direction and orthogonal among themselves" for "any pair of planes orthogonal to the fiber direction and among themselves"
67	At the end of Section 4.1.6, it should read "A transversely isotropic material is described by **five** constants."
67	Section 4.1.6, 3rd line, substitute "if the fibers are randomly distributed in the cross section with a uniform probability density function" for "if the fibers are randomly distributed in the cross section"
67	Section 4.1.7, 5th line, substitute "ratio ν. Any other pair" for "ratio ν, but several other pairs"
69	Fig 4.6, substitute "$\varepsilon 1$" for "$\sigma 1$" in the figure itself.
70	Add http://www.mae.wvu.edu/~barbero/pdf/Addition2Eq4.15.gif after Eq. (4.15) on page 70.
72	4th line, delete "in the direction of loading and perpendicular to it, respectively."
72	Caption to Fig. 4.8 add "[2]" at the end of the caption.
73	Caption of Fig 4.9, replace "interlaminar" for "intralaminar"
75	Caption to Fig. 4.10 add "[2]" at the end of the caption.
78	After Eq. following Eq. (4.44), immediately before Section 4.2.7, add "substituting this into (4.44) results in the CAM formula (4.25)."
78	Section 4.2.7, 1st line, substitute "...containing continuous rovings with orientations that are randomly distributed with a uniform probability density function, held together..." for "...containing randomly placed, continuous rovings held

	together…"
79	Substitute http://www.mae.wvu.edu/~barbero/pdf/Eq4.48.gif for Eq. (4.48) in page 79.
80	Substitute http://www.mae.wvu.edu/~barbero/pdf/Table4.1.gif for Table (4.1) in page 80.
85	Annotation inside Fig. 4.12, replace σ6 for G12. Download updated Fig. 4.12. Corrections are in blue color.
90	Caption of Fig. 4.17, substitute "for the number" for "proportional to the number"
94	In Ex. 4.6 delete (Vf=0.5). The fiber volume fraction is 0.6 as indicated at the end of the line.
95	In Ex. 4.6 replace GPa by MPA twice: for the values of F1c in the last equation of the example and in the paragraph immediately following the last equation.
99	2nd line, 2nd paragraph in sect. 4.4.6, should read "shear acts on a plane parallel to..."
101	In Exercise 4.8, substitute "23C" for "21C" and in the last sentence "Ef and σfa decrease" for "Em decreases"
120	Line 1. Substitute "counterclockwise" for incorrect word "clockwise"
120	Replace (-n) for (n) and (+n) for (-n) in Eqs. 5.38-5.39-5.40-5.41. In other words, transpose all the rotation matrices. The rotation matrices in 5.38-5.39 should have square brackets [], not curly brackets {}.
127	Replace 19.39 for 18.9 in the position 11 of the result matrix QXM2408 of Ex. 5.8
140	Remove superscript "0" from the curvatures "k" in Eq. 6.19.
143	Example 6.2. The applied moment is 1.0 KN. The stresses in pp. 144 are in GPa.
144	Swap D_{11} and D_{22} on the 2nd and 3rd equations. The numerical results are correct.
146	3rd line from top of page, inset "=B66" just before "=0."
147	Last paragraph of section 6.3.4 should read "Balanced laminates having only 45, 0, and 90-layers have identical ..."
154	In denominator of next-to-last Eq., density of Kevlar 49 is (1.45), not (1.44)
159	The Poisson's ratio should not have any units on the graph.
162	The Poisson's ratio should not have any units on the graph.
177	Exercise 6.6, add to question (b). "Take G23=Gm of 9310 Epoxy at room temperature."
179	Exercise 6.22 should reference Exercise 6.21, not 6.2
179	Exercise 6.24 should refer to Table 2.3, not 1.1

186	Next-to-last line before Ex. 7.1, should read "this example for both tensile (P) and compressive (R) transverse load."
186	Ex. 7.1, p. 186, substitute "Use the maximum stress failure criterion" for "Use the maximum failure criterion".
189	"fibers (section 7.1.7)" should read "fibers (section 7.1.6)"
193	In 4th Eq. replace $R^2=$ for $R=$. Then, replace 0.760 for 0.577 twice in the 5th Eq.
195	In Eq. 7.19 replace uppercase F by lowercase f. In Eq. 7.20 replace uppercase F by lowercase f, in $b=1/2(f1*\sigma1+f2*\sigma2)$
210	In Fig. 7.8, between points F & D, replace $\varepsilon_2=-\varepsilon_{1c}$ for $\varepsilon_1=-\varepsilon_{1c}$. <u>Download updated Fig. 7.8.</u> Corrections are in blue color.
214	In 1st Eq., delete R1=, only R2 should be shown.
227	(labeled with solid diamonds) should read (labeled with diamonds).
228	Exercise 7.1. The load is Nx=Ny=1.0 N/mm.
255	After Eq. 8.41 should read "$q_i=N^i_{xs}$ is the shear flow"
283	Eq.(9.17) remove the parenthesis ( ) around the last term in the bracket. Immediately above the Eq., remove "and n". Note: Eq. (9.17) is only a function of m.
316	Footnote "see Exercise 10.6" should read "see Example 10.6"

Please e-mail additional corrections to ebarbero@wvu.edu

Home | Software | Textbooks

Some links require a PDF reader.

Contributors, Users of CADEC and the Book

Corrections have been made thanks to the helpful contributions from:

- California State Polytechnic University, Pomona. Mechanical Engineering Department.

- California State Polytechnic University, Fresno. Civil and Surveying Engineering Department.

- University of Akron, OH. Civil Engineering Department.

- University of Vermont. <u>Mechanical Engineering.</u>

- University of Maine. Department of Civil Engineering.

- University of Puerto Rico. Civil Engineering Department.

- West Virginia University. Civil Engineering Department.

- University of Maryland. Composites Laboratory.

- Composites Group, Montana State University, Chemical Engineering Department, Bozeman, Montana.

- West Virginia University.

- University of Calabria, Italy. Department of Structures.

- University of Trento, Italy. Department of Materials Engineering.

- University of Cordoba, Argentina. Department of Structures.

- NASA Glenn Research Center.

- any others ?

Updated: 10/10/01

Contributors, Users of CADEC and the Book

Corrections have been made thanks to the helpful contributions from:

- California State Polytechnic University, Pomona. Mechanical Engineering Department.

- California State Polytechnic University, Fresno. Civil and Surveying Engineering Department.

- University of Akron, OH. Civil Engineering Department.

- University of Vermont. Mechanical Engineering.

- University of Maine. Department of Civil Engineering.

- University of Puerto Rico. Civil Engineering Department.

- West Virginia University. Civil Engineering Department.

- University of Maryland. Composites Laboratory.

- Composites Group, Montana State University, Chemical Engineering Department, Bozeman, Montana.

- West Virginia University. Mechanical and Aerospace Engineering Department.

- University of Calabria, Italy. Department of Structures.

- University of Trento, Italy. Department of Materials Engineering.

- University of Cordoba, Argentina. Department of Structures.

- any others ?

Solutions Chapter 1
Exercise 1.1

Hoop stress $\sigma = pd/2t \leq F_{1t}/(FS)$

(a) in the hoop direction since $F_{1t} > F_{2t}$

(b) $t = pd(FS)/2F_{1t}$
the largest F_{1t} is for T800/3900-2

$$t = \frac{5(1)4}{2(2698)} = 3.71 \text{ mm}$$

(c) weight per unit length

$$\frac{W}{L} = \rho\pi dt = \frac{\pi pd^2(FS)}{2}\frac{\rho}{F_{1t}} \quad ; \quad t = 4.83 \text{ mm}$$

the lowest ρ/F_{1t} is for AS4/APC2 with W/L = 24 kg/m (density of T800/3900-2 not available)

(d) the cost per unit length is

$$\frac{\$}{L} = \frac{W}{L}(\$/\text{kg}) = \frac{\pi pd^2(FS)}{2}\frac{\rho(\$/\text{kg})}{F_{1t}} \quad ; \quad t = 11.07 \text{ mm}$$

the lowest is for E-Glass/Polyester at $94.6/m

Exercise 1.2

$$t = \frac{5(1)4}{2(270)} = 3.7037 \text{ mm}$$

$$\frac{W}{L} = \rho\pi dt = 314 \text{ kg/m}$$

Exercise 1.3

σ_x = 10 MPa
σ_x = 5 MPa
σ_{xy} = 2.5 MPa

Fig.

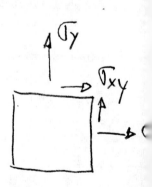

$$p = \frac{\sigma_x + \sigma_y}{2} = \frac{10 + 5}{2} = 7.5 \text{ MPa}$$

$$q = \frac{\sigma_x - \sigma_y}{2} = \frac{10 - 5}{2} = 2.5 \text{ MPa}$$

$$r = \sigma_{xy} = 2.5$$

$$R = \sqrt{q^2 + r^2} = \sqrt{2.5^2 + 2.5^2} = 3.536$$

$$\theta = \frac{1}{2}\tan^{-1}\left(\frac{r}{q}\right) = \frac{1}{2}\tan^{-1}\left(\frac{2.5}{2.5}\right) = 22.5°$$

$$\sigma_I = p + R = 7.5 + 3.536 = 11.04 \text{ MPa}$$

$$\sigma_{II} = p - R = 7.5 - 3.536 = 3.964 \text{ MPa}$$

(a)   Optimum Fiber Orientation
$\theta = 22.5°$ Fig.

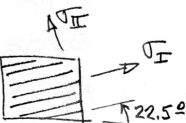

(b)   Check 1-1 Direction

$$\frac{F_{1t}}{\sigma_I} = \frac{1830}{11.04} = 166 \text{ Safety Factor 1-1 Direction}$$

$$\frac{F_{2t}}{\sigma_{II}} = \frac{57 \text{ MPa}}{3.964 \text{ MPa}} = 14.38 \text{ Safety Factor 2-2 Direction}$$

Solutions Chapter 2

Exercise 2.1

See Section 2.1.2, subsection "Strand, tow, end, yarn, and roving"

Exercise 2.2

See Section 2.2.1, subsection "Polyester Resins"

Exercise 2.3

Chlorendic Polyester

Exercise 2.4

Bisphenol-A-Fumarate Polyester

Exercise 2.5

Glass fibers = low cost
Carbon fibers = high stiffness

Exercise 2.6

Polyester = low cost
Epoxy = high mechanical properties

Exercise 2.7

36 K= 36,000 fibers

$$\text{Area} = K\frac{\pi d^2}{4} = 36,000\frac{\pi[7(10^{-6})]^2}{4} = 1.385 \text{ mm}^2$$

Exercise 2.8

Using Eq. 2.1

$$\text{TEX(g/km)} = \frac{496,238}{56 \text{ (yd/lb)}} = 8,861 \text{ g/km}$$

Using Eq. 2.2

$$\text{Area} = \frac{(10^{-5})8,861 \text{ (g/km)}}{2.5 \text{ (g/cc)}} = 0.035 \text{ cm}^2$$

$112/56 = 2$ (Need two 112 yield roving)

Exercise 2.9

$$W = 800 \text{ g/m}^2$$

$$\rho = 2.5 \text{ g/cc} = 2.5(10^6) \text{ g/m}^3$$

$$t = \frac{W}{\rho} = \frac{800}{2.5(10^6)} = 0.32 \text{ mm}$$

Exercise 2.10

Program spreadsheet for all materials in Table 2.1

(a)

$$t_f = \frac{pd/2}{\sigma_{fa}/(F.S.)}; \ t_c = 2t_f$$

where σ_{fa} is the average fiber tensile strength

(b)

$$\frac{W}{L} = \pi d t_c \rho_c \ ; \ \rho_c = V_f \rho_f + V_m \rho_m$$

use t_c from part (a)

(c)

$$\frac{\$}{L} = \frac{\$}{kg} \frac{W}{L}$$

use $/kg values from Exercise 1.1 (d) for the composite and W/L from part (b).
Answers:

(a)   min. thickness: $t_c = 3.116$ mm for IM6
(b)   min. weight: $\frac{W}{L} = 5.75$ kg/m for IM6
(c)   min. cost: $\frac{\$}{m} = 16.2$ $/m for E-glass with $t_c = 4.63$ mm

Exercise 2.11
(a)   Polyester or Chlorendic resins
(b)   Bisphenol
(c)   Epoxy
(d)   Chlorendic

Exercise 2.12
(a)   PEEK
(b)   Phenolic
(c)   Epoxy
(d)   Bismaleimide and Thermoplastics
(e)   Polyester resins with styrene/MMA blends
(f)   Epoxy

Exercise 2.13
Temperature, humidity, and loading rate.

Exercise 2.14
Below

Exercise 2.15
Thermoplastics

Exercise 2.16
From Table 2.4
$T_g = 185\ ^\circ C = T_{gd}$
since there is no moisture change
retention ratio=$\left[\frac{185-149}{185-23} \right]^{1/2} = 0.471$
From Table 2.4
stiffness at 23 °C = 3.12 GPa
strength at 23 °C = 75.8 MPa

stiffness at 149 °C = (0.471)3.12 = 1.47 GPa
strength at 149 °C = (0.471)75.8 = 35.7 MPa
Comparison with values from Table 2.4, 1.4 GPa and 26.2 MPa shows that the retention ratio yields reasonable values.

Exercise 2.17

See Exercise 2.16

Exercise 2.18

It reduces the corrosion resistance of the PMC.

Exercise 2.19

The matrix protects the fibers from chemical attack. Mechanical loads may induce cracking of the matrix, thus accelerating the ingress of chemicals that may attack the fiber.

Exercise 2.20

From Table 2.4, at T_0 = 24 °C (dry), T_{gd} = 260 °C
To account for moisture:

$$T_{gw} = (1 - 0.1m + 0.005m^2)T_{gd}$$

and to add the influence of temperature:

$$\text{retention ratio} = \left[\frac{T_{gw} - T}{T_{gd} - T_0} \right]^{0.5}$$

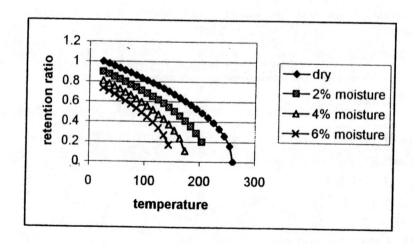

Solutions Chapter 3

Exercise 3.1

In polar winding, the mandrel is fixed and the delivery-eye travels around the mandrel to cover it with fibers. Polar winding is limited to shapes with a length-to-diameter ratio less than 2. It is an ideal process for spherical or quasi-spherical shapes.

Exercise 3.2

BMC is a dough-type mixture. SMC comes in sheets, usually with longer fibers and higher fiber volume fraction.

Exercise 3.3

See Section 3.3

(a)   A release film is used to prevent the laminate from sticking to the breather.

(b)   A peel-ply leaves an imprint on the surface of the laminate to facilitate secondary bonding to other parts.

(c)   The breather is a porous material that allows air and volatiles to reach the vacuum ports.

Exercise 3.4

	Advantage	Disadvantage
Pultrusion	high production rate	only constant cross-sections
Hand lay-up	any shape can be made	low production rate
Compression molding	low cost, high production rate	difficult to orient the fibers
RTM	any fiber orientation possible	costly tooling

Exercise 3.5

(a)   pultrusion

(b)   filament winding

(c)   pultrusion with winding addition

(d)   hand lay-up, compression molding, spray-up

(e)   hand lay-up

Exercise 3.6

(a)   compression molding

(b)   RTM

(c)   hand lay-up/vacuum-bag

Exercise 3.7

(a)   compression molding, but higher temperature and pressure needed

(b)   switch to prepreg lay-up because high density of thermoplastics rules out RTM

(c)   hand lay-up. Switch to autoclave cure to provide higher temperature and pressure.

Solutions Chapter 4

Exercise 4.1

Fig.   RVE

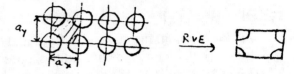

d_f = fiber   diameter

$$V_f = \frac{\text{fiber area}}{\text{total area}}$$

Fiber area = $\frac{\pi}{4}(d_f)^2$ RVE Fiber area = $\frac{\pi}{4}(d_f)^2$

Total area = $a_y a_x$

$$V_f = \frac{\frac{\pi}{4}(d_f)^2}{a_y a_x} = \frac{\pi(d_f)^2}{4 a_y a_x}$$

Exercise 4.2

Representative Volume Element - smallest portion of the material that contains all peculiarities of the material: representative of the material as a whole

Fig.

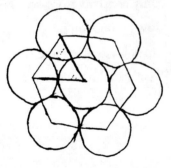

$$V_f = \frac{\text{Fiber area}}{\text{Total area}}$$

All $\angle$'s = $60°$

Area of fiber = $\frac{\pi}{4}d^2$ RVE Fiber Area = $3(\frac{\pi}{4}d^2)(\frac{1}{6}) = \frac{\pi}{8}d^2$

Total area = $\frac{1}{2}$ (base) (height)

Height h = $(d)(\sin(60)) = \frac{\sqrt{3}}{2} d$

Total area = $\frac{1}{2}d(\frac{\sqrt{3}}{2}) d = \frac{\sqrt{3}}{4} d^2$

$$V_f = \frac{\frac{\pi}{8}d^2}{\frac{\sqrt{3}}{4}d^2} = \frac{4\pi}{8\sqrt{3}}$$

$V_f = 0.907$ $V_f = 91\%$

Exercise 4.3

Carbon-Epoxy

$V_f = 70\%$ $E_f = 379$ GPa $v_f = 0.22$ $E_m = 3.3$ GPa $v_m = 0.35$

Compute E_1, E_2, v_{12}, G_{12} using ROM and more accurate formulas.

$$E_1 = E_f V_f + E_m(1 - V_f) = 266 \text{ GPa}$$

$$\frac{1}{E_2} = \frac{V_m}{E_m} + \frac{V_f}{E_f}$$

$$V_m + V_f = 1 \qquad V_m + 0.7 = 1 \qquad V_m = 0.3$$

$E_2 = 10.7$ GPa

$$v_{12} = v_f V_f + v_m V_m \qquad v_{12} = 0.259$$

$$G_{12} = \frac{G_m}{V_m + V_f G_m / G_f}$$

$$G_m = \frac{E_m}{2(1 + v_m)} = \frac{3.3 \times 10^9}{2(1 + 0.35)} = 1.22 \text{ GPa}$$

$$G_f = \frac{E_f}{2(1 + v_f)} = \frac{379 \times 10^9}{2(1 + 0.22)} = 155 \text{ GPa}$$

$$G_{12} = \frac{1.22 \times 10^9}{0.3 + 0.7(1.72 \times 10^9)/155 \times 10^9} = 3.99 \text{ GPa}$$

Now calculate using more accurate formulas.

E_1 has no formula listed for better accuracy.

$$E_2 = E_m \left[\frac{1 + \zeta \eta V_f}{1 - \eta V_f} \right]$$

Assume $\zeta = 2$ (circular or square fiber)

$$\eta = \frac{(E_f/E_m) - 1}{(E_f/E_m) + \zeta} = \frac{(379/3.3) - 1}{(379/3.3) + 2} = 0.974$$

$$E_2 = 3.3 \times 10^9 \left[\frac{1 + 2(0.974)(0.7)}{1 - (0.974)(0.7)} \right] \qquad E_2 = 24.5 \text{ GPa}$$

v_{12} has no formula listed for better accuracy

$$G_{12} = G_m \left[\frac{(1 + V_f) + (1 - V_f)G_m/G_f}{(1 - V_f) + (1 + V_f)G_m/G_f} \right] = 1.22 \times 10^9 \left[\frac{(1 + 0.7) + \frac{1.22}{155}(1 - 0.7)}{(1 - 0.7) + \frac{1.22}{155}(1 + 0.7)} \right] = 6.63 GPa$$

The value for E_2 which was calculated using the ROM equation is a lower bound and is not accurate in the majority of cases. A better prediction was obtained using the semi-empirical Halpin-Tsai formula.

The value for G_{12} which was calculated using the ROM equation is a lower bound and is a simple but not accurate equation for the prediction of the inplane shear modulus. The value obtained using the Cylindrical assemblage model is a better approximation.

Exercise 4.4

Burn-out Test

Sample + container = 50.181 g before burn-out

Sample + container = 49.448 g after burn-out

Container = 47.650 g

Fiber weight = 49.447 g - 47.650 g = 1.797 g

Resin weight = 50.181 g - 49.448 g = 0.733 g

Total weight = 1.797 g + 0.734 g = 2.531 g

$$W_f = \frac{fiber\ weight}{total\ weight} = \frac{1.797}{2.531} = 0.710$$

$$W_m = \frac{matrix\ weight}{total\ weight} = \frac{0.733}{2.531} = 0.29$$

Exercise 4.5

$E_f = 230$ GPa $\qquad\qquad E_m = \frac{E_f}{50} = 4.6$ GPa

$v_f = 0.25$ $\qquad\qquad\qquad v_m = 0.3$

$G_f = \frac{E_f}{2(1+v_f)} = \frac{230}{2.5} = 92$ GPa

$G_m = \frac{E_m}{2.6} = 1.769$ GPa

$V_f = 0.4;\ V_m = 1 - 0.4 = 0.6$

$$E_1 = E_f V_f + E_m(1 - V_f) = 94.76 \text{ GPa}$$

E_2 using Rule of Mixtures

$1/E_2 = V_m/E_m + V_f/E_f \qquad E_2 = 7.566$ GPa

E_2 using Halpin-Tsai

circular fiber: $\xi = 2$

$E_f/E_m = 50;\ \eta = \frac{49}{52} = 0.942$

$$E_2 = 4.6\left[\frac{1 + 2(0.942)0.4}{1 - (0.942)0.4}\right] = 12.944 \text{ GPa}$$

G_{12} using Rule of Mixtures

$$\frac{1}{G_{12}} = \frac{V_m}{E_m} + \frac{V_f}{E_f};\ G_{12} = 2.911 \text{ GPa}$$

Using the Cylindrical Assemblage Model

$$G_{12} = G_m \left[\frac{(1 + V_f) + (1 - V_f)(G_m/G_f)}{(1 - V_f) + (1 + V_f)(G_m/G_f)} \right] = 3.966 \text{ GPa}$$

Rules of Mixtures

$$v_{12} = 0.25(0.4) + 0.3(0.6) = 0.28$$

Exercise 4.6

Compute elastic properties for unidirectional lamina of the following material combinations:
(a) E-glass Polyester (isophthalic)

$V_f = 0.55$ $\quad\quad E_f = 72.345 \text{ GPa}$ $\quad v_f = 0.22$
$E_m = 3.4 \text{ GPa}$ $\quad\quad v_m = 0.38$

$E_1 = E_m(1 - V_f) + E_f V_f = 41.32 \text{ GPa}$
Using inverse rule of mixtures

$\frac{1}{E_2} = \frac{V_m}{E_m} + \frac{V_f}{E_f}$; $E_2 = 7.145 \text{ GPa}$
E_2 using Halpin Tsai

$$\eta = \frac{\frac{E_f}{E_m} - 1}{\frac{E_f}{E_m} + 2} = 0.8711$$

$$E_2 = E_m \left[\frac{1 + \zeta \eta V_f}{1 - \eta V_f} \right] = 12.78 \text{ GPa}$$

$$v_{12} = v_m V_m + v_f V_f = 0.292$$

G_{12} using rule of mixtures and the CAM method

$G_m = E_m/2(1 + v_m) = 1.232 \text{ GPa}$ $\quad\quad G_f = E_f/2(1 + v_f) = 29.65 \text{ GPa}$

$\frac{1}{G_{12}} = \frac{V_m}{E_m} + \frac{V_f}{E_f}$ $\rightarrow$ $G_{12} = 2.605 \text{ GPa}$ $\quad\quad$ Inverse Rule of Mixtures

$$G_{12} = G_m \left[\frac{(1 + V_f) + (1 - V_f)\,G_m/G_f}{(1 - V_f) + (1 + V_f)G_m/G_f} \right] = 3.757 \text{ GPa} \quad\quad \text{CAM Method}$$

$$G_{23} = G_m \left[\frac{V_f + \eta_{23}(1 - V_f)}{\eta_{23}(1 - V_f) + V_f G_m/G_f} \right] = 3.404 \text{ GPa} \quad \eta_{23} = \frac{3 - 4v_f + G_m/G_f}{4(1 - v_m)} = 0.6181 \text{ SPP Method}$$

(b) S-glass Epoxy (9310) at 23 °C
$E_f = 85 \text{ GPa}$ $\quad\quad E_m = 3.12 \text{ GPa}$ $\quad V_f = 0.55$
$v_f = 0.22$ $\quad\quad\quad v_m = 0.38$

$E_1 = 48.154 \text{ GPa}$
$E_2 = 6.636 \text{ GPa}$ $\quad\quad\quad$ Using IROM
$E_2 = 12.24 \text{ GPa}$ $\quad\quad\quad$ Using Halpin-Tsai

$v_{12} = 0.292$

$G_{12} = 2.4153$ GPa   Using IROM

$G_{12} = 3.534$ GPa   Using CAM

$G_{23} = 3.189$ GPa   Using SPP

(c) Carbon (T300)/vinyl ester

$E_f = 230$ GPa $v_m = 0.38$ $V_f = 0.55$

$E_m = 3.4$ GPa $v_f = 0.2$

$E_1 = 128.03$ GPa   Using ROM

$E_2 = 7.421$ GPa   Using IROM

$E_2 = 14.732$ GPa   Using Halpin-Tsai

$v_{12} = 0.281$

$G_{12} = 2.691$ GPa   Using IROM

$G_{12} = 4.073$ GPa   Using CAM

$G_{23} = 3.633$ GPa   Using SPP

Exercise 4.7

Estimate F_{1t} and F_{2t} for AS4-D/Peek with $V_f = 0.5$

$E_f = 241$ GPa $v_m = 0.4$

$E_m = 3.24$ GPa $\sigma_{fa} = 4.134$ GPa

$F_{1t} = \sigma_{fa}V_f + (\sigma_{fa}E_m/E_f)(1 - V_f) = 2.095$ GPa

F_{2t} using Nielson

$F_{2t} = \sigma_{mu}C_v(1 - V_f^{1/3})E_2/E_m$

Using Halpin-Tsai for the E_2 calculation

$$\tau = 2$$

$$\eta = \frac{(E_f/E_m) - 1}{(E_f/E_m) + \tau}$$

$$E_2 = E_m \left[\frac{1 + \tau\eta V_f}{1 - \eta V_f} \right] = 12.224 \text{ GPa}$$

$\sigma_{mu} = 0.100$ GPa

$F_{2t} = 77.87$ MPa

F_{2t} using Chamis

$F_{2t} = \sigma_{mu}C_v \left[1 + (V_f - \sqrt{V_f})(1 - E_m/E_f) \right]$

$C_v = 1$

$F_{2t} = 79.6$ MPa

Exercise 4.8

Estimate the tensile strength of Kevlar-epoxy with $V_f = 0.5$ at two temperatures, at 21 °C and 149 °C using the strength of materials approach. Assume that the E_m decreases linearly with temperature up to 75% reduction at 171 °C. The void content is negligible.

At 23 °C	At 149 °C	At 171 °C
$E_f = 131$ GPa	$E_f = ?$	$E_f = 98.25$ GPa
$E_m = 3.12$ GPa	$E_m = 1.4$ GPa	$\sigma_{fa} = 2.715$ GPa
$\sigma_{fa} = 3.62$ GPa	$\sigma_{mu} = 26.2$ MPa	
$\sigma_{mu} = 75.8$ MPa		

The linear relationship between E_f and T based on the above data is the following:
$E_f(T) = -0.221(T) + 136.09$
$E_f(149°C) = 103.161$ GPa
$E_f(21°C) = 131.449$ GPa

The linear relationship between E_m and T is the following:
$E_m(T) = -0.014(T) + 3.434$
$E_m(21°C) = 3.147$ GPa

The linear relationship between σ_{fa} and T is the following:
$\sigma_{fa}(T) = -0.00718(T) + 3.785$
$\sigma_{fa}(21°C) = 3.632$ GPa
$\sigma_{fa}(149°C) = 2.85$ GPa

The linear relationship between σ_{mu} and T is the following:
$\sigma_{mu}(T) = -0.39365(T) + 84.85397$
$\sigma_{mu}(21°C) = 76.587$ MPa
$F_{1t} = \sigma_{fa}V_f + (\sigma_{fa}\frac{E_m}{E_f})(1 - V_f)$

At 21 °C	At 149 °C
$F_{1t} = 1.8533$ GPa	$F_{1t} = 1.444$ GPa

F_{2t} using Nielson: $C_v = 1$ $\zeta = 2$

$$E_2 = E_m \left[\frac{1 + \zeta\eta V_f}{1 - \eta V_f} \right]$$

$$\eta = \frac{\frac{E_f}{E_m} - 1}{\frac{E_f}{E_m} + \zeta}$$

$E_2(21°C) = 11.38$ GPa
$E_2(149°C) = 5.28$ GPa

$$F_{2t} = \sigma_{mu}C_v\left[1 - V_f^{1/3} \right]\frac{E_2}{E_m}$$

At 21 °C   At 149 °C

$F_{2t} = 57.08$ MPa $\qquad$ $F_{2t} = 20.37$ MPa

F_{2t} using Chamis $\qquad\qquad$ $C_v = 1$

$$F_{2t} = \sigma_{mu} C_v \left[1 + (V_f - \sqrt{V_f})(1 - \frac{E_m}{E_f}) \right]$$

At 21 °C $\qquad\qquad$ At 149 °C

$F_{2t} = 61.11$ MPa $\qquad$ $F_{2t} = 20.85$ MPa

Exercise 4.9

Reference to Example 4.6

Select a different matrix that will handle 1,100 MPa of compressive strength.

$F_{1c} \geq 1.1$ GPa

$E_f = 241$ GPa $\qquad v_f = 0.2$

$V_f = 0.6$

$$F_{1c} = (\frac{\chi}{0.21} + 1)^{-0.69}(G_{12}) \qquad\qquad \chi = \frac{G_{12}\Omega}{F_6}$$

G_{12} found by using the CAM method.

$$G_{12} = G_m \left[\frac{(1 + V_f) + (1 - V_f)G_m/G_f}{(1 - V_f) + (1 + V_f)G_m/G_f} \right]$$

F_6 found by assuming that $\tau_{mu} = \sigma_{mu}$, $C_v = 1$, and using the SPP method

$$F_6 = \tau_{mu} C_v \left[1 + (V_f - \sqrt{V_f})(1 - G_m/G_f) \right]$$

Calculations	PEEK	PPS	PSUL	PEI	PAI	K-III	LARC-TPI	Epoxy	units
E_m	3.24	3.30	2.48	3.00	2.76	3.76	3.72	3.12	GPa
v_m	0.4	0.37	0.37	0.37	0.37	0.365	0.36	0.38	-
σ_{mu}	0.1	0.0827	0.0703	0.105	0.08957	0.1020	0.1192	0.119	GPa
G_f	100.4	100.4	100.4	100.4	100.4	100.4	100.4	100.4	GPa
G_m	1.157	1.204	0.9051	1.095	1.006	1.377	1.368	1.130	GPa
G_{12}	4.437	4.611	3.502	4.208	3.878	5.241	5.206	4.339	GPa
V_v	0	0	0	0	0	0	0	0	-
C_v	1	1	1	1	1	1	1	1	-
τ_{mu}	0.1	0.0827	.07030	0.1050	0.08957	0.1020	0.1192	0.0758	GPa
ζ	2	2	2	2	2	2	2	2	-
η	0.9607	0.9600	0.9698	0.9636	0.9665	0.9546	0.9551	0.9621	-
E_2	16.47	16.75	12.83	15.33	14.17	18.88	18.70	15.90	GPa
F_6	0.0827	0.06843	0.05814	0.08687	0.07409	0.08444	0.09867	0.06271	GPa
χ	1.320	1.658	1.483	1.192	1.288	1.527	1.298	1.703	-
F_{1c}	1.127	1.021	0.8298	1.135	0.996	1.219	1.336	0.9449	GPa

Exercise 4.10

$$\frac{v_{12}}{E_1} = \frac{v_{21}}{E_2}$$

$v_{12}/E_1 = 0.00225$

$v_{21}/E_2 = 0.00205$ $2.25(10^{-3}) \approx 2.05(10^{-3})$

Exercise 4.11

Select a matrix, fiber, and fiber volume fraction to obtain a material with the following properties:

$E_1 \geq 30$ GPa $\frac{E_1}{E_2} \leq 3.5$

Pick the following materials:

D-glass $E_f = 55$ GPa
Vinyl ester $E_m = 3.6$ GPa
$V_f = 0.55$
$E_1 = E_f V_f + E_m(1 - V_f)$ Using ROM
$E_1 = 31.87$ GPa

Using Halpin-Tsai

$$E_2 = E_m \left[\frac{1 + \zeta \eta V_f}{1 - \eta V_f} \right] \qquad \zeta = 2; \eta = \frac{\frac{E_f}{E_m} - 1}{\frac{E_f}{E_m} + \zeta} = 0.826$$

$E_2 = 13.85$ GPa

$\frac{31.87}{13.85} = 2.301 \leq 3.5$

Exercise 4.12

Give an approximate value for the compressive strength of and Al-Boron composite with $V_f = 0.4$.

Aluminum 2024:   Boron fibers:
$E_m = 71$ GPa $E_f = 400$ GPa
$v_m = 0.334$ $\sigma_{ult} = 3.4$GPa
$G = 26.6$ GPa $v_f = 0.25$
$\sigma_y = 76$ MPa
fatigue limit = 90 MPa

Since Boron fibers are straight, use the same analysis of tensile strength.
$F_{1c} = 3.4 \left[0.4 + \frac{71}{400}(1 - 0.4) \right] = 1.7221$ GPa

the strain at failure is
$$\epsilon_{1c} = F_{1c}/E_1 = 0.0085$$

$$E_1 = (0.4)(400) + (1 - 0.4)(71) = 202.6 \text{ GPa}$$
the strain at yield of Aluminum is
$$\epsilon_y = \frac{76}{71,000} = 0.00107$$
Since the strain at yield of aluminum is much lower that the strain at failure computed above, the value of F_{1c} is recomputed assuming that the matrix is elasto-plastic.
$$F_{1c} = 3.4(0.4) + 0.076(1 - 0.4) = 1.4056 \text{ GPa}$$
$$\epsilon_{1c} \approx \frac{3.4}{400} \approx 0.0085$$

Exercise 4.13

Compute the approximate material properties for the material in Exercise 4.12

(a) Using ROM: $\quad E_1 = 400(0.4) + 71(0.6) = 202.6 \text{ GPa}$

(b) Using Inverse ROM: $\qquad 1/E_2 = V_m/E_m + V_f/E_f \qquad E_2 = 105.8 \text{ GPa}$

Using Halpin-Tsai:
$$E_2 = E_m \left[\frac{1 + \zeta \eta V_f}{1 - \eta V_f} \right] \qquad \eta = 0.607$$

$$E_2 = 139.3 \text{ GPa}$$

(c) Using ROM:
$$1/G_{12} = V_m/G_m + V_f/G_f \qquad G_m = E_m/2(1 + v_m) \qquad G_f = E_f/2(1 + v_f)$$

$$G_{12} = 39.92 \text{ GPa}$$

Using Cylindrical Assemblage Model:
$$G_{12} = G_m \left[\frac{(1 + V_f) + (1 - V_f) G_m/G_f}{(1 - V_f) + (1 + V_f)G_m/G_f} \right] = 47.92 \text{ GPa}$$

(d) Using ROM: $\quad v_{12} = 0.25 (0.4) + 0.334 (0.6) = 0.3$

Comments: Both inverse rule of mixtures for E_2 and G_{12} underestimate the properties significantly.

Exercise 4.14

Compute the tensile strength of the composite in 4.12

Similarly to Exercise 4.9, the matrix yields early and may be considered to be elasto-plastic.

$$F_{1t} = (0.4)3.4 + (1 - 0.4)0.076 = 1.4056 \text{ GPa}$$

Solutions Chapter 5
Exercise 5.1

Compute the reduced stiffness and compliance matrices, and the intralaminar shear stiffness and compliance matrices for a layer with:

$E_1 = 35$ GPa $G_{12} = 1.75$ GPa $v_{12} = 0.3$
$E_2 = 3.5$ GPa $G_{23} = 0.35$ GPa.

Matrix Definitions:
Reduced stiffness $[Q]$ Compliance$[S]$
Shear stiffness$[Q^*]$ Compliance $[S^*]$

$$v_{21} = v_{12}E_2/E_1 = \frac{0.3(3.5)}{35} = 0.03$$

$\Delta = 1 - v_{12}v_{21} = 1 - (0.3)(0.03) = 0.991$
$Q_{11} = E_1/\Delta = 35/0.991 = 35.318$ GPa
$Q_{12} = Q_{21} = v_{12}E_2/\Delta = (0.3)(3.5)/0.991 = 1.0595$ GPa
$Q_{22} = E_2/\Delta = 3.5/0.991 = 3.5318$ GPa
$Q_{66} = G_{12} = 1.75$ GPa

$$[Q] = \begin{bmatrix} \frac{E_1}{\Delta} & \frac{v_{12}E_2}{\Delta} & 0 \\ \frac{v_{12}E_2}{\Delta} & \frac{E_2}{\Delta} & 0 \\ 0 & 0 & G_{12} \end{bmatrix} = \begin{bmatrix} 35.318 & 1.06 & 0 \\ 1.06 & 3.532 & 0 \\ 0 & 0 & 1.75 \end{bmatrix} \text{GPa}$$

$Q_{44}^* = G_{23} = 0.35$ GPa
$Q_{55}^* = G_{13} = 1.75$ GPa

$$[Q^*] = \begin{bmatrix} 0.35 & 0 \\ 0 & 1.75 \end{bmatrix} \text{GPa}$$

$S_{11} = 1/E_1 = 1/35 = 0.0286$ GPa^{-1}
$S_{12} = S_{21} = -v_{12}/E_1 = -(0.3)/35 = -0.00857$ GPa^{-1}
$S_{22} = 1/E_2 = 1/3.5 = 0.2857$ GPa^{-1}
$S_{66} = 1/G_{12} = 1/1.75 = 0.5714$ GPa^{-1}

$$[S] = \begin{bmatrix} 1/E_1 & -v_{12}/E_1 & 0 \\ -v_{12}/E_1 & 1/E_2 & 0 \\ 0 & 0 & 1/G_{12} \end{bmatrix} = \begin{bmatrix} 0.0286 & -0.00857 & 0 \\ -0.00857 & 0.2857 & 0 \\ 0 & 0 & 0.5714 \end{bmatrix} \text{GPa}^{-1}$$

$S_{44}^* = 1/G_{23} = 1/0.35 = 2.8571 \text{ GPa}^{-1}$
$S_{55}^* = 1/G_{13} = 1/1.75 = 0.5714 \text{ GPa}^{-1}$

$$[S^*] = \begin{bmatrix} 2.857 & 0 \\ 0 & 0.571 \end{bmatrix} \text{GPa}^{-1}$$

Exercise 5.2

Compute the safety factor using stress transformation for carbon-epoxy (AS4/3501-6). The composite layer has a fiber orientation of $\theta = 60°$. The state of stress is as follows:

$\sigma_x = 10$ MPa $\qquad \sigma_y = 5$ MPa $\qquad \sigma_{xy} = 2.5$ MPa

$$\begin{Bmatrix} \sigma_1 \\ \sigma_2 \\ \sigma_6 \end{Bmatrix} = \begin{bmatrix} m^2 & n^2 & 2mn \\ n^2 & m^2 & -2mn \\ -mn & mn & m^2 - n^2 \end{bmatrix} \begin{Bmatrix} \sigma_x \\ \sigma_y \\ \sigma_{xy} \end{Bmatrix}$$

where $m = \cos(\theta)$ and $n = \sin(\theta)$

$$\begin{Bmatrix} \sigma_1 \\ \sigma_2 \\ \sigma_6 \end{Bmatrix} = \begin{Bmatrix} 8.4151 \\ 6.585 \\ -3.4151 \end{Bmatrix} \text{MPa}$$

From the material property data in Table 1.1:

$F_{1t} = 1830$ MPa
$F_{2t} = 57$ MPa
$F_6 = 71$ MPa

F.S. (Longitudinal) = $F_{1t}/\sigma_1 = 1830/8.4151 = 217.5$
F.S. (Transverse) = $F_{2t}/\sigma_2 = 57/6.585 = 8.66$
F.S. (Shear) = $F_6/\sigma_6 = 71/-3.4151 = 20.79$

F.S. = Min(F.S.$_L$, F.S.$_T$, F.S.$_S$) = 8.66

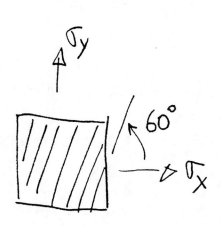

Exercise 5.3

Compute the rotation matrix $[T]$ for a layer oriented at 45° with respect to the x axis. Use it to find $[\bar{Q}]$ in global coordinates, corresponding to $[Q]$ in material coordinates computed in 5.1.

$$m = \cos(\theta) = \sqrt{2}/2 \qquad n = \sin(\theta) = \sqrt{2}/2$$

$$R = \begin{bmatrix} 1 & 0 & 0 \\ 0 & 1 & 0 \\ 0 & 0 & 2 \end{bmatrix} \qquad [T] = \begin{bmatrix} 0.5 & 0.5 & 1.0 \\ 0.5 & 0.5 & -1.0 \\ -0.5 & 0.5 & 0 \end{bmatrix}$$

$$[\bar{Q}] = [T]^{-1}[Q][R][T][R]^{-1}$$

Using a spred sheet to multiply the matrices:

$$[\bar{Q}] = \begin{bmatrix} 11.99 & 8.492 & 7.947 \\ 8.492 & 11.99 & 7.947 \\ 7.947 & 7.947 & 9.183 \end{bmatrix} \text{GPa}$$

Exercise 5.4

Plot the variation of $\bar{Q}_{11}$ as a function of θ between 0 and 90° using the material properties of Exercise 4.10. When does $\bar{Q}_{11}$ become numerically equal to $\bar{Q}_{22}$.

$$E_1 = 156.403 \text{ GPa} \qquad E_2 = 7.786 \text{ GPa}$$
$$\nu_{12} = 0.352 \qquad \nu_{21} = 0.016$$
$$G_{12} = 3.762 \text{ GPa}$$

$$\bar{Q}_{11} = Q_{11}\cos^4(\theta) + 2(Q_{12} + Q_{66})\sin^2(\theta)\cos^2(\theta) + Q_{22}\sin^4(\theta)$$
$$\bar{Q}_{22} = Q_{11}\sin^4(\theta) + 2(Q_{12} + Q_{66})\sin^2(\theta)\cos^2(\theta) + Q_{22}\cos^4(\theta)$$

From the two equations above, $\bar{Q}_{11} = \bar{Q}_{22}$ when $\theta = 45°$.

$$\Delta = 1 - \nu_{12}\nu_{21}$$
$$Q_{11} = E_1/\Delta = 157.29 \text{ GPa}$$
$$Q_{12} = \nu_{12}/\Delta = 2.757 \text{ GPa}$$

$Q_{66} = G_{12} = 3.762$ GPa
$Q_{22} = E_2/\Delta = 7.830$ GPa

$$\overline{Q}_{11}(\theta) = 157.29\cos^4(\theta) + 20.562\sin^2(\theta)\cos^2(\theta) + 7.830\sin^4(\theta)$$
$$\overline{Q}_{22}(\theta) = 157.29\cos^4(\theta) + 20.562\sin^2(\theta)\cos^2(\theta) + 7.830\cos^4(\theta)$$
$$\overline{Q}_{11} = \overline{Q}_{22} \text{ for } \theta = 45°$$

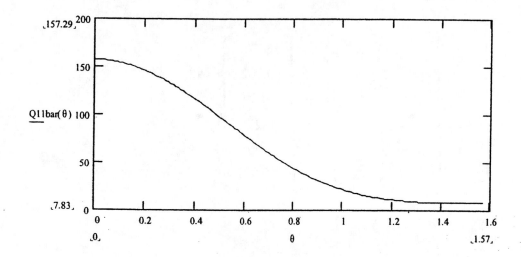

Exercise 5.5

A tensile test specimen of the material in Exercise 5.1 is 10 mm wide, 2 mm thick, and 150 mm and is subjected to an axial force of 200 N. The fibers are oriented at $\theta = 10°$ with respect to the loading axis.

a) Compute to inplane and shear strains in the material coordinate system.
b) Compute the inplane and shear strains in the global coordinate system.
c) Is the cross-section under a uniaxial state of strain in the global coordinates?
     Explain the origin of each nonzero strain.

(a) $E_1 = 35$ GPa $\nu_{12} = 0.3$ $G_{23} = 0.35$ GPa
          $E_2 = 3.5$ GPa $G_{12} = 1.75$ GPa

$\sigma_x = F/A = 200$ N$/0.00002$ m$^2 = 10$ MPa
$m = \cos(10°) = 0.985$ $n = \sin(10°) = 0.174$

$$[T] = \begin{bmatrix} 0.970 & 0.030 & 0.343 \\ 0.030 & 0.970 & -0.343 \\ -0.171 & 0.171 & 0.940 \end{bmatrix}$$

$$[S] = \begin{bmatrix} 0.0286 & -0.00857 & 0 \\ -0.00857 & 0.286 & 0 \\ 0 & 0 & 0.5714 \end{bmatrix}$$

$$\begin{Bmatrix} \sigma_1 \\ \sigma_2 \\ \sigma_6 \end{Bmatrix} = [T] \begin{Bmatrix} \sigma_x \\ \sigma_y \\ \sigma_{xy} \end{Bmatrix} = \begin{Bmatrix} 9.70 \\ 0.302 \\ -1.71 \end{Bmatrix} \text{ MPa}$$

$$\begin{Bmatrix} \epsilon_1 \\ \epsilon_2 \\ \gamma_6 \end{Bmatrix} = [S] \begin{Bmatrix} \sigma_1 \\ \sigma_2 \\ \sigma_6 \end{Bmatrix} = \begin{Bmatrix} 0.275 \\ 0.003 \\ -0.977 \end{Bmatrix}$$

(b)

$$\begin{Bmatrix} \epsilon_x \\ \epsilon_y \\ \gamma_{xy} \end{Bmatrix} = [T]^{-1} \begin{Bmatrix} \epsilon_1 \\ \epsilon_2 \\ \gamma_6 \end{Bmatrix} = \begin{Bmatrix} 0.433 \\ -0.156 \\ -0.825 \end{Bmatrix}$$

(c)   No. The axial load produces an axial strain $\epsilon_x = 0.825$, a transverse strain (Poisson effect) $\epsilon_y = -0.156$, and a shear $\gamma = 0.825$. The shear is produced by the fibers trying to get oriented along the load direction.

Exercise 5.6

Compute the transformed reduced stiffness matrix for the material in Exercise 5.1 at $\theta = 45°$

and 90°.

$E_1 = 35$ GPa	$G_{12} = 1.75$ GPa	$v_{12} = 0.3$
$E_2 = 3.5$ GPa	$G_{23} = 0.35$ GPa.	

From Exercise 5.3 with $\theta = 45°$.

$$m = \cos(\theta) = \sqrt{2}/2 \qquad n = \cos(\theta) = \sqrt{2}/2$$

$$R = \begin{bmatrix} 1 & 0 & 0 \\ 0 & 1 & 0 \\ 0 & 0 & 2 \end{bmatrix} \qquad [T] = \begin{bmatrix} 0.5 & 0.5 & 1.0 \\ 0.5 & 0.5 & -1.0 \\ -0.5 & 0.5 & 0 \end{bmatrix}$$

$$[\bar{Q}] = [T]^{-1}[Q][R][T][R]^{-1}$$

$$[\bar{Q}] = \begin{bmatrix} 11.992 & 8.492 & 7.947 \\ 8.492 & 11.992 & 7.947 \\ 7.947 & 7.947 & 9.183 \end{bmatrix}$$

for $\theta = 90°$

$$[T] = \begin{bmatrix} 0 & 1 & 0 \\ 1 & 0 & 0 \\ 0 & 0 & -1 \end{bmatrix} \qquad [Q] = \begin{bmatrix} 35.318 & 1.060 & 0 \\ 1.060 & 3.532 & 0 \\ 0 & 0 & 1.75 \end{bmatrix} \text{GPa}$$

$$[\bar{Q}] = \begin{bmatrix} 3.532 & 1.060 & 0 \\ 1.060 & 35.32 & 0 \\ 0 & 0 & 1.75 \end{bmatrix} \text{GPa}$$

(c) In Exercise 5.4 at 45°, $\bar{Q}_{16} \neq 0$ and $\bar{Q}_{26} \neq 0$. At 90° $Q_{16} = Q_{26} = 0$, thus the laminate is specially orthotropic. In other words, at 90° the material axes coincide with the global axes. Also at 90°, $\bar{Q}_{11} = Q_{22}$ and $\bar{Q}_{22} = Q_{11}$.

Exercise 5.7

Contracted notation is a simplified way of writing stress and strain symbols with one subscript instead of 2.

Exercise 5.8

The plane stress assumption is when the height and width of an object are much greater than the thickness, making $\sigma_3 = 0$.

Exercise 5.9

Plane defined in parametric form:

$F(x,y,z) = x/3 + y/4 + z/5 - 1 = 0$. Then the stresses are:

$$[\sigma] = \begin{bmatrix} 100 & -5 & 2 \\ -5 & 10 & 1 \\ 2 & 1 & -20 \end{bmatrix}$$

Compute the stress vector at $x = y = z = \frac{60}{47}$ using Cauchy's law ($t_i = \sigma_{ij}n_j$).
Need to find the normal to the plane.

$\frac{\partial F}{\partial x} = \frac{1}{3}$ $\frac{\partial F}{\partial y} = \frac{1}{4}$ $\frac{\partial F}{\partial z} = \frac{1}{5}$ $n_x = \frac{\partial F}{\partial x} / \sqrt{(\frac{\partial F}{\partial x})^2 + (\frac{\partial F}{\partial y})^2 + (\frac{\partial F}{\partial z})^2} = 0.7212$

n_y and n_z computed the same way as n_x
$n_y = 0.5409$ $n_z = 0.4327$

Using Cauchy's Law (Eq. 5.3)

$$\begin{Bmatrix} t_x \\ t_y \\ t_z \end{Bmatrix} = \begin{bmatrix} 100 & -5 & 2 \\ -5 & 10 & 1 \\ 2 & 1 & -20 \end{bmatrix} \begin{Bmatrix} 0.7212 \\ 0.5409 \\ 0.7327 \end{Bmatrix} = \begin{Bmatrix} 70.28 \\ 2.236 \\ -6.67 \end{Bmatrix} \text{ GPa}$$

$$n_y = \frac{\partial F}{\partial y} / \sqrt{\left(\frac{\partial F}{\partial x}\right)^2 + \left(\frac{\partial F}{\partial y}\right)^2 + \left(\frac{\partial F}{\partial z}\right)^2} = 0.5409$$

$$n_z = \frac{\partial F}{\partial z} / \sqrt{\left(\frac{\partial F}{\partial x}\right)^2 + \left(\frac{\partial F}{\partial y}\right)^2 + \left(\frac{\partial F}{\partial z}\right)^2} = 0.4327$$

Solutions Chapter 6
Exercise 6.1

Compute the A, B, and D matrices for a bimetallic strip with each layer 2 mm thick.
Material properties:

Aluminum
$E = 71$ GPa
$\alpha = 23(10^6)1/°C$

Copper
$E = 119$ GPa
$\alpha = 17(10^6)1/°C$

assume $v = 0.3$ for both materials
Since both copper and aluminum are isotropic, $\overline{Q} = Q$.
For aluminum:
$Q_{11} = Q_{22} = E/(1 - v^2) = 78.02$ GPa
$Q_{12} = vE/(1 - v^2) = 23.41$ GPa
$Q_{66} = E/2(1 + v) = 27.31$ GPa
For copper:
$Q_{11} = Q_{22} = E/(1 - v^2) = 130.8$ GPa
$Q_{12} = vE/(1 - v^2) = 39.24$ GPa
$Q_{66} = E/2(1 + v) = 45.77$ GPa
$A_{ij} = \sum[(Q_{ij})_k t_k]$

$A_{11} = \sum[(Q_{11})_k t_k] = 78.02(10^9)(0.002) + 130.8(10^6)(0.002) = 4.176(10^8)$ Pa m
$A_{12} = 1.253(10^8)$ Pa m
$A_{16} = 0$
$A_{22} = 4.176(10^8)$ Pa m
$A_{26} = 0$
$A_{66} = 1.462(10^8)$ Pa m
$B_{ij} = \sum[(Q_{ij})_k t_k \overline{z}_k]$
$B_{11} = \sum[(Q_{11})_k t_k \overline{z}_k] = 78.02(10^9)(0.002)(-0.001) + 130.8(10^9)(0.002)(0.001) = -1.056(10^5)$
Pa m^2

$B_{12} = -3.166(10^4)$ Pa m^2
$B_{22} = -1.056(10^5)$ Pa m^2
$B_{16} = 0$
$B_{26} = 0$
$B_{66} = -3.692(10^4)$ Pa m^2
$D_{ij} = \sum[(Q_{ij})_k [t_k(\overline{z}_k)^2 + (t_k)^3/12]]$
$D_{11} = \sum[(Q_{11})_k [t_k(\overline{z}_k)^2 + (t_k)^3/12]] = (78.02(10^9) + 130.9(10^9))[(0.002)(0.001)^2 + (t_k)^3/12]$
$= 5.57(10^2)$ Pa m^3
$D_{12} = 1.671(10^2)$ Pa m^3
$D_{22} = 5.569(10^2)$ Pa m^3
$D_{16} = D_{26} = 0$
$D_{26} = 1.949(10^2)$ Pa m^3

$$[A] = \begin{bmatrix} 4.176 & 1.253 & 0 \\ 1.253 & 4.176 & 0 \\ 0 & 0 & 1.462 \end{bmatrix} (10^8) \text{ Pa m}$$

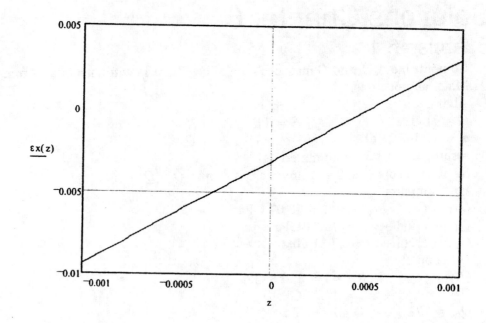

Exercise 6.3

Compute the stresses in global coordinates at the top and bottom of each layer for the material in Example 6.1.

$\epsilon_x^0 = 765(10^{-6})$

$\epsilon_y^0 = -280(10^{-6})$

$\kappa_{xy} = 34.2(10^{-6})1/mm.$

All remaining strains and curvatures equal to zero.

From Example 6.1, the bottom surface has $\theta_1 = 55°$ and $\theta_2 = -55°$, with $t_1 = t_2 = 0.635$ mm. The material properties are given in Example 5.1.

At the top and bottom of the laminate

$$\left\{ \begin{array}{c} \epsilon_x \\ \epsilon_y \\ \gamma_{xy} \end{array} \right\} = \left\{ \begin{array}{c} \epsilon_x^0 \\ \epsilon_y^0 \\ \gamma_{xy}^0 \end{array} \right\} + z \left\{ \begin{array}{c} \kappa_x \\ \kappa_y \\ \kappa_{xy} \end{array} \right\} = \left\{ \begin{array}{c} 765 \\ -280 \\ 0 \end{array} \right\} + 0.635 \left\{ \begin{array}{c} 0 \\ 0 \\ 34.2 \end{array} \right\} = \left\{ \begin{array}{c} 765 \\ -280 \\ 21.72 \end{array} \right\}$$

$$\left\{ \begin{array}{c} \sigma_x \\ \sigma_y \\ \sigma_{xy} \end{array} \right\} = [\overline{Q}] \left\{ \begin{array}{c} \epsilon_x \\ \epsilon_y \\ \gamma_{xy} \end{array} \right\}$$

For the bottom surface ($\theta = 55°$)

$$\left\{ \begin{array}{c} \sigma_x \\ \sigma_y \\ \sigma_{xy} \end{array} \right\} = \left[\begin{array}{ccc} 12.402 & 5.710 & 1.217 \\ 5.710 & 15.472 & 3.000 \\ 1.217 & 3.000 & 6.239 \end{array} \right] \left\{ \begin{array}{c} 765 \\ -280 \\ -21.72 \end{array} \right\} (10^{-6}) = \left\{ \begin{array}{c} 7.861 \\ -0.0292 \\ 0.0445 \end{array} \right\} \text{MPa}$$

For the top surface ($\theta = -55°$)

$$\left\{\begin{array}{c} \sigma_x \\ \sigma_y \\ \sigma_{xy} \end{array}\right\} = \left[\begin{array}{ccc} 12.402 & 5.710 & -1.217 \\ 5.710 & 15.472 & -3.000 \\ -1.217 & -3.000 & 6.239 \end{array}\right] \left\{\begin{array}{c} 765 \\ -280 \\ 21.72 \end{array}\right\}(10^{-6}) = \left\{\begin{array}{c} 7.861 \\ -0.0292 \\ -0.0445 \end{array}\right\} \text{MPa}$$

For the middle surface:

$$\left\{\begin{array}{c} \sigma_x \\ \sigma_y \\ \sigma_{xy} \end{array}\right\} = \left[\begin{array}{ccc} 12.402 & 5.710 & \pm1.217 \\ 5.710 & 15.472 & \pm3.00 \\ \pm1.217 & \pm3.00 & 6.239 \end{array}\right] \left\{\begin{array}{c} 765 \\ -280 \\ 0 \end{array}\right\}(10^{-6}) = \left\{\begin{array}{c} 7.89 \\ 0.0360 \\ \pm0.0910 \end{array}\right\} \text{MPa}$$

Layer	Z	σ_x (MPa)	σ_y (MPa)	σ_{xy} (MPa)
2	0.635	7.86	-0.0292	0.0445
2	0	7.89	0.0360	-0.0910
1	0	7.89	0.0360	0.0910
1	-0.635	7.86	-0.0292	-0.0445

Exercise 6.4

Derive the seven stress resultants by integrating over the thickness of a laminated plate.

$$\sigma_x = A^{(k)}z\sin(\alpha x)\cos(\beta y) \qquad N = 3 \text{ and } t_1 = t_2 = t_3$$
$$\sigma_y = B^{(k)}z\sin(\alpha x)\sin(\beta y)$$
$$\sigma_{xy} = 0$$
$$\sigma_{yz} = Q^{(k)}z\sin(\alpha x)\cos(\beta y)$$
$$\sigma_{xz} = R^{(k)}z\sin(\alpha x)\sin(\beta y)$$

$$\left\{\begin{array}{c} N_x \\ N_y \\ N_{xy} \end{array}\right\} = \sum_{k=1}^{N}\left[\int_{z_{k-1}}^{z_k}\left\{\begin{array}{c} \sigma_x \\ \sigma_y \\ \sigma_{xy} \end{array}\right\}^{(k)} dz\right]$$

$$N_x = \int_{-1.5t}^{-0.5t}\sigma_x dz + \int_{-0.5t}^{0.5t}\sigma_x dz + \int_{0.5t}^{1.5t}\sigma_x dz \qquad \text{Same equation for } \sigma_y \text{ and } \sigma_{xy}.$$

$$N_x = -t^2(A^{(1)} - A^{(3)})\sin(\alpha x)\sin(\beta y)$$
$$N_y = -t^2(B^{(1)} - B^{(3)})\sin(\alpha x)\sin(\beta y)$$
$$N_{xy} = 0$$

$$\left\{\begin{array}{c} V_x \\ V_y \end{array}\right\} = \sum_{k=1}^{N}\left[\int_{z_{k-1}}^{z_k}\begin{array}{c} \sigma_{yz} \\ \sigma_{xz} \end{array} dz\right]$$

$$V_x = \int_{-1.5t}^{-0.5t}\sigma_{xz} dz + \int_{-0.5t}^{0.5t}\sigma_{xz} dz + \int_{0.5t}^{1.5t}\sigma_{xz} dz \qquad \text{Same equation for } \sigma_{yz}.$$
$$V_x = (R^{(1)} + R^{(2)} + R^{(3)})t\cos(\alpha x)\sin(\beta y)$$
$$V_y = (Q^{(1)} + Q^{(2)} + Q^{(3)})t\sin(\alpha x)\cos(\beta y)$$

$$\left\{\begin{array}{c} M_x \\ M_y \\ M_{xy} \end{array}\right\} = \sum_{k=1}^{N}\left[\int_{z_{k-1}}^{z_k}\left[\begin{array}{c} \sigma_x \\ \sigma_y \\ \sigma_{xy} \end{array}\right]^{(k)} zdz\right]$$

$$M_x = \int_{-1.5t}^{-0.5t}\sigma_x zdz + \int_{-0.5t}^{0.5t}\sigma_x zdz + \int_{0.5t}^{1.5t}\sigma_x zdz \qquad \text{Same equation for } M_y \text{ and } M_{xy}.$$

$$M_x = (13A^{(1)} + A^{(2)} + 13A^{(3)})(t^3/12)\sin(\alpha x)\sin(\beta y)$$
$$M_y = (13B^{(1)} + B^{(2)} + 13B^{(3)})(t^3/12)\sin(\alpha x)\sin(\beta y)$$
$$M_{xy} = 0$$

Exercise 6.5

Find σ_1, σ_2, and σ_6 at $z = 0$ on both layers of a laminate $[\pm 30]_T$ subject to $N_x = 1\ \frac{N}{m}$. All remaining forces and moments equal zero.

$E_1 = 137.8$ GPa $\qquad E_2 = 9.6$ GPa

$G_{12} = 5.2$ GPa $\qquad v_{12} = 0.3$

$t_1 = t_2 = 1.27$ mm $\qquad N_x = \frac{1}{1000}\ \frac{N}{mm}$

$$[Q] = \begin{bmatrix} 138.67 & 2.898 & 0 \\ 2.898 & 9.66 & 0 \\ 0 & 0 & 5.2 \end{bmatrix}\ \text{GPa}$$

$[\bar{Q}]$ for the 1st layer ($\theta = 30°$) $\qquad\qquad$ $[\bar{Q}]$ for the 2nd layer ($\theta = -30°$)

$$[\bar{Q}] = \begin{bmatrix} 83.592 & 25.723 & 41.110 \\ 25.723 & 19.09 & 14.753 \\ 41.110 & 14.753 & 28.025 \end{bmatrix}\ \text{GPa} \qquad [\bar{Q}] = \begin{bmatrix} 83.592 & 25.723 & -41.110 \\ 25.723 & 19.09 & -14.753 \\ -41.110 & -14.753 & 28.025 \end{bmatrix}$$

GPa

$$[A] = \begin{bmatrix} 212.33 & 65.34 & 0 \\ 65.34 & 48.484 & 0 \\ 0 & 0 & 71.182 \end{bmatrix}\ \text{GPa mm} \qquad [B] = \begin{bmatrix} 0 & 0 & -66.305 \\ 0 & 0 & -23.796 \\ -66.305 & -23.796 & 0 \end{bmatrix}$$

GPa mm^2

$$[D] = \begin{bmatrix} 114.15 & 35.129 & 0 \\ 35.129 & 26.067 & 0 \\ 0 & 0 & 38.270 \end{bmatrix}\ \text{GPa mm}^3$$

$$\begin{Bmatrix} 0.001 \\ 0 \\ 0 \\ 0 \\ 0 \\ 0 \end{Bmatrix} = \begin{bmatrix} 212.33 & 65.34 & 0 & 0 & 0 & -66.305 \\ 65.34 & 48.484 & 0 & 0 & 0 & -23.796 \\ 0 & 0 & 71.182 & -66.305 & -23.796 & 0 \\ 0 & 0 & -66.305 & 114.15 & 35.129 & 0 \\ 0 & 0 & -23.796 & 35.129 & 26.067 & 0 \\ -66.305 & -23.796 & 0 & 0 & 0 & 38.270 \end{bmatrix} \begin{Bmatrix} \epsilon_x^0 \\ \epsilon_y^0 \\ \gamma_{xy}^0 \\ \kappa_x \\ \kappa_y \\ \kappa_{xy} \end{Bmatrix}$$

$$\begin{Bmatrix} \epsilon_x^0 \\ \epsilon_y^0 \\ \gamma_{xy}^0 \\ \kappa_x \\ \kappa_y \\ \kappa_{xy} \end{Bmatrix} = \begin{Bmatrix} 0.125 \\ -0.0893 \\ 0 \\ 0 \\ 0 \\ 0.161 \end{Bmatrix} (10^{-4})$$

Table of answers	ϵ_x	ϵ_y	γ_{xy}	σ_x	σ_y	σ_{xy}	σ_1	σ_2	σ_6	
-30 top	0.125	-0.0893	0.204	-0.254	-1.50	1.90	-2.22	0.458	1.49	
-30 bottom	0.125	-0.0893	0	8.13	1.50	-3.81	9.77	-0.139	0.963	(10^{-4})
+30 top	0.125	-0.0893	0	8.13	1.50	3.81	9.77	-0.139	-0.963	
+30 bottom	0.125	-0.0893	-0.204	-0.254	-1.50	-1.90	-2.22	0.458	-1.49	

Exercises 6.6 and 6.7

Use CADEC to find the values for the A, B, D, and H matrices for the listed laminates. See CADEC output.

Exercise 6.8

For a [0/90]$_s$ laminate made from Kevlar49-epoxy with each layer 1.0 mm thick, compute the stresses in material coordinates at the top and bottom of each layer when $N_x = 1 \frac{N}{m}$.

Since the laminate is symmetric $[B] = 0$. Since only an inplane load is applied, only the $[A]$ matrix is needed.

$$[Q] = \begin{bmatrix} 76.441 & 1.885 & 0 \\ 1.885 & 5.546 & 0 \\ 0 & 0 & 2.07 \end{bmatrix} (10^9) \text{ N/m}^2$$

For the 0° layer, $[Q] = [\overline{Q}]$. For the 90° layer simply interchange Q_{11} and Q_{22}

$$[\overline{Q}] = \begin{bmatrix} 5.546 & 1.885 & 0 \\ 1.885 & 76.441 & 0 \\ 0 & 0 & 2.07 \end{bmatrix} (10^9) \text{ N/m}^2$$

The laminate can be analyzed as a 3-layer laminate with the center layer 0.002 m thick at 90° and two outer layers at 0°, each 0.001 m thick.

$$[A] = \begin{bmatrix} 163.97 & 7.54 & 0 \\ 7.54 & 163.97 & 0 \\ 0 & 0 & 8.28 \end{bmatrix} (10^6) \text{ N/m}$$

Then compute the inplane strains (they are constant throughout the laminate because there are no curvatures).

$$\begin{Bmatrix} \epsilon_x^0 \\ \epsilon_y^0 \\ \gamma_{xy}^0 \end{Bmatrix} = [A]^{-1} \begin{Bmatrix} 1 \\ 0 \\ 0 \end{Bmatrix} = \begin{Bmatrix} 6.14 \\ -.2827 \\ 0 \end{Bmatrix} (10^{-9}) \text{m/m}$$

For the bottom and top of the laminate.

$$\begin{Bmatrix} \sigma_x \\ \sigma_y \\ \sigma_{xy} \end{Bmatrix} = [Q] \begin{Bmatrix} \epsilon_x^0 \\ \epsilon_y^0 \\ \gamma_{xy}^0 \end{Bmatrix} = \begin{bmatrix} 76.441 & 1.885 & 0 \\ 1.885 & 5.546 & 0 \\ 0 & 0 & 2.07 \end{bmatrix} \begin{Bmatrix} 6.14 \\ -.2827 \\ 0 \end{Bmatrix} = \begin{Bmatrix} 469 \\ 10.06 \\ 0 \end{Bmatrix}$$

N/m^2

For the middle two layers.

$$\begin{Bmatrix} \sigma_x \\ \sigma_y \\ \sigma_{xy} \end{Bmatrix} = [Q] \begin{Bmatrix} \epsilon_x^0 \\ \epsilon_y^0 \\ \gamma_{xy}^0 \end{Bmatrix} = \begin{bmatrix} 5.546 & 1.885 & 0 \\ 1.885 & 76.441 & 0 \\ 0 & 0 & 2.07 \end{bmatrix} \begin{Bmatrix} 6.14 \\ -.2827 \\ 0 \end{Bmatrix} = \begin{Bmatrix} 33.52 \\ -10.04 \\ 0 \end{Bmatrix}$$

N/m^2

For $\theta = 0°$, $\sigma_1 = \sigma_x$ and $\sigma_2 = \sigma_y$.
For $\theta = 90°$, $\sigma_1 = \sigma_y$ and $\sigma_2 = \sigma_x$.

Layer	θ	σ_1 (Pa)	σ_2 (Pa)	σ_6 (Pa)
3	0	469	10.06	0
2	90	-10.04	33.52	0
1	0	469	10.06	0

Exercise 6.9

Consider a $[45/-45]_s$ laminate with the material properties of Exercise 6.8. Compute the mid-surface strains and the curvatures for the following load cases.

(a) $N_x = 1000$ N/m

(b) $M_{xy} = 1$ Nm/m

(c)   Comment on the coupling effects observed.

From Exercise 6.8:

$$[Q] = \begin{bmatrix} 76.441 & 1.885 & 0 \\ 1.885 & 5.546 & 0 \\ 0 & 0 & 2.07 \end{bmatrix} 10^9 \text{N/m}^2$$

For $\theta = 45°$:

$$[\bar{Q}] = \begin{bmatrix} 23.509 & 19.369 & 17.723 \\ 19.369 & 23.509 & 17.723 \\ 17.723 & 17.723 & 19.554 \end{bmatrix} 10^9 \text{ N/m}^2$$

For $\theta = -45°$:

$$[\bar{Q}] = \begin{bmatrix} 23.509 & 19.369 & -17.723 \\ 19.369 & 23.509 & -17.723 \\ -17.723 & -17.723 & 19.554 \end{bmatrix} 10^9 \text{ N/m}^2$$

$$[A] = \begin{bmatrix} 94.036 & 77.476 & 0 \\ 77.476 & 94.036 & 0 \\ 0 & 0 & 78.216 \end{bmatrix} 10^6 \text{ N/m} \qquad [B] = 0 \qquad \text{Because of symmetry}$$

$$[D] = \begin{bmatrix} 125.386 & 103.306 & 70.8947 \\ 103.306 & 125.386 & 70.8947 \\ 70.8947 & 70.8947 & 104.288 \end{bmatrix} \text{Nm}$$

(a) See also CADEC output

$$\left\{ \begin{array}{c} \epsilon_x^0 \\ \epsilon_y^0 \\ \gamma_{xy}^0 \end{array} \right\} = [A]^{-1} \left\{ \begin{array}{c} 1 \\ 0 \\ 0 \end{array} \right\} = \left\{ \begin{array}{c} 0.0331 \\ -0.0273 \\ 0 \end{array} \right\} 10^{-3} \text{ m/m}$$

(b) See also CADEC output

$1 \ 0$

ebarbero

From: Dr. Pizhong Qiao [qiao@uakron.edu]
Sent: Friday, March 10, 2000 5:31 PM
To: barbero@cemr.wvu.edu
Subject: Correction on Solution

Dr. Barbero:

FYI.
When I graded the homework assignments, I found an error in "Solutions Manual for Intr. to Comp. Matls. Design". For page 6.9 or Prob. 6.9 (b), the solution curvature for kappa x, kappa y, and kappa xy are -0.00514, -0.00514, and 0.0166, respectively (unit: 1/mm).

Talk to you soon. Pizhong

Pizhong Qiao (Chiao), Ph.D.,P.E.
Assistant Professor
Department of Civil Engineering
244 Sumner Street, ASEC, Rm.210
The University of Akron
Akron, OH 44325-3905
Phone: (330)972-5226; Fax:(330)972-6020
Email: Qiao@uakron.edu

$$\left\{ \begin{array}{c} \kappa_x \\ \kappa_y \\ \kappa_{xy} \end{array} \right\} = [D]^{-1} \left\{ \begin{array}{c} 0 \\ 0 \\ 1 \end{array} \right\} = \left\{ \begin{array}{c} -0.0264 \\ -0.0189 \\ -0.0051 \end{array} \right\}$$

(c) The inplane loads only produce mid-surface strains and no curvatures because $[B] = 0$. Bending moment produces only curvatures.

Exercises 6.10-6.11

Use CADEC to solve these Exercises.

6.9

Laminate Definition   Filename: [C:\BOOK\PROBLEMS\PROB6_8.DEF]   [Plot]   (1/2)

NOTE: Layer #1 is the bottom layer, therefore the first value should represent the bottom layer, then a space and then the value for the next layer up, and so on.

Laminate Name: [Problem 6.8, 6.9, 6.10, m, N, Pa, Kevlar-Epoxy Table 1.1]

Number of Layers: [3]   Number of Materials: [1]   Total Thickness [4.000E-3]

Layer Thicknesses: [0.001 0.002 0.001]

Layup Angles: [45 -45 45]

Layer Materials: [1 1 1]

Loading: Nx [0]   Mx [1]   Qx [0]
           Ny [0]   My [0]   Qy [0]
           Nxy [0]   Mxy [0]

Temperature Change: [0]

Moisture Concentration Change: [0]   Use Current Laminate Properties ➡

Safety Factor: [1]   Set Up Laminate Mat. Properties

6.9
  cont.

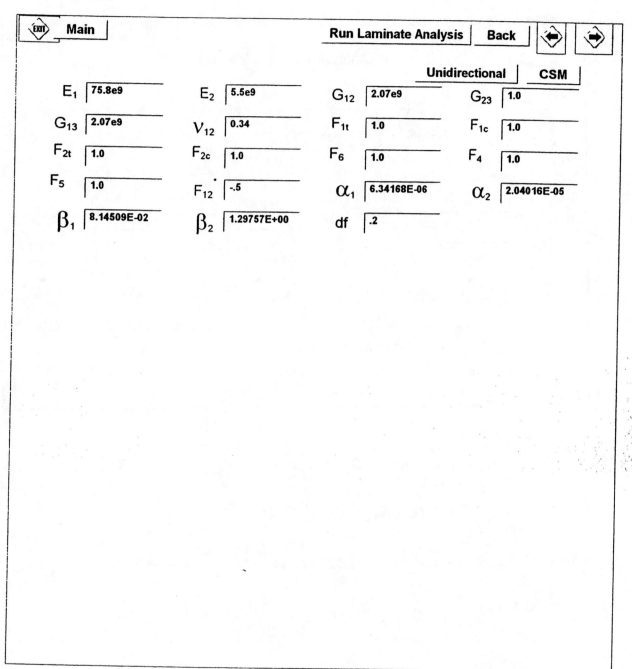

			Unidirectional	**CSM**

E_1 `75.8e9` E_2 `5.5e9` G_{12} `2.07e9` G_{23} `1.0`

G_{13} `2.07e9` ν_{12} `0.34` F_{1t} `1.0` F_{1c} `1.0`

F_{2t} `1.0` F_{2c} `1.0` F_6 `1.0` F_4 `1.0`

F_5 `1.0` F_{12} `-.5` α_1 `6.34168E-06` α_2 `2.04016E-05`

β_1 `8.14509E-02` β_2 `1.29757E+00`   df `.2`

Main   **Run Laminate Analysis**   **Back**

6.9

(a)

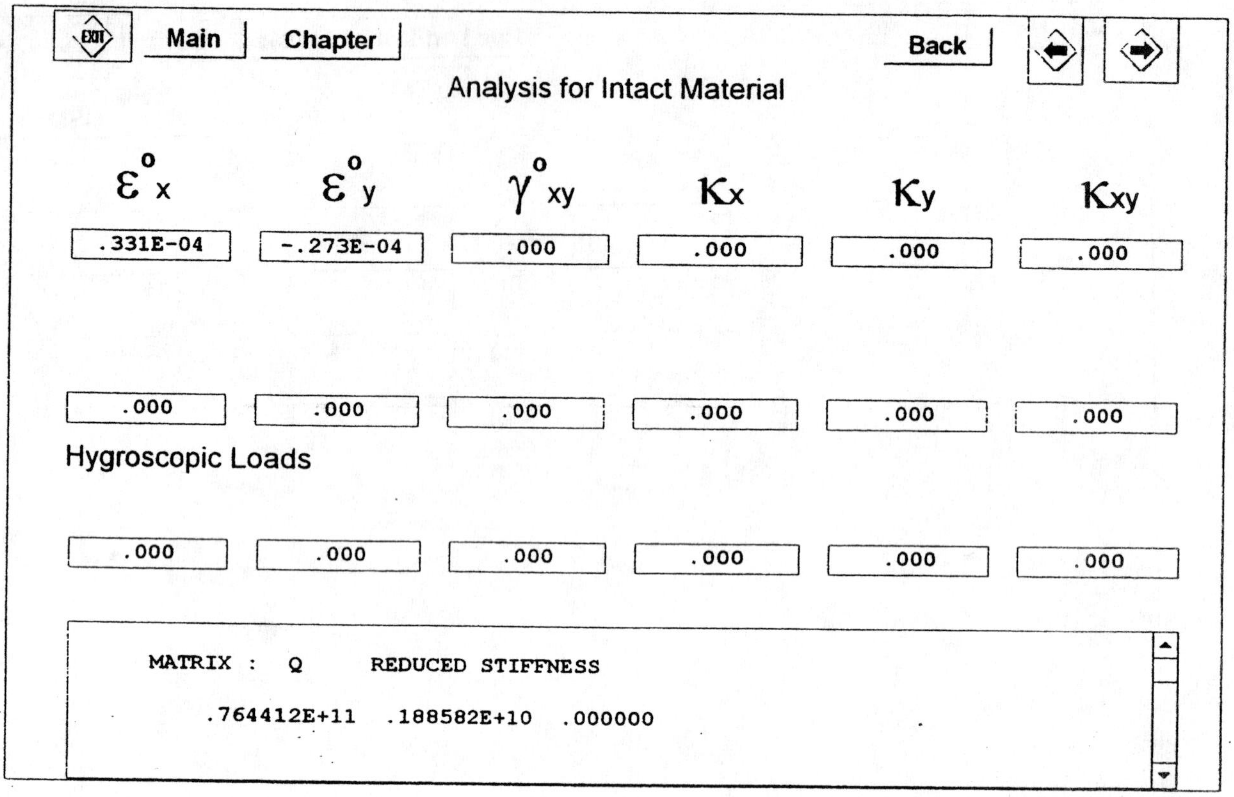

EXIT
Main
Chapter
Back
Analysis for Intact Material
ε°x
ε°y
γ°xy
Kx
Ky
Kxy
.331E-04
-.273E-04
.000
.000
.000
.000
.000
.000
.000
.000
.000
.000
Hygroscopic Loads
.000
.000
.000
.000
.000
.000
MATRIX : Q REDUCED STIFFNESS
.764412E+11 .188582E+10 .000000

Inplane load produces only mid-surface strains because $[B] = 0$

Also, $\gamma^{o}_{xy} = 0$ because the laminate is balanced, so $A_{16} = 0$

6.9

(b)

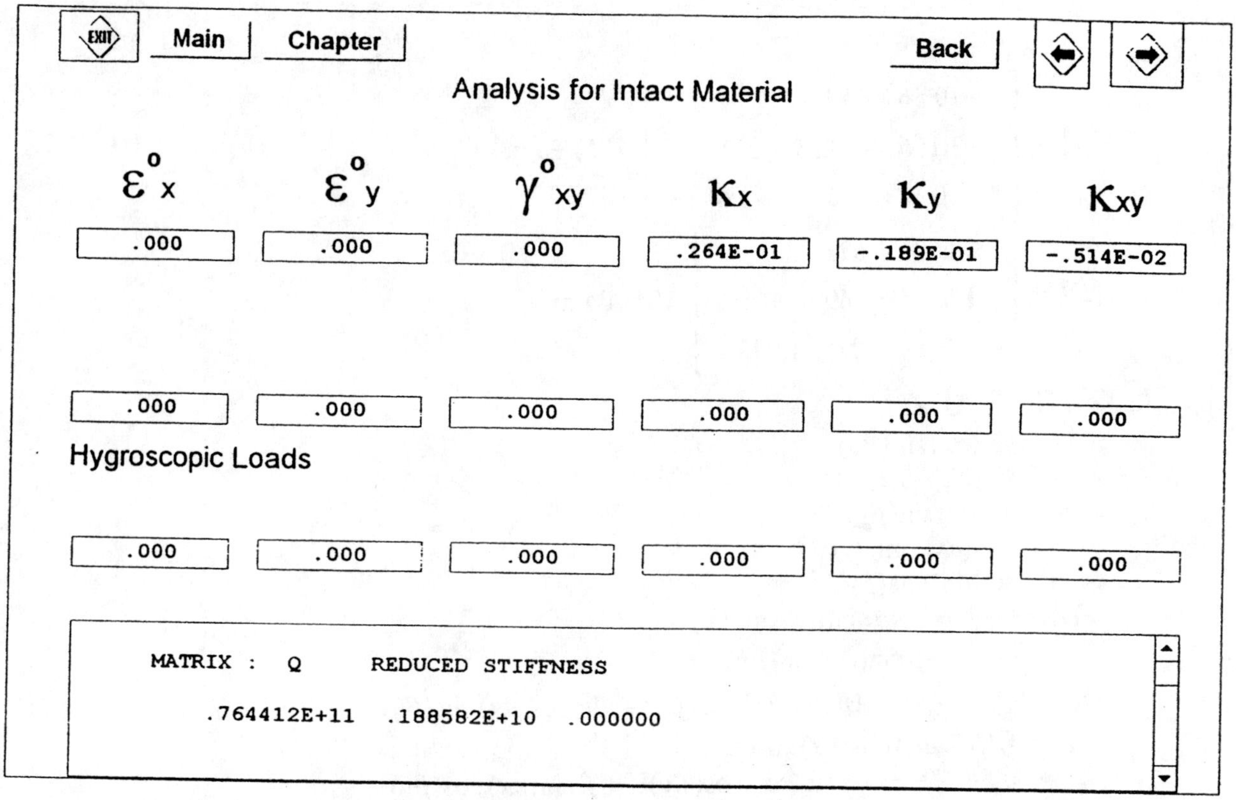

Analysis for Intact Material

ε^o_x	ε^o_y	γ^o_{xy}	K_x	K_y	K_{xy}
.000	.000	.000	.264E-01	-.189E-01	-.514E-02

.000	.000	.000	.000	.000	.000

Hygroscopic Loads

.000	.000	.000	.000	.000	.000

```
    MATRIX :  Q      REDUCED STIFFNESS

      .764412E+11   .188582E+10   .000000
```

Bending moment produces only
curvature because [B] = 0.

$K_{xy} \neq 0$ because $D_{16} \neq 0$.

$$[B] = \begin{bmatrix} -10.56 & -3.166 & 0 \\ -3.166 & -10.56 & 0 \\ 0 & 0 & -3.692 \end{bmatrix} (10^4) \ \text{Pa m}^2$$

$$[D] = \begin{bmatrix} 5.569 & 1.671 & 0 \\ 1.671 & 5.569 & 0 \\ 0 & 0 & 1.949 \end{bmatrix} (10^2) \ \text{Pa m}^3$$

Exercise 6.2

$u_0 = A \sin(\alpha x) \sin(\beta y)$

$v_0 = B \sin(\alpha x) \cos(\beta y)$

$w = C \sin(\alpha x) \sin(\beta y)$

$\phi_x = Q \cos(\alpha x) \sin(\beta y)$

$\phi_y = R \sin(\alpha x) \cos(\beta y)$

(a)$\epsilon_x^0 = \frac{\partial u_0}{\partial x} = -A\alpha \sin(\alpha x) \sin(\beta y)$

$\epsilon_y^0 = \frac{\partial v_0}{\partial y} = -B\beta \sin(\alpha x) \sin(\beta y)$

$\gamma_{xy}^0 = \frac{\partial u_0}{\partial y} + \frac{\partial v_0}{\partial x} = A\beta \cos(\alpha x)\cos(\beta y) + B\alpha \cos(\alpha x)\cos(\beta y)$

$\quad = (A\beta + B\alpha)\cos(\alpha x)\cos(\beta y)$

$\gamma_{yz}^0 = -\phi_y + \frac{\partial w_0}{\partial y} = -R\sin(\alpha x)\cos(\beta y) + C\beta \sin(\alpha x)\cos(\beta y)$

$\quad = (-R + C\beta)\sin(\alpha x)\cos(\beta y)$

$\gamma_{xz}^0 = -\phi_x + \frac{\partial w_0}{\partial x} = -Q\cos(\alpha x)\sin(\beta y) + C\alpha \cos(\alpha x)\sin(\beta y)$

$\quad = (-Q + C\alpha)\cos(\alpha x)\sin(\beta y)$

$\kappa_x = -\frac{\partial \phi_x}{\partial x} = \alpha Q \sin(\alpha x)\sin(\beta y)$

$\kappa_y = -\frac{\partial \phi_y}{\partial y} = \beta R \sin(\alpha x)\sin(\beta y)$

$\kappa_{xy} = -(\frac{\partial \phi_x}{\partial y} + \frac{\partial \phi_y}{\partial x}) = -[\beta Q \cos(\alpha x)\cos(\beta y) + \alpha R \cos(\alpha x)\cos(\beta y)]$

$\quad = (Q\beta + R\alpha)[-\cos(\alpha x)\cos(\beta y)]$

(b) $\quad \epsilon_x = \epsilon_x^0 - z\frac{\partial \phi_y}{\partial y} = (-A + Qz)\alpha \sin(\alpha x)\sin(\beta y)$

$\epsilon_y = \epsilon_y^0 - z\frac{\partial \phi_x}{\partial x} = (-B + Rz)\beta \sin(\alpha x)\sin(\beta y)$

$\epsilon_z = 0$

$\gamma_{xy} = \gamma_{xy}^0 - z(\frac{\partial \phi_x}{\partial y} + \frac{\partial \phi_y}{\partial x}) = [(A\beta + B\alpha) - z(Q\beta + R\alpha)]\cos(\alpha x)\cos(\beta y)$

$\gamma_{yz} = \gamma_{yz}^0 = (-R + C\beta)\sin(\alpha x)\cos(\beta y)$

$\gamma_{xz} = \gamma_{xz}^0 = (-Q + C\alpha)\cos(\alpha x)\sin(\beta y)$

(c) For $A = B = C = 0.001$, $Q = R = 2$, $\alpha = \beta = \pi, x = y = \frac{1}{2}$, and $t = 0.002$, evaluate and plot the strain as a function of the thickness coordinate z.

$\epsilon_x(z) = -(0.001 + 2z)\pi$

$\epsilon_y(z) = -(0.001 + 2z)\pi$

$\gamma_{xy}(z) = 0$

$\gamma_{yz}(z) = 0$

$\gamma_{xz}(z) = 0$

- 6.10

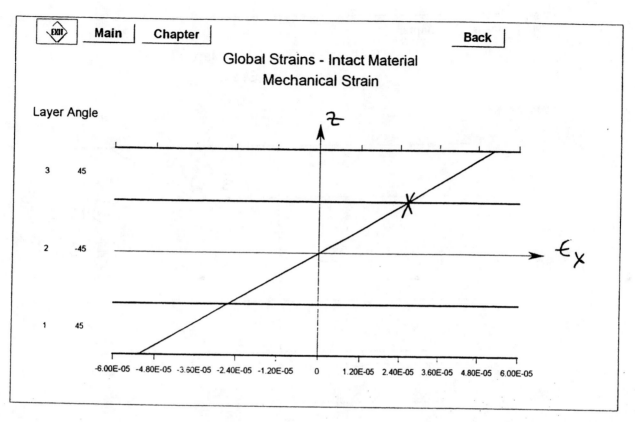

$\varepsilon_x = 26.4 \ (10)^{-6}$

$\varepsilon_y = -18.9 \ (10)^{-6}$

$\gamma_{xy} = -5.14 \ (10)^{-6}$

strains are continuous across interfaces.

5.11

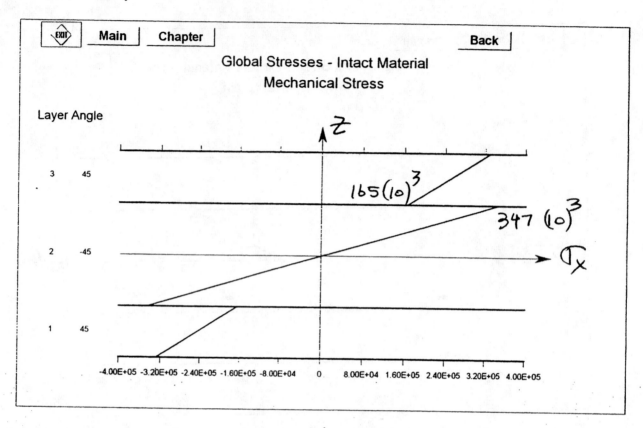

σ_x jumps at the interface because the material changes from -45 to $+45$.

Exercise 6.12

Write down those terms that are equal to zero in the constitutive equations.

(a)   Symmetric laminate
$[B] = 0$

(b)   Balanced anti-symmetric laminate
$A_{16} = A_{26} = D_{16} = D_{26} = 0$

(c)   Cross Ply laminate
$A_{16} = A_{26} = B_{16} = B_{26} = D_{16} = D_{26} = 0$

(d)   Specially Orthotropic laminate
$A_{16} = A_{26} = B_{16} = B_{26} = D_{16} = D_{26} = 0$

Exercise 6.13

Define and explain the following: (a) Regular laminate, (b) Balanced laminate, and (c) Specially orthotropic laminate

(a) Regular: All layers have the same thickness.

(b) Balanced: Each layer at $+\theta$ has a complementary at $-\theta$ with same material and thickness. If $\theta = 0°$, then the complementary is at $90°$.

(c) Specially orthotropic laminate: All layers are at $0°$ or $90°$, that is $Q_{16} = Q_{26} = 0$. This may include $\pm\theta$ fabrics.

Exercise 6.14

Indicate whether the following are continuous or discontinuous functions through the thickness. Also, indicate why some values are continuous or discontinuous and state which locations through the thickness this occurs.

(a) The strains are continuous across interfaces because the layers are perfectly bonded.

(b) The stresses are discontinuous across interfaces because the stiffnesses, represented by $[Q]$, change from one layer to another even if the same material (but different angle) is used.

Exercise 6.15

Demonstrate that an angle-ply laminate has $A_{16} = A_{26} = 0$ using an example laminate of your choice. In which case is $D_{16} = 0$? In which case is $D_{16} \neq 0$?

$D_{16} = 0$ for anti-symmetric angle-ply

$D_{16} \neq 0$ for symmetric angle ply

Exercise 6.16

Given that all of the mid-surface strains are zero, and:

$\kappa_x = 1.0 \ 1/m$

$\kappa_y = 0.5 \ 1/m$

$\kappa_{xy} = 0.25 \ 1/m$

Compute the global strains at $z = -1.27$ mm for the laminate in Exercise 6.5.

$$\left\{ \begin{array}{c} \epsilon_x \\ \epsilon_y \\ \gamma_{xy} \end{array} \right\} = \left\{ \begin{array}{c} \epsilon_x^0 \\ \epsilon_y^0 \\ \gamma_{xy}^0 \end{array} \right\} + z \left\{ \begin{array}{c} \kappa_x \\ \kappa_y \\ \kappa_{xy} \end{array} \right\} = -1.27(10^{-3}) \left\{ \begin{array}{c} 1.0 \\ 0.5 \\ 0.25 \end{array} \right\} = \left\{ \begin{array}{c} -0.00127 \\ -0.000635 \\ -0.0003175 \end{array} \right\}$$

Exercise 6.17

Compute the stresses in global coordinates at $z = -1.27$ mm using the results of exercise 6.16. Note that $\theta = 30°$.

$$\begin{Bmatrix} \sigma_x \\ \sigma_y \\ \sigma_{xy} \end{Bmatrix} = [\overline{Q}]\begin{Bmatrix} \epsilon_x \\ \epsilon_y \\ \gamma_{xy} \end{Bmatrix} = \begin{bmatrix} 83.592 & 25.723 & 41.110 \\ 25.723 & 19.09 & 14.753 \\ 41.110 & 14.753 & 28.025 \end{bmatrix} 10^9 \begin{Bmatrix} -0.00127 \\ -0.000635 \\ -0.0003175 \end{Bmatrix}$$

$$= -\begin{Bmatrix} 0.135 \\ 0.049 \\ 0.070 \end{Bmatrix} \text{GPa} = -\begin{Bmatrix} 135 \\ 49 \\ 70.4 \end{Bmatrix} \text{MPa}$$

Exercise 6.18

Show that a balanced laminate, not necessarily symmetric, has $A_{16} = A_{26} = 0$. In which case is $D_{16} = 0$.

Balanced: For every layer at θ, there is another layer of the same thickness at $-\theta$. If $\theta = 0$, there is layer at $90°$.

Since $Q_{16}(\theta) = -Q_{16}(-\theta)$ it is simple to show that $A_{16} = 0$.

$D_{16} = 0$ only if the laminate is antisymmetric.

Exercise 6.19

Compute the $[Q]$ matrix of a layer reinforced of a balanced bidirectional fabric from Table 2.3. Assume $V_f = \text{0.30}$ 0.37 for E-glass fibers and isophthalic polyester resin.

Use TH27 (from Table 2.3)

$w = 450$ g/m^2 in the $\pm 45°$ direction ($225 \frac{g}{m^2}$ in each direction)

$w = 450$ g/m^2 in the $90°$ direction

From Eq. 4.9, with $\rho_f = 2.5$ g/cc for E-glass fibers:

$t_c = w/(1000\rho_f V_f)$

$t_{45°} = 0.3$ mm $= t_{-45°}$

$t_{90°} = 0.6$ mm

$t_{total} = 1.2$ mm

and Isophthalic from 2.1 and 2.4 and using CAD to perform the micromechanics computat the micromechanics computat we obtain

Material properties for E-Glass/Polyester in Table 1.1.

$E_1 = \text{37.9}$ GPa 28.91 $\quad \nu_{12} = 0.32$

$E_2 = \text{11.3}$ GPa 8.25 $\quad G_{12} = \text{3.3}$ GPa 2.50

For the unidirectional layer:

$$[Q]_{0°} = \begin{bmatrix} \text{38.95} \ 29.78 & \text{3.483} \ 2.72 & 0 \\ \text{3.483} \ 2.72 & \text{11.61} \ 8.50 & 0 \\ 0 & 0 & \text{3.30} \ 2.50 \end{bmatrix} \text{GPa} = [\overline{Q}]$$

← copy #'s from sheet w/ 3 decimal places

$[\overline{Q}]_{0°}$

```
MATRIX : Q        REDUCED STIFFNESS
   29.7846        2.72710          .000000
   2.72710        8.50095          .000000
    .000000        .000000        2.50381
MATRIX : Q*       INTERLAMINAR SHEAR STIFFNESS
   2.31888         .000000
    .000000       2.50381
ANGLE =  45.
```

$\overline{Q}]_{45°}$

```
MATRIX :  Qbar       TRANSFORMED REDUCED STIFFNESS
   13.4387        8.43112         5.32090
    8.43112       13.4387         5.32090
    5.32090        5.32090        8.20782
ANGLE =  45.
MATRIX :  Q*bar      TRANSFORMED REDUCED STIFFNESS
   2.41134         .924650E-01
    .924650E-01   2.41134
MATRIX :  Q        REDUCED STIFFNESS
   29.7846        2.72710          .000000
   2.72710        8.50095          .000000
    .000000        .000000        2.50381
MATRIX : Q*       INTERLAMINAR SHEAR STIFFNESS
   2.31888         .000000
    .000000       2.50381
ANGLE = -45.
```

$\overline{Q}]_{-45°}$

```
MATRIX :  Qbar       TRANSFORMED REDUCED STIFFNESS
   13.4387        8.43112        -5.32090
    8.43112       13.4387        -5.32090
   -5.32090       -5.32090        8.20782
ANGLE = -45.
MATRIX :  Q*bar      TRANSFORMED REDUCED STIFFNESS
   2.41134        -.924650E-01
   -.924650E-01   2.41134
MATRIX :  Q        REDUCED STIFFNESS
   29.7846        2.72710          .000000
   2.72710        8.50095          .000000
    .000000        .000000        2.50381
MATRIX :  Q*      INTERLAMINAR SHEAR STIFFNESS
   2.31888         .000000
    .000000       2.50381
ANGLE =  90.
```

$\overline{Q}]_{90°}$

```
MATRIX :  Qbar       TRANSFORMED REDUCED STIFFNESS
    8.50095        2.72710         -.469161E-16
    2.72710       29.7846          .135012E-14
    -.469161E-16   .135012E-14    2.50381
ANGLE =  90.
MATRIX :  Q*bar      TRANSFORMED REDUCED STIFFNESS
   2.50381         .113233E-16
    .113233E-16   2.31888
```

$$[\overline{Q}]_{45°} = \begin{bmatrix} \text{17.681} & \text{11.081} & 6.833 \\ \text{11.081} & \text{17.681} & 6.833 \\ 6.833 & 6.833 & 10.897 \end{bmatrix} \text{GPa}$$

copy from sheet

$$[\overline{Q}]_{-45°} = \begin{bmatrix} \text{17.681} & \text{11.081} & -6.833 \\ \text{11.081} & \text{17.681} & -6.833 \\ -6.833 & -6.833 & \text{10.897} \end{bmatrix} \text{GPa}$$

copy from sheet

For the ±45 fabric, since both fiber orientations are intertwined:

$$[\overline{Q}]_{\pm 45°} = \tfrac{1}{2}\left[[\overline{Q}]_{45°} + [\overline{Q}]_{-45°}\right] = \begin{bmatrix} 12.438 & 8.431 & 0 \\ 8.431 & 13.438 & 0 \\ 0 & 0 & 8.207 \end{bmatrix} \text{GPa}$$

For the 90° layer:

$$[\overline{Q}]_{90°} = \begin{bmatrix} \text{11.61} & 3.483 & 0 \\ -3.483 & 38.95 & 0 \\ 0 & 0 & \text{3.30} \end{bmatrix} \text{GPa}$$

copy from sheet

For the whole fabric, since the three types of fiber orientation can be assumed to be intertwined through the thickness

$$[\overline{Q}]_{TH27} = (1/t_{90} + t_{45} + t_{-45})\left[t_{90}[\overline{Q}]_{90} + t_{45}[\overline{Q}]_{45} + t_{-45}[\overline{Q}]_{-45}\right]$$

$$[\overline{Q}]_{TH27} = \begin{bmatrix} 10.910 & 5.579 & 0 \\ 5.579 & 21.612 & 0 \\ 0 & 0 & 5.356 \end{bmatrix} \text{GPa}$$

This result can also be obtained modeling the laminate as a [45/-45/90₂]ₛ with a layer thickness of 1.2/8 mm and dividing the entries in the [A] matrix by the laminate thickness $t = 1.2$ mm.

Exercise 6.20

Once $[\overline{Q}]_{TH27}$ is known from Exercise 6.19, the usual computations can be used to get the A, B, and D matrices.

$$[A] = t[\overline{Q}]_{TH27}$$
$$[B] = 0$$
$$[D] = \tfrac{t^3}{12}[\overline{Q}]_{TH27} = (t^3/12)[\overline{Q}]_{TH27}$$

with $t = 1.2$ mm.

This result can also be obtained modeling the laminate as a [45/-45/90₂]ₛ with a layer thickness of 0.15 mm to get the laminate moduli.

$E_x = \text{12.78 GPa}\ 9.535$ $\nu_{xy} = 0.258\ 9$

$E_y = \text{24.7 GPa}\ 18.77$ $G_{xy} = \text{7.123 GPa}\ 5.342$

Note that a [45/-45/90₂]ₛ models perfectly the inplane behavior of the TH27 as evidenced by

$r_N = r_B = 0$. Then, use these moduli to model TH27 as a $0°$ single layer 1.2 mm thick in CADEC.

Exercise 6.21

Temperature load for Aluminum: $\quad P_{Al} = \alpha \Delta T A E = 23(10^{-6})(10)(0.002)71(10^9) = 32660$ $\frac{Nm}{m}$

Temperature load for Copper: $\quad P_{Cu} = \alpha \Delta T A E = 17(10^{-6})(10)(0.002)119(10^9) = 40460$ $\frac{Nm}{m}$

$N_x = P_{Al} + P_{Cu} = 32660 + 40460 = 73120 \, \frac{Nm}{m} = N_y$

$M_x = P_{Cu}(0.001) + P_{Al}(0.001) = 32660(0.001) + 40460(0.001) = 7.8 \, \frac{N}{m} = M_y$

Equations (6.15) $\quad [N] = [A][\epsilon^0] + [B][\kappa]$
$\qquad\qquad\qquad [M] = [B][\epsilon^0] + [D][\kappa]$

$$
\begin{Bmatrix} 73120 \\ 73120 \\ 0 \end{Bmatrix} = \begin{bmatrix} 4.176 & 1.235 & 0 \\ 1.253 & 5.569 & 0 \\ 0 & 0 & 1.949 \end{bmatrix}(10^8)\begin{Bmatrix} \epsilon_x^0 \\ \epsilon_y^0 \\ \gamma_{xy}^0 \end{Bmatrix}\begin{bmatrix} -10.56 & -3.166 & 0 \\ -3.166 & -10.56 & 0 \\ 0 & 0 & -3.692 \end{bmatrix}(10
$$

$$
\begin{Bmatrix} 7.8 \\ 7.8 \\ 0 \end{Bmatrix} = \begin{bmatrix} -10.56 & -3.166 & 0 \\ -3.166 & -10.56 & 0 \\ 0 & 0 & -3.692 \end{bmatrix}(10^4)\begin{Bmatrix} \epsilon_x^0 \\ \epsilon_y^0 \\ \gamma_{xy}^0 \end{Bmatrix} + \begin{bmatrix} 5.569 & 1.671 & 0 \\ 1.671 & 5.569 & 0 \\ 0 & 0 & 1.949 \end{bmatrix}(10
$$

solving: $\quad \begin{Bmatrix} \epsilon_x^0 \\ \epsilon_y^0 \\ \gamma_{xy}^0 \end{Bmatrix} = \begin{Bmatrix} 0.0001443 \\ 0.0001443 \\ 0 \end{Bmatrix} \qquad \begin{Bmatrix} \kappa_x \\ \kappa_y \\ \kappa_{xy} \end{Bmatrix} = \begin{Bmatrix} 0.03814 \\ 0.03814 \\ 0 \end{Bmatrix}\frac{1}{m}$

Exercise 6.22

$\epsilon_x = \epsilon_x^0 + z\kappa_x = 0.0001443 + z(0.03814) = \epsilon_y$

At $\quad z = 0.002 \qquad\qquad \epsilon_x = \epsilon_y = 0.0001443 + (0.002)(0.03814) = 2.206(10^{-4})$

$\qquad z = 0 \qquad\qquad\quad \epsilon_x = \epsilon_y = 0.0001443 = 1.443(10^{-4})$

$\qquad z = -0.002 \qquad\quad \epsilon_x = \epsilon_y = 0.0001443 - (0.002)(0.03814) = 0.6802(10^{-4})$

Stresses for Aluminum:

At $z = 0.002$

$$
\begin{Bmatrix} \sigma_x \\ \sigma_y \\ \sigma_{xy} \end{Bmatrix} = \begin{bmatrix} 78.02 & 2.341 & 0 \\ 2.341 & 78.02 & 0 \\ 0 & 0 & 27.31 \end{bmatrix}(10^9)\begin{Bmatrix} \epsilon_x - \epsilon_T \\ \epsilon_y - \epsilon_T \\ 0 \end{Bmatrix}
$$

$\epsilon_T = \alpha \Delta T = 23(10^{-6})(10) = 2.3(10^{-4})$

$$\left\{ \begin{array}{c} \sigma_x \\ \sigma_y \\ \sigma_{xy} \end{array} \right\} = \left\{ \begin{array}{c} -9.534 \\ -9.534 \\ 0 \end{array} \right\} (10^5) \text{ Pa}$$

At $z = 0$

$$\left\{ \begin{array}{c} \sigma_x \\ \sigma_y \\ \sigma_{xy} \end{array} \right\} = \left[\begin{array}{ccc} 78.02 & 2.341 & 0 \\ 2.341 & 78.02 & 0 \\ 0 & 0 & 27.31 \end{array} \right] (10^9) \left\{ \begin{array}{c} \epsilon_x^0 - \epsilon_T \\ \epsilon_y^0 - \epsilon_T \\ 0 \end{array} \right\}$$

$$\left\{ \begin{array}{c} \sigma_x \\ \sigma_y \\ \sigma_{xy} \end{array} \right\} = \left\{ \begin{array}{c} -86.93 \\ -86.93 \\ 0 \end{array} \right\} (10^5) \text{ Pa}$$

Stresses for Copper:

At $z = 0$

$$\left\{ \begin{array}{c} \sigma_x \\ \sigma_y \\ \sigma_{xy} \end{array} \right\} = \left[\begin{array}{ccc} 130.8 & 39.24 & 0 \\ 39.24 & 130.8 & 0 \\ 0 & 0 & 45.77 \end{array} \right] (10^9) \left\{ \begin{array}{c} \epsilon_x - \epsilon_T \\ \epsilon_y - \epsilon_T \\ 0 \end{array} \right\}$$

$$\left\{ \begin{array}{c} \sigma_x \\ \sigma_y \\ \sigma_{xy} \end{array} \right\} = \left\{ \begin{array}{c} -43.70 \\ -43.70 \\ 0 \end{array} \right\} (10^5) \text{ Pa}$$

At $z = -0.002$

$$\left\{ \begin{array}{c} \sigma_x \\ \sigma_y \\ \sigma_{xy} \end{array} \right\} = \left[\begin{array}{ccc} 13.08 & 3.924 & 0 \\ 3.924 & 13.08 & 0 \\ 0 & 0 & 4.577 \end{array} \right] (10^9) \left\{ \begin{array}{c} \epsilon_x - \epsilon_T \\ \epsilon_y - \epsilon_T \\ 0 \end{array} \right\}$$

$$\left\{ \begin{array}{c} \sigma_x \\ \sigma_y \\ \sigma_{xy} \end{array} \right\} = \left\{ \begin{array}{c} -275.4 \\ -275.4 \\ 0 \end{array} \right\} (10^5) \text{ Pa}$$

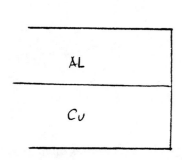

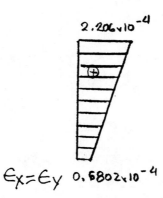

$$2.206 \times 10^{-4}$$

$E_x = E_y \quad 0.6802 \times 10^{-4}$

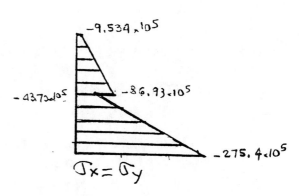

-9.534×10^5

$-4373 \times 10^5 \quad -86.93 \times 10^5$

-275.4×10^5

$\sigma_x = \sigma_y$

Exercise 6.23

[30/ − 45/45/ − 30]	balanced anti-symmetric
[30/0/ − 45/45/90/ − 30]	balanced anti-symmetric
[30/0/ − 30/ − 30/90/30]	balanced
[±30/ ± 45]	angle-ply
[0/90/ ± 45]	quasi-isotropic

Exercise 6.24 (See Table 2.3)

M1500-CSM	isotropic
CM1808 [CSM/±45]	balanced and specially orthotropic
TVM3408 [CSM/±45/0]	specially orthotropic
XM2408 [CSM/±45]	balanced and specially orthotropic
UM1608 [CSM/0]	specially orthotropic

A FEW MORE

They all have $A_{16} = B_{16} = D_{16} = 0$ because the ±45 and 0/90 layers are fabrics with $Q_{16} = Q_{26} = 0$. See Example 6.5.

Exercise 6.25

Laminate	Structure	Description
M1500	[CSM]	isotropic
C24	[0/90]	balanced, specially orthotropic
CM1808	[CSM/0/90]	balanced, specially orthotropic
Q30	[±45/0/90]	specially orthotropic
CM1810	[CSM/0/90]	balanced, specially orthotropic
QM5620	[CSM/0/90/±45]	balanced, specially orthotropic
TH27	[±45/90]	specially orthotropic
TVM3408	[CSM/±45/0]	specially orthotropic
UM1810	[CSM/0]	specially orthotropic
UM1608	[CSM/0]	specially orthotropic
XM2408	[CSM/±45]	balanced, specially orthotropic

All have $A_{16} = B_{16} = D_{16} = 0$ because all are specially orthotropic.

Exercise 6.26

Use the properties from the laminate in Exercise 6.8 and input into CADEC to find the laminate moduli. See CADEC output.

6.26

LaminateModuli: Inplane   Bending

$E_x =$ `4.092E+10` $E_x^b =$ `6.745E+10`

$E_y =$ `4.092E+10` $E_y^b =$ `1.431E+10`

$G_{xy} =$ `2.066E+9` $G_{xy}^b =$ `2.070E+9`

$\nu_{xy} =$ `0.046` $\nu_{xy}^b =$ `0.131`

$r_N =$ `0.000` $r_M =$ `0.000`

$r_B =$ `0.`

Material Stresses - Intact Material
Mechanical Stress

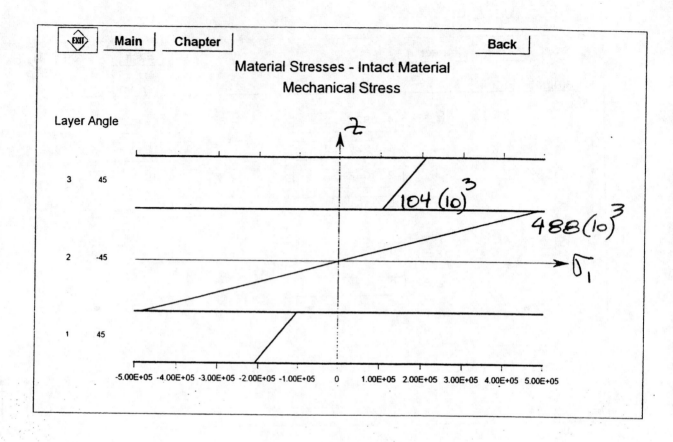

Exercise 6.27

$[0_2/90/\pm 45]_s$ laminate of E-glass polyester with properties given in Table 1.1

10 layers total

4 layers at 0°: $\alpha = 0.4$

2 layers 90°: $\beta = 0.2$

2 layers at 45° and 2 at -45°: $\gamma = 0.4$

From carpet plots

     $E_x = 23$ GPa   (CADEC yields 22.78 GPa)

To get E_y use: $\alpha = 0.2, \beta = 0.4, \gamma = 0.4$

     $E_y = 17.75$ GPa   (CADEC = 17.79 GPa)

     $v_{xy} = 0.35$ (CADEC = 0.335 GPa)

     $G_{xy} = 6.4$ GPa   (CADEC = 6.329 GPa)

See also CADEC output.

6.27

EXIT | Main | Chapter | Load Laminate | Save Laminate | Back

(1/2)

Filename: C:\BOOK\PROBLEMS\P6_26.DEF

Laminate Definition for: Problem 6.26, mm, MPa, glass-epoxy of carpet plots (carpet.dat

Plot

Number of Layers: 7 Number of Materials: 1 Total Thickness 1.000E+1

Layer Thicknesses: 2 1 1 2 1 1 2

Layup Angles: 0 90 45 -45 45 90 0

Layer Materials: 1 1 1 1 1 1 1

Loading:

Nx	1	Mx	0	Qx	0
Ny	0	My	0	Qy	0
Nxy	0	Mxy	0		

Temperature Change: 0

Moisture Concentration Change: 0

Safety Factor: 1

Use Current Laminate Properties ➡

Set Up Laminate Mat. Properties

6.27

EXIT **Main**

Run Laminate Analysis | **Back** | ← | →

Unidirectional | **CSM**

E_1 `3.78725E+01` E_2 `1.12711E+01` G_{12} `3.33156E+00` G_{23} `3.03426E+00`

G_{13} `3.33156E+00` V_{12} `0.30000` F_{1t} `9.03035E-01` F_{1c} `0.3788109689`

F_{2t} `4.36240E-02` F_{2c} `6.73614E-02` F_6 `4.36240E-02` F_4 `7.59000E-02`

F_5 `4.36240E-02` F_{12} `-0.5` α_1 `6.50423E-06` α_2 `2.20427E-05`

β_1 `8.30418E-02` β_2 `1.25159E+00`   df `0.2`

6.27

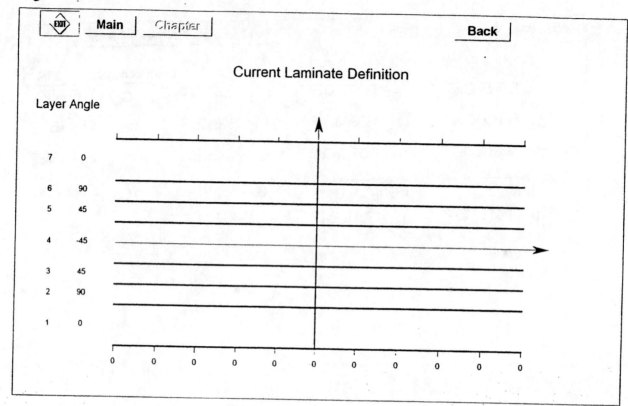

Current Laminate Definition

Layer Angle

7	0
6	90
5	45
4	-45
3	45
2	90
1	0

0 0 0 0 0 0 0 0 0 0 0

6.27

LaminateModuli: Inplane   Bending

$E_x =$ 2.278E+1 $E^b_x =$ 3.243E+1

$E_y =$ 1.779E+1 $E^b_y =$ 1.565E+1

$G_{xy} =$ 6.369E+0 G^b_{xy} 3.810E+0

$\nu_{xy} =$ 0.333 $\nu^b_{xy} =$ 0.244

$r_N =$ 0.000 $r_M =$ 0.1

$r_B =$ 0.000

6.27

Filename: C:\BOOK\PROBLEMS\CARPET.DAT (1/2)

<u>Layer Material Properties for:</u> E-GLASS/POLYESTER(ISOPHTALIC)

Fiber Properties:

E_f = 72.345

V_f = 0.22

G_f = 2.96496E+01

α_f = 5.4e-6

k_f = 1.05

V_f = .5

a/b = 1

ρ_f = 2.5

σ_{fa} = 1.725

ρ_c = 1.85000E+00

Ω[deg] = 3.53

Matrix Properties:

E_m = 3.4

V_m = 0.38

G_m = 1.23188E+00

α_m = 30.e-6

k_m = .2

β_m = 0.6

ρ_m = 1.2

σ_{mu} = 75.9e-3

σ_{muc} = 117.2e-3

τ_{mu} = 75.9e-3

V_v = .01

Elastic Properties:

E_1 = 3.78725E+01

E_2 = 1.12711E+01

G_{12} = 3.33156E+00

G_{23} = 3.03426E+00

V_{12} = 0.30000

V_{23} = 0.57335

Transport Properties:

k_1 = 6.25000E-01

k_2 = 3.36000E-01

6.30

6.27

Thermal and Hygroscopic Expansion Coefficients:

α_1 = 6.50423E-06

α_2 = 2.20427E-05

β_1 = 8.30418E-02

β_2 = 1.25159E+00

Continuous Strand Mat Properties:

E = 2.12466E+01

G = 7.55184E+00

0.406719

1.02334E-05

3.89306E-01

F_{csm-t} = F_{csm-c} = 1.39697E-01

F_{csm-6} = 6.98485E-02

F_{csm-4} = F_{csm-5} = 7.59000E-02

Unidirectional Layer Strengths:

F_{1t} = 9.03035E-01

F_{1c} = 0.37881096899

F_{2t} = 4.36240E-02

F_{2c} = 6.73614E-02

F_6 = F_5 = 4.36240E-02

F_4 = 7.59000E-02

Solutions Chapter 7

Exercise 7.1

Consider a $[\pm 45]_T$ laminate with

$E_1 = 137.8$ GPa $\qquad E_2 = 10$ GPa

$G_{12} = 5.236$ GPa $\qquad v_{12} = 0.3$

$F_{1t} = 826.8$ MPa $\qquad F_{1c} = 482.3$ MPa

$F_{2t} = F_{2c} = 68.9$ MPa $\quad F_6 = 241.15$ MPa

$t_k = 0.127$ mm subject to $N_x = N_y = 1.0$ N/mm.

Determine the strength ratio at the bottom surface of the laminate using:

(a) Maximum stress

(b) Maximum strain

(c) Tsai-Wu

Using CADEC to find the stresses and strains:

Layer	σ_1	σ_2
2 top	-4.84	2.62
2 bottom	16.4	1.52
1 top	16.4	1.52
1 bottom	-4.84	2.62

MPa

Layer	ϵ_1	ϵ_2
2 top	$-.408(10^{-4})$	$.273(10^{-3})$
2 bottom	$.116(10^{-3})$	$.116(10^{-3})$
1 top	$.116(10^{-3})$	$.116(10^{-3})$
1 bottom	$-.408(10^{-4})$	$.273(10^{-3})$

(a) Using maximum stress Eq. 7.3

$R_1 = 99.6, R_2 = 26.3, R_6 = \infty$

$R = \min(R_i) = F_{2t}/\sigma_2 = 68.9/2.62 = 26.3$

(b) First estimate strains to failure using Eq. 7.5. Using maximum strain criterion Eq. 7.4

$R = \min(R_i) = \epsilon_{2t}/\epsilon_2 = 0.00689/0.000273 = 25.2$

(c) Using Eq. 7.16

$f_1 = 1/F_{1t} - 1/F_c = -8.639(10^{-4}) \; \frac{1}{\text{MPa}}$

$f_2 = 1/F_{2t} - 1/F_{2c} = 0$

$f_{11} = 1/F_{1t}F_{1c} = 2.508(10^{-6}) \; \frac{1}{(\text{MPa})^2}$

$f_{22} = 1/F_{2t}F_{2c} = 2.11(10^{-4}) \; \frac{1}{(\text{MPa})^2}$

Use Eq. 7.17 and $f^*_{12} = -1/2$ to find f_{12}.

$f_{12} = (-1/2)1/(F_{1t})^2 = f^*_{12}/(F_{1t})^2 = -7.314(10^{-7}) \; \frac{1}{(\text{MPa})^2}$

Then use Eq. 7.20-7.21

$$a = f_{11}\sigma_1^2 + f_{22}\sigma_2^2 + 2f_{12}\sigma_1\sigma_2 = 0.001526$$
$$b = \frac{1}{2}(f_1\sigma_1) = 0.002091$$
$$R = \frac{1}{a}\left[-b + \sqrt{b^2 + a}\right] = 24.3$$

See also CADEC output.

Exercise 7.2

$[0/45/-45]_s$ laminate subject to $N_x = N_y = 1000 \frac{N}{mm}$. Compute $[A]$ and $[D]$ matrices for the intact laminate. Each layer is 2.5 mm thick.

$E_1 = 204.6$ GPa $\qquad$ $E_2 = 20.46$ GPa
$G_{12} = 6.89$ GPa $\qquad$ $v_{12} = 0.3$
$F_{1t} = 826.8$ MPa $\qquad$ $F_{1c} = 482.3$ MPa
$F_{2t} = F_{2c} = 68.9$ MPa $\quad$ $F_6 = 241.15$ MPa
Using CADEC:

$$[A] = \begin{bmatrix} 1.70(10^6) & 5.61(10^5) & 0 \\ 5.61(10^5) & 7.71(10^5) & 0 \\ 0 & 0 & 5.71(10^5) \end{bmatrix} \text{GPa mm}$$

$$[D] = \begin{bmatrix} 4.64(10^7) & 5.641(10^6) & 2.903(10^6) \\ 5.641(10^6) & 9.650(10^6) & 2.903(10^6) \\ 2.903(10^6) & 2.903(10^6) & 5.836(10^6) \end{bmatrix} \text{GPa mm}^3$$

See CADEC output.

7.1/1

Maximum Stress Failure Criterion

Ply	Angle	Rint-Top	Rint-Bot	Rdeg-Top	Rdeg-Bot
2	-45.	26.3(2)	45.4(2)	39.8(1)	31.2(1)
1	45.	45.4(2)	26.3(2)	31.2(1)	39.8(1)

FPF	FF
26.3	31.2

7.1/2

Maximum Strain Failure Criterion

Ply	Angle	Rint-Top	Rint-Bot	Rdeg-Top	Rdeg-Bot
2	-45.	25.2(2)	51.7(1)	39.6(1)	31.2(1)
1	45.	51.7(1)	25.2(2)	31.2(1)	39.6(1)

FPF	FF
25.2	31.2

7.1/3

Tsai-Wu Failure Criterion

Ply	Angle	Rint-Top	Rint-Bot	Rdeg-Top	Rdeg-Bot
2	-45.	24.3	36.7	39.8	31.2
1	45.	36.7	24.3	31.2	39.8

FPF	FF
24.3	31.2

7.2 1/3

Filename: C:\BOOK\PROBLEMS\P7_2.DEF

Laminate Definition for: Problem 7.2, mm, MPa, Nx=Ny=1000[N/mm]

Plot

Number of Layers: 5   Number of Materials: 1   Total Thickness | 1.500E+1

Layer Thicknesses: 2.5 2.5 5 2.5 2.5

Layup Angles: 0 45 -45 45 0

Layer Materials: 1 1 1 1 1

Loading: Nx 1000   Mx 0   Qx 0
            Ny 1000   My 0   Qy 0
            Nxy 0   Mxy 0

Temperature Change: 0

Moisture Concentration Change: 0   Use Current Laminate Properties ➡

Safety Factor: 1   Set Up Laminate Mat. Properties

| EXIT   Main | | | Run Laminate Analysis | Back | ◄ | ► |

Unidirectional   **CSM**

E_1	204.6e3	E_2	20.46e3	G_{12}	6.89e3	G_{23}	1.0
G_{13}	6.89e3	ν_{12}	0.30000	F_{1t}	826.8	F_{1c}	482.3
F_{2t}	68.9	F_{2c}	68.9	F_6	241.15	F_4	241.15
F_5	241.15	F_{12}	-0.5	α_1	6.50423E-06	α_2	2.20427E-05
β_1	8.30418E-02	β_2	1.25159E+00	df	0.2		

Analysis for Intact Material

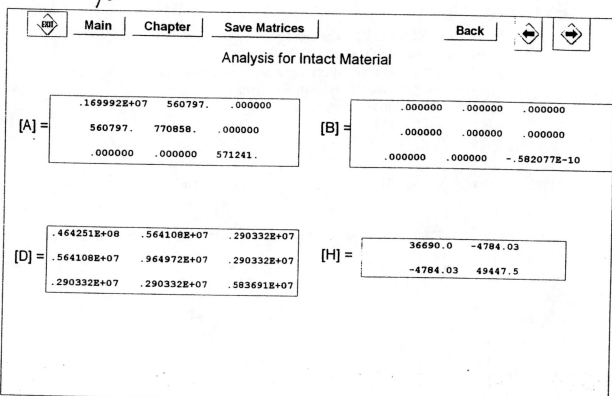

$$[A] = \begin{bmatrix} .169992E+07 & 560797. & .000000 \\ 560797. & 770858. & .000000 \\ .000000 & .000000 & 571241. \end{bmatrix}$$

$$[B] = \begin{bmatrix} .000000 & .000000 & .000000 \\ .000000 & .000000 & .000000 \\ .000000 & .000000 & -.582077E-10 \end{bmatrix}$$

$$[D] = \begin{bmatrix} .464251E+08 & .564108E+07 & .290332E+07 \\ .564108E+07 & .964972E+07 & .290332E+07 \\ .290332E+07 & .290332E+07 & .583691E+07 \end{bmatrix}$$

$$[H] = \begin{bmatrix} 36690.0 & -4784.03 \\ -4784.03 & 49447.5 \end{bmatrix}$$

Exercise 7.3

Using the results of Exercise 7.2 to compute:
- (a) $\epsilon_x, \epsilon_y, \gamma_{xy}$ at the interface between layers
- (b)   stresses $\sigma_x, \sigma_y, \sigma_{xy}$ on all layers, right next to the interface
- (c) $\epsilon_1, \epsilon_2, \gamma_6$ using the transformation equations
- (d)   stresses in the material coordinates

(a)
$$\left\{ \begin{array}{c} \epsilon_x^0 \\ \epsilon_y^0 \\ \gamma_{xy}^0 \end{array} \right\} = [A]^{-1} \left\{ \begin{array}{c} N_x \\ N_y \\ N_{xy} \end{array} \right\} = \left\{ \begin{array}{c} 0.211 \\ 1.144 \\ 0 \end{array} \right\} (10^{-3})$$

The mid-surface strains are the same throughout because there is no curvature in the laminate.

(b)
$$\left\{ \begin{array}{c} \sigma_x \\ \sigma_y \\ \sigma_{xy} \end{array} \right\} = [\bar{Q}] \left\{ \begin{array}{c} \epsilon_x \\ \epsilon_y \\ \gamma_{xy} \end{array} \right\} \qquad [Q] = \begin{bmatrix} 2.06(10^5) & 6193 & 0 \\ 6193 & 2.06(10^4) & 0 \\ 0 & 0 & 6890 \end{bmatrix}$$

for $\theta = 0, \bar{Q} = Q$

$$\left\{ \begin{array}{c} \sigma_x \\ \sigma_y \\ \sigma_{xy} \end{array} \right\} = \left\{ \begin{array}{c} 50.63 \\ 24.92 \\ 0 \end{array} \right\} \text{MPa}$$

for $\theta = +45$

$$[\bar{Q}] = \begin{bmatrix} 66785 & 53005 & 46475 \\ 53005 & 66785 & 46475 \\ 46475 & 46475 & 53701 \end{bmatrix} \qquad \left\{ \begin{array}{c} \sigma_x \\ \sigma_y \\ \sigma_{xy} \end{array} \right\} = \left\{ \begin{array}{c} 74.71 \\ 87.575 \\ 62.965 \end{array} \right\} \text{MPa}$$

For the $\theta = -45$ layers, the answers are the same except for σ_{xy}, which is -62.965 MPa.

The final distribution is:

Layer	ϵ_1	ϵ_2	γ_6
5 top	0.211	1.14	0
5 bottom	0.211	1.14	0
4 top	0.677	0.677	0.933
4 bottom	0.677	0.677	0.933
3 top	0.677	0.677	-0.933
3 bottom	0.677	0.677	-0.933
2 top	0.677	0.677	0.933
2 bottom	0.677	0.677	0.933
1 top	0.211	1.14	0
1 bottom	0.211	1.14	0

(10^{-3})

(d)
$$\left\{ \begin{array}{c} \sigma_1 \\ \sigma_2 \\ \sigma_6 \end{array} \right\} = [T] \left\{ \begin{array}{c} \sigma_x \\ \sigma_y \\ \sigma_{xy} \end{array} \right\}$$

For $\theta = 0$, the values of the two sets of stresses are equal.

For $\theta = \pm 45$:
$$\left\{ \begin{array}{c} \sigma_1 \\ \sigma_2 \\ \sigma_6 \end{array} \right\} = \left\{ \begin{array}{c} 144.1 \\ 18.18 \\ \pm 6.43 \end{array} \right\} \text{MPa}$$

7.3. (d) continues after CADEC results

1.3 (a)

		ϵ_x	ϵ_y			
5	TOP	.211E-03	.114E-02	.000	.000	.000
5	BOT	.211E-03	.114E-02	.000	.000	.000
4	TOP	.211E-03	.114E-02	.000	.000	.000
4	BOT	.211E-03	.114E-02	.000	.000	.000
3	TOP	.211E-03	.114E-02	.000	.000	.000
3	BOT	.211E-03	.114E-02	.000	.000	.000
2	TOP	.211E-03	.114E-02	.000	.000	.000
2	BOT	.211E-03	.114E-02	.000	.000	.000
1	TOP	.211E-03	.114E-02	.000	.000	.000
1	BOT	.211E-03	.114E-02	.000	.000	.000

1.5 (b)

		σ_x	σ_y			
5	TOP	50.6	24.9	.000	.000	.000
5	BOT	50.6	24.9	.000	.000	.000
4	TOP	74.7	87.5	62.9	.000	.000
4	BOT	74.7	87.5	62.9	.000	.000
3	TOP	74.7	87.5	-62.9	.000	.000
3	BOT	74.7	87.5	-62.9	.000	.000
2	TOP	74.7	87.5	62.9	.000	.000
2	BOT	74.7	87.5	62.9	.000	.000
1	TOP	50.6	24.9	.000	.000	.000
1	BOT	50.6	24.9	.000	.000	.000

7.3(c)

		ε_1	ε_2	γ_6		
5	TOP	.211E-03	.114E-02	.000	.000	.000
5	BOT	.211E-03	.114E-02	.000	.000	.000
4	TOP	.677E-03	.677E-03	.933E-03	.000	.000
4	BOT	.677E-03	.677E-03	.933E-03	.000	.000
3	TOP	.677E-03	.677E-03	-.933E-03	.000	.000
3	BOT	.677E-03	.677E-03	-.933E-03	.000	.000
2	TOP	.677E-03	.677E-03	.933E-03	.000	.000
2	BOT	.677E-03	.677E-03	.933E-03	.000	.000
1	TOP	.211E-03	.114E-02	.000	.000	.000
1	BOT	.211E-03	.114E-02	.000	.000	.000

7.3(d)

		σ_1	σ_2	σ_6		
5	TOP	50.6	24.9	.000	.000	.000
5	BOT	50.6	24.9	.000	.000	.000
4	TOP	144.	18.2	6.43	.000	.000
4	BOT	144.	18.2	6.43	.000	.000
3	TOP	144.	18.2	-6.43	.000	.000
3	BOT	144.	18.2	-6.43	.000	.000
2	TOP	144.	18.2	6.43	.000	.000
2	BOT	144.	18.2	6.43	.000	.000
1	TOP	50.6	24.9	.000	.000	.000
1	BOT	50.6	24.9	.000	.000	.000

5	TOP	50.6	24.9

7.15

7.3.(d) continued.

The final distribution is:

Layer	σ_1	σ_2	σ_6
5 top	50.6	24.9	0
5 bottom	50.6	24.9	0
4 top	144.1	18.18	6.43
4 bottom	144.1	18.18	6.43
3 top	144.1	18.18	-6.43
3 bottom	144.1	18.18	-6.43
2 top	144.1	18.18	6.43
2 bottom	144.1	18.18	6.43
1 top	50.6	24.9	0
1 bottom	50.6	24.9	0

(MPa)

Exercise 7.4

Using the stresses found in Exercise 7.3, evaluate the strength ratio R on top and bottom of each layer using:

(a)   Tsai-Wu
(b)   Maximum strain
(c)   Maximum stress

Minimum value of R computed through the thickness corresponds to the first ply failure strength ratio.

(a)   for $\theta = 0°$

$$\left\{ \begin{array}{c} \sigma_1 \\ \sigma_2 \\ \sigma_6 \end{array} \right\} = \left\{ \begin{array}{c} 50.6 \\ 24.92 \\ 0 \end{array} \right\} \text{ MPa}$$

for $\theta = 45°$

$$\left\{ \begin{array}{c} \sigma_1 \\ \sigma_2 \\ \sigma_6 \end{array} \right\} = \left\{ \begin{array}{c} 144.1 \\ 18.18 \\ 6.43 \end{array} \right\} \text{ MPa}$$

for $\theta = -45°$

$$\left\{ \begin{array}{c} \sigma_1 \\ \sigma_2 \\ \sigma_6 \end{array} \right\} = \left\{ \begin{array}{c} 144.1 \\ 18.18 \\ -6.43 \end{array} \right\} \text{ MPa}$$

7.16

$$f_1 = 1/F_{1t} - 1/F_{1c} = -8.639(10^{-4}) \frac{1}{MPa}$$
$$f_2 = 1/F_{2t} - 1/F_{2c} = 0$$
$$f_{11} = 1/(F_{1t}F_{1c}) = 2.508(10^{-6}) \frac{1}{(MPa)^2}$$
$$f_{22} = 1/(F_{2t}F_{2c}) = 2.11(10^{-4}) \frac{1}{(MPa)^2}$$
$$f_{66} = 1/(F_6)^2 = 1.720(10^{-5}) \frac{1}{(MPa)^2}$$

$$f_{12} = -1/2(F_{1t})^2 = f_{12}^*/(F_{1t})^2 = -7.314(10^{-7}) \frac{1}{(MPa)^2}$$

for $\theta = 0°$
$$a = f_{11}\sigma_1^2 + f_{22}\sigma_2^2 + 2f_{12}\sigma_1\sigma_2 = 0.1083$$
$$b = \tfrac{1}{2}(f_1\sigma_1) = -0.0219$$
$$R = \tfrac{1}{a}\left[-b + \sqrt{b^2 + a}\,\right] = 2.884$$

for $\theta = 45°$
$$a = f_{11}\sigma_1^2 + f_{22}\sigma_2^2 + 2f_{12}\sigma_1\sigma_2 = 0.1180$$
$$b = \tfrac{1}{2}(f_1\sigma_1) = -0.0622$$
$$R = \tfrac{1}{a}\left[-b + \sqrt{b^2 + a}\,\right] = 3.486$$

for $\theta = -45°$, $R = 3.486$ because σ_6 is squared.

(b)   Maximum strain criterion
for $\theta = 0°$
$$\left\{ \begin{array}{c} \epsilon_1 \\ \epsilon_2 \\ \gamma_6 \end{array} \right\} = \left\{ \begin{array}{c} 0.211 \\ 1.144 \\ 0 \end{array} \right\} (10^{-3})$$

for $\theta = \pm 45°$
$$\left\{ \begin{array}{c} \epsilon_1 \\ \epsilon_2 \\ \gamma_6 \end{array} \right\} = \left\{ \begin{array}{c} 0.677 \\ 0.677 \\ \pm 0.933 \end{array} \right\} (10^{-3})$$

for $\theta = 0°$
$$R = \epsilon_{2t}/\epsilon_2 = 2.944$$

for $\theta = \pm 45°$
$$R = \epsilon_{2t}/\epsilon_2 = 4.972$$

(c)   Maximum Stress Criterion

for $\theta = 0°$

$R = F_{2t}/\sigma_2 = 2.765$

for $\theta = \pm45°$

$R = F_{2t}/\sigma_2 = 3.79$

Exercise 7.5

Recompute all results from Exercise 7.4 for a material which has been loaded beyond the FPF load.

$f_d = 0.2$

Refer to CADEC for the results.

The final distribution is:

Layer	σ_x	σ_y	σ_{xy}
5 top	50.6	24.9	0
5 bottom	50.6	24.9	0
4 top	74.7	87.6	62.9
4 bottom	74.7	87.6	62.9
3 top	74.7	87.6	-62.9
3 bottom	74.7	87.6	-62.9
2 top	74.7	87.6	62.9
2 bottom	74.7	87.6	62.9
1 top	50.6	24.9	0
1 bottom	50.6	24.9	0

(MPa)

(c)
$$\left\{ \begin{array}{c} \epsilon_1 \\ \epsilon_2 \\ \gamma_6 \end{array} \right\} = [T] \left\{ \begin{array}{c} \epsilon_x \\ \epsilon_y \\ \gamma_{xy} \end{array} \right\}$$

For $\theta = 0$ the two sets of strains are equal.

For $\theta = 45$:
$$\left\{ \begin{array}{c} \epsilon_1 \\ \epsilon_2 \\ \gamma_6 \end{array} \right\} = \left\{ \begin{array}{c} 6.774 \\ 6.774 \\ 9.32 \end{array} \right\} (10^{-4})$$

For $\theta = -45$:
$$\left\{ \begin{array}{c} \epsilon_1 \\ \epsilon_2 \\ \gamma_6 \end{array} \right\} = \left\{ \begin{array}{c} 6.774 \\ 6.774 \\ -9.32 \end{array} \right\} (10^{-4})$$

7.4 & 7.5

Tsai-Wu Failure Criterion

Ply	Angle	Intact Laminate		Degraded Laminate	
		Rint-Top	Rint-Bot	Rdeg-Top	Rdeg-Bot
5	0.	2.88	2.88	51.2	51.2
4	45.	3.48	3.48	4.48	4.48
3	-45.	3.48	3.48	4.48	4.48
2	45.	3.48	3.48	4.48	4.48
1	0.	2.88	2.88	51.2	51.2

Print   FPF [2.88]   FF [4.48]

7.4 & 7.5

Maximum Strain Failure Criterion

Ply	Angle	Intact Laminate		Degraded Laminate	
		Rint-Top	Rint-Bot	Rdeg-Top	Rdeg-Bot
5	0.	2.94(2)	2.94(2)	52.6(1)	52.6(1)
4	45.	4.97(2)	4.97(2)	4.49(1)	4.49(1)
3	-45.	4.97(2)	4.97(2)	4.49(1)	4.49(1)
2	45.	4.97(2)	4.97(2)	4.49(1)	4.49(1)
1	0.	2.94(2)	2.94(2)	52.6(1)	52.6(1)

Print   FPF [2.94]   FF [4.49]

7.4 & 7.5

EXIT | Main | Chapter | Back

Maximum Stress Failure Criterion

Ply	Angle	Intact Laminate Rint-Top	Intact Laminate Rint-Bot	Degraded Laminate Rdeg-Top	Degraded Laminate Rdeg-Bot
5	0.	2.76(2)	2.76(2)	51.2(1)	51.2(1)
4	45.	3.79(2)	3.79(2)	4.48(1)	4.48(1)
3	-45.	3.79(2)	3.79(2)	4.48(1)	4.48(1)
2	45.	3.79(2)	3.79(2)	4.48(1)	4.48(1)
1	0.	2.76(2)	2.76(2)	51.2(1)	51.2(1)

Print FPF 2.76 FF 4.48

7.4 & 7.5

Truncated Maximum Strain Failure Criterion

Ply	Angle	R with matrix cutoff		R without matrix cutoff	
		Top	Bottom	Top	Bottom
5	0.	2.94(2)	2.94(2)	19.2(1)	19.2(1)
4	45.	4.97(2)	4.97(2)	5.97(1)	5.97(1)
3	-45.	4.97(2)	4.97(2)	5.97(1)	5.97(1)
2	45.	4.97(2)	4.97(2)	5.97(1)	5.97(1)
1	0.	2.94(2)	2.94(2)	19.2(1)	19.2(1)

Print | Laminate R with matrix cutoff [2.94]   without matrix cutoff [5.97]

Exercise 7.6

A $[\pm 45]_T$ laminate with the same material properties as Exercise 7.2.
Determine the values of $N_x, N_y, N_{xy}, M_x, M_y, M_{xy}$ required to produce a curvature of:

$\kappa_x = 0.00545$

$\kappa_y = -0.00486$

$E_1 = 204.6$ GPa $\qquad E_2 = 20.46$ GPa

$G_{12} = 6.89$ GPa $\qquad v_{12} = 0.3$

$F_{1t} = 826.8$ MPa $\qquad F_{1c} = 482.3$ MPa

$F_{2t} = F_{2c} = 68.9$ MPa $\quad F_6 = 241.15$ MPa

By solving for the A, B, and D matrices:

$$
\begin{Bmatrix} N_x \\ N_y \\ N_{xy} \\ M_x \\ M_y \\ M_{xy} \end{Bmatrix}
=
\begin{bmatrix}
3.34 & 2.65 & 0 & 0 & 0 & -2.90 \\
2.65 & 3.34 & 0 & 0 & 0 & -2.90 \\
0 & 0 & 2.68 & -2.90 & -2.90 & 0 \\
0 & 0 & -2.90 & 6.95 & 5.52 & 0 \\
0 & 0 & -2.90 & 5.52 & 6.95 & 0 \\
-2.90 & -2.90 & 0 & 0 & 0 & 5.59
\end{bmatrix}
\begin{Bmatrix} 0 \\ 0 \\ 0 \\ 0.00545 \\ -0.00486 \\ 0 \end{Bmatrix}
$$

$$
\begin{Bmatrix} N_x \\ N_y \\ N_{xy} \\ M_x \\ M_y \\ M_{xy} \end{Bmatrix}
=
\begin{Bmatrix} 0 \\ 0 \\ -171.3 \\ 1108 \\ -372.0 \\ 0 \end{Bmatrix} \frac{N}{m}
$$

See also CADEC output.

7.6

Plate Stiffness Equations

$N_x =$ 0.00000E+00

$N_y =$ 0.00000E+00

$N_{xy} =$ -1.71296E+02

$M_x =$ 1.10792E+03

$M_y =$ -3.71985E+02

$M_{xy} =$ 0.00000E+00

$=$

3.34E+05	2.65E+05	0.00E+00	0.00E+00	0.00E+00	-2.90E+05
2.65E+05	3.34E+05	0.00E+00	0.00E+00	0.00E+00	-2.90E+05
0.00E+00	0.00E+00	2.68E+05	-2.90E+05	-2.90E+05	0.00E+00
0.00E+00	0.00E+00	-2.90E+05	6.95E+05	5.52E+05	0.00E+00
0.00E+00	0.00E+00	-2.90E+05	5.52E+05	6.95E+05	0.00E+00
-2.90E+05	-2.90E+05	0.00E+00	0.00E+00	0.00E+00	5.59E+05

$\varepsilon^o_x =$ 0

$\varepsilon^o_y =$ 0

$\gamma^o_{xy} =$ 0

$K_x =$ 0.00545

$K_y =$ -0.00486

$K_{xy} =$ 0

$Q_y =$ 0.00000E+00

$Q_x =$ 0.00000E+00

$=$

| 1.44E+04 | 0.00E+00 |
| 0.00E+00 | 1.44E+04 |

$\gamma_{yz} =$ 0.00000E+00

$\gamma_{xz} =$ 0.00000E+00

Matrices [ABD] and [H] are based on the current intact material, or last matrix file loaded

Exercise 7.7

If all the forces determined in Exercise 7.6 are increased proportionally, use the Tsai-Wu criterion to determine the following:

(a)   First Ply Failure load

(b)   Fiber failure load with degradation factor $f_d = 0.5$

$$f_1 = 1/F_{1t} - 1/F_{1c} = -8.639(10^{-4})\ \frac{1}{\text{MPa}}$$

$$f_2 = 1/F_{2t} - 1/F_{2c} = 0$$

$$f_{11} = 1/F_{1t}F_{1c} = 2.508(10^{-6})\ \frac{1}{(\text{MPa})^2}$$

$$f_{22} = 1/F_{2t}F_{2c} = 2.11(10^{-4})\ \frac{1}{(\text{MPa})^2}$$

$$f_{66} = 1/(F_6)^2 = 1.720(10^{-5})\frac{1}{(\text{MPa})^2}$$

$$f_{12} = -1/2(F_{1t})^2 = f_{12}^*/(F_{1t})^2 = -7.314(10^{-7})\ \frac{1}{(\text{MPa})^2}$$

From CADEC:

Layers	σ_1	σ_2	σ_6
2 top	157	19.8	178
2 bottom	-.00777	.561(10^{-3})	0
1 top	.00777	-.561(10^{-3})	0
1 bottom	-157	-19.8	178

Using the Tsai-Wu criterion for the 1 bottom layer:

$a = 0.685$

$b = 0.06782$

Solving this the $R = 1.12$, which happens to be the lowest value.
The same procedure is used for the other layers to compute the other R values.

FPF occurs due to a shear failure at $\sigma_6 = (1.12)(178) = 199$ MPa

FF occurs at $M_x = (2.73)(1108) = 3024\ \frac{\text{Nm}}{\text{m}}$

See also CADEC output.

7.7

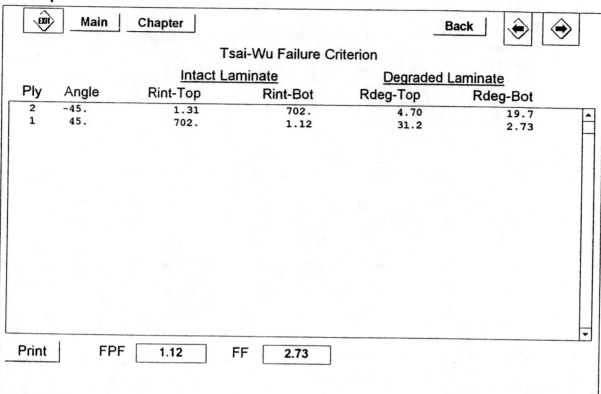

Ply	Angle	Intact Laminate		Degraded Laminate	
		Rint-Top	Rint-Bot	Rdeg-Top	Rdeg-Bot
2	-45.	1.31	702.	4.70	19.7
1	45.	702.	1.12	31.2	2.73

Tsai-Wu Failure Criterion

Print | FPF 1.12 | FF 2.73

Exercise 7.8

Asymmetric Quasi-Isotropic laminate with a $[0/90/\pm45]_s$ geometry and is 8 mm thick. It is made of carbon-epoxy AS4/3510-6.

Explain the mode of failure under the following loading conditions:

(a) $N_x = 1000 \ \frac{N}{mm}$

(b) $N_y = -1000 \ \frac{N}{mm}$

(c) $N_{xy} = 1000 \ \frac{N}{mm}$

(d) $N_{xy} = -1000 \ \frac{N}{mm}$

By using CADEC:

(a) $N_x = 1000$, Fiber failure on the $0°$ layer.

(b) $N_y = -1000$, Fiber failure on the $90°$ layer.

(c) $N_{xy} = 1000$, Fiber failure on the $-45°$, because $F_{1c} < F_{1t}$.

 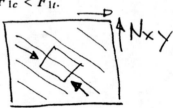

Since $F_{1c} < F_{1t}$, the $-45°$ Fails First.

(d) $N_{xy} = -1000$, Fiber failure on the $+45°$.

Exercise 7.9

Same as Exercise 7.8, use CADEC.

7.8 KT.9

Exit | Main | Chapter | Back

Truncated Maximum Strain Failure Criterion

Ply	Angle	R with matrix cutoff Top	R with matrix cutoff Bottom	R without matrix cutoff Top	R without matrix cutoff Bottom
7	0.	5.84(1)	5.84(1)	5.84(1)	5.84(1)
6	90.	2.51(2)	2.51(2)	12.2(1)	12.2(1)
5	45.	7.03(2)	7.03(2)	16.4(1)	16.4(1)
4	-45.	7.03(2)	7.03(2)	16.4(1)	16.4(1)
3	45.	7.03(2)	7.03(2)	16.4(1)	16.4(1)
2	90.	2.51(2)	2.51(2)	12.2(1)	12.2(1)
1	0.	5.84(1)	5.84(1)	5.84(1)	5.84(1)

Print | Laminate R with matrix cutoff **2.51** | without matrix cutoff **5.84**

$N_x = 1000$

7.8 & 7.9

| | | EXIT | Main | Chapter | | | | | Back | | ⬅ | ◇ |

Truncated Maximum Strain Failure Criterion

Ply	Angle	R with matrix cutoff		R without matrix cutoff	
		Top	Bottom	Top	Bottom
7	0.	20.4(1)	20.4(1)	20.4(1)	20.4(1)
6	90.	3.50(1)	3.50(1)	3.50(1)	3.50(1)
5	45.	9.80(1)	9.80(1)	9.80(1)	9.80(1)
4	-45.	9.80(1)	9.80(1)	9.80(1)	9.80(1)
3	45.	9.80(1)	9.80(1)	9.80(1)	9.80(1)
2	90.	3.50(1)	3.50(1)	3.50(1)	3.50(1)
1	0.	20.4(1)	20.4(1)	20.4(1)	20.4(1)

Print Laminate R with matrix cutoff   3.50   without matrix cutoff   3.50

$N_y = -1000$

7.8&7.9

EXIT | Main | Chapter | Back

Truncated Maximum Strain Failure Criterion

Ply	Angle	R with matrix cutoff Top	R with matrix cutoff Bottom	R without matrix cutoff Top	R without matrix cutoff Bottom
7	0.	1000. (2)	1000. (2)	1000. (6)	1000. (6)
6	90.	1000. (2)	1000. (2)	1000. (6)	1000. (6)
5	45.	4.54 (1)	4.54 (1)	4.54 (1)	4.54 (1)
4	-45.	1.95 (2)	1.95 (2)	2.72 (1)	2.72 (1)
3	45.	4.54 (1)	4.54 (1)	4.54 (1)	4.54 (1)
2	90.	1000. (2)	1000. (2)	1000. (6)	1000. (6)
1	0.	1000. (2)	1000. (2)	1000. (6)	1000. (6)

Print | Laminate R with matrix cutoff **1.95** without matrix cutoff **2.72**

$N_{xy} = 1000$

7.8 & 7.9

Truncated Maximum Strain Failure Criterion

Ply	Angle	R with matrix cutoff		R without matrix cutoff	
		Top	Bottom	Top	Bottom
7	0.	1000.(6)	1000.(6)	1000.(6)	1000.(6)
6	90.	1000.(6)	1000.(6)	1000.(6)	1000.(6)
5	45.	1.95(2)	1.95(2)	2.72(1)	2.72(1)
4	-45.	4.54(1)	4.54(1)	4.54(1)	4.54(1)
3	45.	1.95(2)	1.95(2)	2.72(1)	2.72(1)
2	90.	1000.(6)	1000.(6)	1000.(6)	1000.(6)
1	0.	1000.(6)	1000.(6)	1000.(6)	1000.(6)

Print Laminate R with matrix cutoff [1.95]   without matrix cutoff [2.72]

$N_{xy} = -1000$

731

Exercise 7.10

Use the material properties described in Exercise 4.10 and Example 7.1 to verify that the experimental values of strain to failure (listed below) are higher than those computed assuming linear elastic behavior.

Experimental data:

$\epsilon_{1t} = 11900(10^{-6})$ $\epsilon_{1c} = 8180(10^{-6})$

$\epsilon_{2t} = 2480(10^{-6})$ $\epsilon_{2c} = 22100(10^{-6})$

$\gamma_{6u} = 30000(10^{-6})$

Experimental data:

$F_{1t} = 1.826 \text{ GPa}$ $F_{1c} = 1.134 \text{ GPa}$

$F_{2t} = 19 \text{ MPa}$ $F_{2c} = 131 \text{ MPa}$

$F_6 = 75 \text{ MPa}$

Eq. 7.1:

$\epsilon_{1t} = F_{1t}/E_1 = 0.0117$ $\epsilon_{1c} = F_{1c}/E_1 = 0.00725$

$\epsilon_{2t} = F_{2t}/E_2 = 0.00244$ $\epsilon_{2c} = F_{2c}/E_2 = 0.0168$

$\gamma_{6u} = F_6/G_{12} = 0.0199$

All values computed are less than the experimentally found results.

ϵ_{2c} and γ_{6u} are much greater because of the non-linearity of the material.

Solutions Chapter 8
Exercise 8.1

Design an I beam of length $L = 4$ m to be supported at the ends and loaded at the mid-span by a load $P = 1000$ N. The thickness of the flange and web is t; the depth is h; the width is b; all measured at the midsurface of the panels (web and flanges).

The maximum deflection shouldn't exceed 1/800 of the span.
Use carpet plots for E glass polyester with $V_f = 0.5$.
The height to width ratio is $h/b = 2$.
The width to thickness ratio is $b/t = 10$.

For a simply supported beam with a load at the midspan, the maximum deflection is:

$$\delta_{max} = \frac{PL^3}{48EI} + \frac{PL}{4GA}$$

Select $\alpha = 0.8$ and $\gamma = 0.2$
From Figures 6.8-6.10 we obtain:
$E_x = 32.8$ GPa
$G_{xy} = 4.8$ GPa
If neglect shear deformation at this point, the moment of inertia can be found.

$$I = \frac{PL^3}{48E\delta_{max}} = \frac{(1000)(4.0)^3(800)}{48(32.8(10^9))(4)} = 8.13(10^{-6}) \text{ m}^4$$

By using this approximation for I, the beams dimensions can be found by assuming that the beam is thin-walled.

$$I = \frac{th^3}{12} + 2tb\left(\frac{h}{2}\right)^2 = \frac{h^4}{60}$$

$h = 148$ mm
To account for the additional shear deformation, round to
$h = 150$ mm
$b = 75$ mm
$t = 1.5$ mm
With these new dimensions we compute the maximum deflection.

$$I \approx \frac{h^4}{60} = 8.4375(10^{-6}) \text{ m}^4$$

$$A_w \approx th = 1125(10^{-6}) \text{ m}^2$$

$$\delta_b = \frac{1000(4)^3}{(48)32.8(10)^9 8.4375(10)^{-6}} = 4.818 \text{ mm}$$

$$\delta_S = \frac{1000(4)}{(48)4.8(10)^9 125(10)^{-6}} = 0.185 \text{ mm}$$

$$\delta_T = \delta_b + \delta_S = 5.003 \text{ mm} \approx L/800$$

Exercise 8.2

Determine the factor of safety at first ply failure of the beam designed in Exercise 8.1.

From the carpet plots in Figures 7.10-7.12, $\alpha = 0.8$ and $\beta = 0.2$.

$F_{xt} = 235$ MPa

$F_{xc} = 205$ MPa

$F_{xy} = 45$ MPa

The maximum moment at the midspan and the shear force are

$$M = \frac{PL}{4} = \frac{1000(4)}{4} = 1000 \text{ Nm}$$

$$Q = \frac{P}{2} = 500 \text{ N}$$

The average compressive stress in the flange is approximately

$$\sigma_x = \frac{Mc}{I} = \frac{M}{I}\left(\frac{h-t}{2}\right) = \frac{1000}{8.4375(10)^{-6}}\left(\frac{150 - 7.5}{2000}\right)$$

$$\sigma_x = 8.45(10)^6 \text{ N/m}^2$$

$$FS = \frac{S_{xc}}{\sigma_x} = \frac{205}{8.45} = 24.2$$

The first moment of area at the neutral axis of the beam is approximately

$$S = tb\left(\frac{h-t}{2}\right) + t\left(\frac{h-t}{2}\right)\left(\frac{h-t}{4}\right) = 59(10)^{-6} \text{ m}^3$$

Therefore, the maximum shear stress is

$$\tau_{max} = \frac{SQ}{It} = \frac{59(10)^{-6}500}{8.4375(10)^{-6}0.0075} = 0.466 \text{ MPa}$$

$$F.S. = \frac{S_{xy}}{\tau_{max}} = \frac{45}{0.466} = 96$$

it can be seen that the deflection constraint in Exercise 8.1 controls the design.

Exercise 8.3

Determine the maximum tip load P that can be applied to the beam of Example 6.6 before the bottom flange buckles.

Assume the flange to be simply supported on the webs, which gives a conservative answer.

Box beam

$h = 188$ mm $t = 4.7$ mm

$b = 94$ mm

From Example 6.6:

$I \cong \frac{h^4}{96} = 13.01(10^{-6}) \text{ m}^4$

$A_w = 2ht = 1.767(10^{-3}) \text{ m}^2$

Cantilever beam with $L = 2$ m subject to a tip load of $P = 1000$ N. Limit of tip deflection to $\frac{1}{300}$ of the span. The unsupported length of the compression flange along the length of the beam is $L = 2000$ mm. Assume the flange simply supported on the webs.

$N_x = M/bh$

$M_{max} = PL$

$N_x^{CR} = (2\pi^2/b^2)\left[\sqrt{D_{11}D_{22}} + D_{12} + 2D_{66}\right]$

$P = N_x^{CR}(bh)/L$

Using the carpet plots in Figures 6.10-6.12 (inplane moduli could be used as an approximation Fig. 6.8-6.10) with $\alpha = 0.8$, $\gamma = 0.2$:

$E_x^b = 25$ GPa

$E_y^b = 15$ GPa

$G_{xy}^b = 7$ GPa

$$P_{CR} = 2(188 + 94)201 = 113,364 \text{ N}$$

which is lower than the Euler (global) buckling load in Exercise 8.4.

Exercise 8.6

A simply supported rectangular beam:
L = 100 mm
b = 10 mm
Made of $[\pm45/0_2]_s$ glass-polyester with each layer 0.125 mm thick
Uniform load of q = 1000 N/m applied
Compute the maximum deflection

From the carpet plots, with $\alpha = \gamma = 0.5$
E_x = 24.8 GPa
G_{xy} = 7.1 GPa
Since the laminate has 8 layers at 0.125 mm each, the height h = 1 mm

$$I = \frac{bh^3}{12} = 8.33(10^{-11}) \text{ m}^4$$

$$A = hb = 1.0(10^{-5}) \text{ m}^2$$

From Table 8.1

$$\delta_{max} = \frac{5}{384}\frac{qL^4}{EI} + \frac{qL^2}{8GA} = 6.479(10^{-4}) \text{ m}$$

Exercise 8.7

Same beam dimensions as Exercise 8.6, but the material is made of carbon-epoxy AS4/3501-6.
A transverse load P is applied at the midspan.
(a)   compute the deflection as a function of P
(b)   compute the maximum load that can be applied to the beam for first ply failure using the Tsai-Wu criterion

(a) I =bh^3/12= 8.33(10^{-13}) m^4 A = 1.0(10^{-5}) m^2
Using CADEC the laminate moduli are
E_x^b = 38.96 GPa G_{xy}^b = 3.093 GPa
δ_{max} =PL^3/(48EI)+PL/(4GA)= 0.00642(P)
(b) Maximum Moment: M=PL/4
Shear: $Q = \frac{P}{2} = V$
Fig.

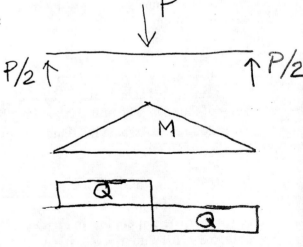

The above are absolute moment [Nm] and shear force [N]. To load the laminate, compute the values per unit length. Since the width of the beam is b = 10 mm

$$M_x = \frac{M}{b} = \frac{PL}{4b} = \frac{100P}{(4)(10)} = 2.5P$$

$$V_x = Q_x = \frac{Q}{b} = \frac{P}{2b} = \frac{P}{2(10)} = 0.05P$$

or V_x

Using a reference load $P = 1$, input M_x and Q_x into CADEC (Chapter 6). The strength properties are given in Table 1.1. Using the Tsai-Wu criterion, FPF occurs at $P = 14.6$ and FF at $P = 23.4$.

Exercise 8.8

Compute the deflection of the beam in Example 6.6 using the formulation of Section 8.2.

$L = 2$ m

$P = 1000$ N loaded at the tip

From CADEC

$(EI)_\eta = (EI)_{max} = 427$ Nm^2

$(GA)_\eta = 2.85(10^6)$ N

From Table 8.1

$$\delta = \frac{PL^3}{3EI} + \frac{PL}{GA} = (0.006255 + 0.000702)P$$

$\delta = 6.957$ mm

The slight difference with Example 6.6 is due to the improved computation of the shear stiffness of the box section.

See CADEC output.

8.8

File name: `c:\book\examples\exmp6_6.tws`

Beam Name: Example 6.6 and Problem 8.8

Number of Nodes 4   Number of Segments 4   Number of Forces 1

Select all that apply:

☐ Symmetric Sections
☑ Closed Section
☑ Compute Deformations

Build Laminate Data

Build Isotropic Data

Build Concentrated Stiffnes

Strength Computations:

○ No
○ FPF
◉ FF Long
○ FF Short

Set Up Sections ➡

3

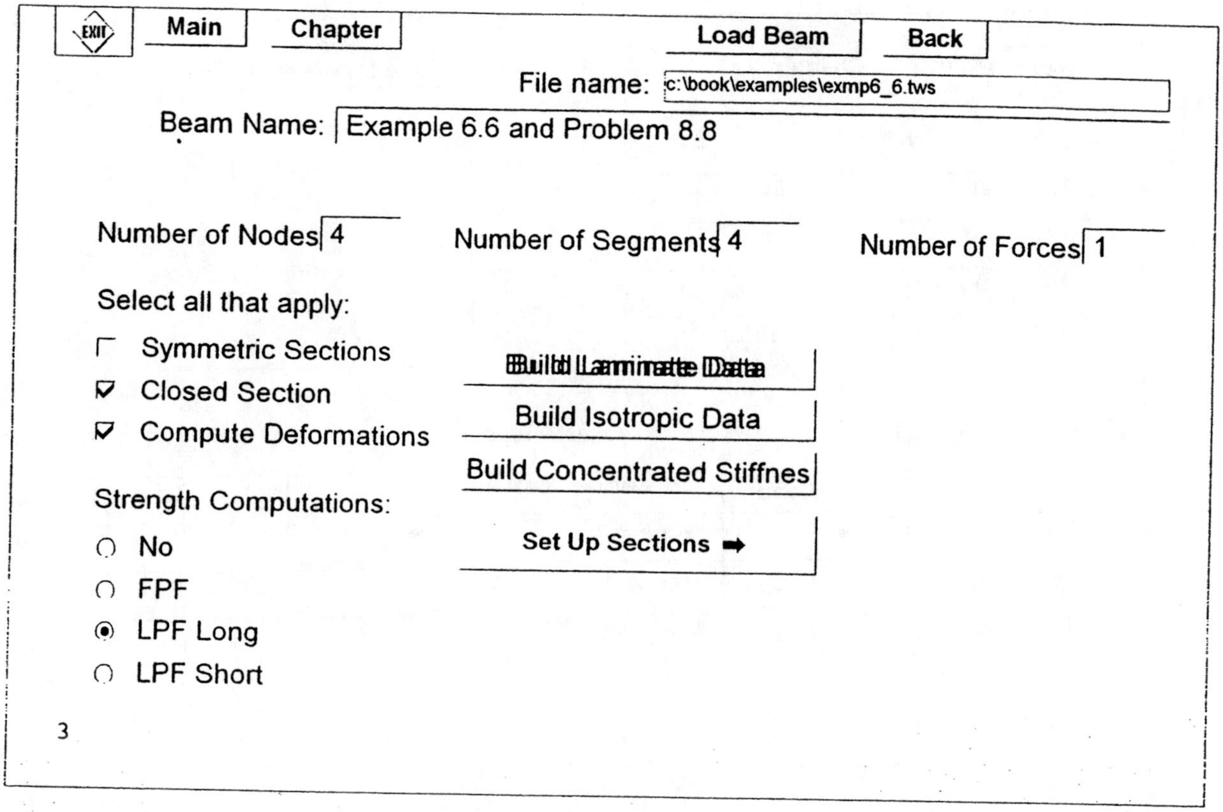

| EXIT | Main | Chapter | | Load Beam | Back |

File name: c:\book\examples\exmp6_6.tws

Beam Name: Example 6.6 and Problem 8.8

Number of Nodes 4   Number of Segments 4   Number of Forces 1

Select all that apply:

☐ Symmetric Sections

☑ Closed Section

☑ Compute Deformations

| Build Laminate Data |
| Build Isotropic Data |
| Build Concentrated Stiffnes |
| Set Up Sections ➡ |

Strength Computations:

◯ No

◯ FPF

◉ LPF Long

◯ LPF Short

3

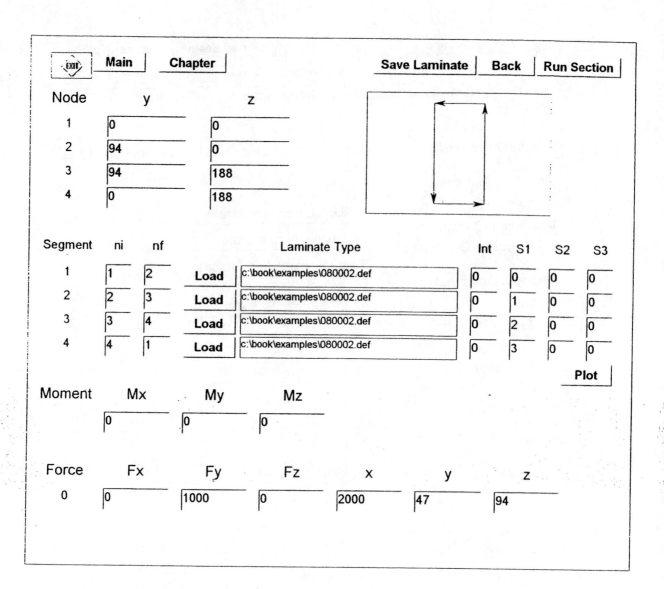

Node	y	z
1	0	0
2	94	0
3	94	188
4	0	188

Segment	ni	nf		Laminate Type	Int	S1	S2	S3
1	1	2	Load	c:\book\examples\080002.def	0	0	0	0
2	2	3	Load	c:\book\examples\080002.def	0	1	0	0
3	3	4	Load	c:\book\examples\080002.def	0	2	0	0
4	4	1	Load	c:\book\examples\080002.def	0	3	0	0

Plot

Moment	Mx	My	Mz
	0	0	0

Force	Fx	Fy	Fz	x	y	z
0	0	1000	0	2000	47	94

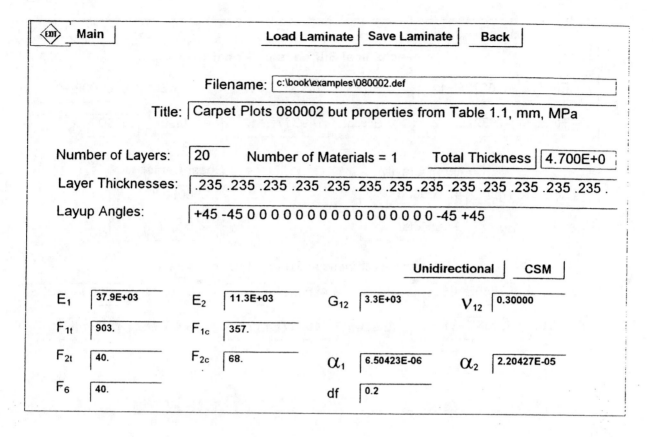

Main | Load Laminate | Save Laminate | Back

Filename: c:\book\examples\080002.def

Title: Carpet Plots 080002 but properties from Table 1.1, mm, MPa

Number of Layers: 20   Number of Materials = 1   Total Thickness | 4.700E+0

Layer Thicknesses: .235 .235 .235 .235 .235 .235 .235 .235 .235 .235 .235 .235 .235 .

Layup Angles: +45 -45 0 0 0 0 0 0 0 0 0 0 0 0 0 0 -45 +45

Unidirectional | CSM

E_1 | 37.9E+03 E_2 | 11.3E+03 G_{12} | 3.3E+03 ν_{12} | 0.30000

F_{1t} | 903. F_{1c} | 357.

F_{2t} | 40. F_{2c} | 68. α_1 | 6.50423E-06 α_2 | 2.20427E-05

F_6 | 40.   df | 0.2

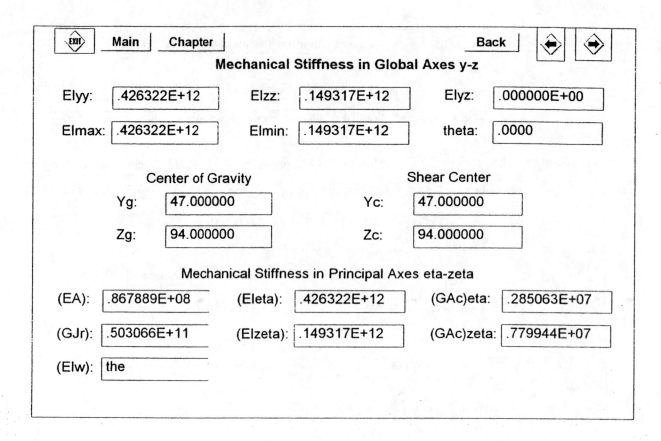

Mechanical Stiffness in Global Axes y-z

Elyy: `.426322E+12`   Elzz: `.149317E+12`   Elyz: `.000000E+00`

Elmax: `.426322E+12`   Elmin: `.149317E+12`   theta: `.0000`

Center of Gravity		Shear Center	
Yg:	47.000000	Yc:	47.000000
Zg:	94.000000	Zc:	94.000000

Mechanical Stiffness in Principal Axes eta-zeta

(EA): `.867889E+08`   (Eleta): `.426322E+12`   (GAc)eta: `.285063E+07`

(GJr): `.503066E+11`   (Elzeta): `.149317E+12`   (GAc)zeta: `.779944E+07`

(Elw): `the`

Exercise 8.9

Compute the torsional stiffness of the beam designed in Example 6.6.
Use CADEC printout for Exercise 8.8 and see Example 8.5.
$(GJ_R) = 50.3(10^9)$ Nmm^2

Exercise 8.10

Compute the torsional stiffness of the beam in Example 8.4.
Using Eq. 8.45

$$GJ_R = 4 \sum_{i=1}^{n} \overline{H}_i b_i$$

For the beam in Example 8.4, there are 5 segments, all with the same segment stiffness H_i given in Example 8.4.

$H_i = 4.0$ Nm
Since $e_q = 0$, $\overline{H}_i = H_i$, then:

$$GJ_R = 4\overline{H}_i\left[(h - \tfrac{h}{16}) + 2\tfrac{h}{2}\right] \text{ with } h = 0.025$$

$GJ_R = 0.775$ Nm^2
CADEC results:

$H_i = 4.323$ Nm
$GJ_R = 0.820$ Nm^2

The discrepancy is explained as follows. The laminate has $\delta_{16} \neq 0$ and $\delta_{26} \neq 0$. H_i was computed in Example 8.4 using Eq. 8.14-8.15 which assume $\delta_{16} = \delta_{26} = 0$. CADEC calculates H_i by matrix condensation on Eq. 8.13 to refine the result by incorporating the influence of $\delta_{16} \neq 0$ and $\delta_{26} \neq 0$ (see reference [11]). Also, CADEC corrects the perimeter to avoid counting twice the contribution of the overlapping areas at the corners where several segments converge.

Exercise 8.11

Consider a box beam with a constant area A and thickness t. The depth h and width b are variable.

(a) Derive an expression for the moment of inertia per unit area in terms of the ratio $r = h/b$.

(b) Derive an expression for the failure moment per unit area in terms of the ratio $r = h/b$, for a constant material strength F_x.

(c) Comment on the significance of your results with regard to thin-walled construction.

(a)

$$\frac{I}{A} = r^2 \frac{(1 + r/3)}{4(1 + r)}$$

(b)

$$\frac{M}{A} = F_x r \frac{(1 + r/3)}{2(1 + r)}$$

(c) The moment of inertia grows with r^2, thus increasing the stiffness for constant area. The moment capacity grows linearly with r. However, the depth of a section will be limited by local buckling of the panels when the thickness reduces too much.

Solutions Chapter 9

Exercise 9.1

Select a material, laminate stacking sequence, and thickness to carry a uniformly distributed load $p = 172$ kPa.

Plate dimensions: 15 cm wide by 2 m long and simply supported around the boundary

Base design on the analysis of a plate strip 15 cm long and 1 cm wide, which can be analyzed as a beam.

System can be analyzed as a wide beam. See Section 9.3.1

Glass-Polyester is the material with $\alpha = 1$ with fibers oriented along the shorter dimension.

$F_x^b = 379$ MPa

$F_{xy}^b = 44$ MPa

For simply supported beam under uniformly distributed load

$$M_{max} = \frac{WL^2}{8} = \frac{(172)(0.15)}{8} = 3.23 \text{ kNm}$$

Using Equation 9.1:

$$M_x = \frac{F_x^b(t)^2}{6}$$

yields $t = 7.2$ mm.

The M_y is ignored because $a/b = 200/15 \gg 1$, then $M_y \ll M_x$ (x direction along the width of the plate)

Exercise 9.2

Compute the critical edge load N_x^{CR} for a plate simply supported around the boundary.

length $a = 2$ m

width $b = 1$ m

thickness $t = 1$ cm

The laminate is made of E-glass-polyester with $V_f = 0.5$ with a symmetric configuration defined by $\alpha = \beta = 0.2$, $\gamma = 0.6$.

From carpet plots in Figure 6.11:

$E_x^b = 17$ GPa $v_{xy}^b = 0.42$

$E_y^b = 17$ GPa $G_{xy}^b = 8.2$ GPa

From Eq. 6.42

$\Delta = 1 - v_{xy}^b v_{yx}^b = 0.8236$

$D_{11} = t^3 E_x^b/(12\Delta) = 1720$ $D_{12} = t^3 E_x^b v_{yx}^b/(12\Delta) = 722.4$

$D_{22} = t^3 E_y^b/(12\Delta) = 1720$ $D_{66} = t^3 G_{xy}^b/12 = 683.3$

Using Equation 9.4:

$$N_x^{CR} = \frac{\pi^2 D_{22}}{b^2}\left[m^2 \frac{D_{11}}{D_{22}}\left(\frac{b}{a}\right)^2 + 2\frac{D_{12} + 2D_{66}}{D_{22}} + \frac{1}{m^2}\left(\frac{a}{b}\right)^2\right] = 75.19\frac{\text{kN}}{\text{m}}$$

$R_m = (a/b)(D_{22}/D_{11})^{0.25} = 2$

$m = 2$

$v_{xy}^b = 0.49$

From Eq. 6.42

$D_{11} = t^3 E_x^b/12\Delta = 252$ Nm

$\Delta = 1 - v_{xy}v_{yx} = 0.856$

$D_{22} = t^3 E_y^b/12\Delta = 152$ Nm

$D_{12} = t^3 E_x^b E_y^b/12\Delta = 74.29$ Nm

$D_{66} = t^3 G_{xy}^b/12 = 60.563$ Nm

Then

$N_x^{CR} = 8.74(10^5) \frac{N}{m}$

$P = 7720$ N

You can use CADEC to check the D values. Using the properties of glass-polyester from Table 1.1 for a $[\pm45/0_8]_S$, CADEC yields

$D_{11} = 247.169$ Nm

$D_{12} = 62.2161$ Nm

$D_{22} = 126.088$ Nm

$D_{66} = 10.6287$ Nm

$D_{26} = D_{16} = 3.19257$ Nm

Note that $D_{16} \neq 0$, $D_{26} \neq 0$ but they are neglected in the computation of the buckling load, which is

$P = 7106$ N.

Exercise 8.4

Determine the Euler buckling load of a column clamped at the base and free at the loaded end. The column has the same dimensions and material of the beam in Example 6.6.

$P_{CR} = \pi^2 EI/(L_e)^2$ $L_e = KL = (2.0)(2.0) = 4$ for a clamped-free column

$E = 32.8$GPa

$I = 13.01(10^{-6})$ m^4

$A = 2t(h + b) = 2.65(10^{-3})$ m^2

$$P_{CR} = \frac{\pi^2(13.01(10^{-6}))(32.8(10^9))}{16} = 263 \text{ kN}$$

Because the beam is relatively short, local buckling should be checked. To check the strength of the material

$\sigma_x = P_{CR}/A = 99.25$ MPa $< F_{xc}$ for this beam. From carpet plot for $\alpha = 0.8$, $\beta = 0.2$, $S_{xc} = 205$ MPa. The column will not crush but buckle under load.

Exercise 8.5

Determine an approximate estimate for the load P that causes local buckling to a column with the dimensions and material of Example 6.6.

A conservative estimate of N_x^{CR} is found the same way as in Exercise 8.3, except for the value of b. All panels are assumed simply supported. In this case b is the width of the wider wall of the box.

$N_x^{CR} = 2.18(10^5)$ N/m

Since all four walls are of the same material and laminate configuration, they are all subjected to the same force per unit length N_x^{CR}. Therefore, the total load at buckling is:

$P_{CR} = 2(0.188 + 0.094)N_x^{CR} = 123$ kN

Using CADEC's values from Exercise 8.3,

$$N_x^{CR} = \frac{2\pi^2}{(188)^2}\left(\sqrt{(247,169)(126,088)} + 62,216 + 2(60,628)\right) = 201 \text{ N/mm}$$

The axial load that produces buckling of the walls is

$$\sigma_{CR} = \frac{N_x^{CR}}{t} = 7.52 \text{ MPa}$$

From Figure 9.5, with $a/b = 2$ the dimensionless buckling stress $= 4.4$

$$4.45 = \frac{\sigma_{CR}12\Delta b^2}{\pi^2 E_y^b t^2} \qquad \sigma_{CR} = 7.55 \text{ MPa}$$

Note: to get E_y^b interchanging the value of α by value of β in the carpet plot Fig. 6.11.

Exercise 9.3

Repeat Exercise 9.2 with $a = 5$, and $b = 2$. Since $a \gg b$:

$$N_x^{CR} = \frac{2\pi^2}{b^2} \left[\sqrt{D_{11}D_{22}} + D_{12} + 2D_{66} \right] = 75.182 \frac{\text{kN}}{\text{m}}$$

There is very little change from Exercise 9.2.

Exercise 9.4

Repeat the computations of exercise 9.3 for various values of γ between 0 and 1, keeping $\alpha = \beta = (1 - \gamma)/2$.

Mathcad used for these computations.

$$t := .01 \quad b := 1 \quad a := 5 \quad i := 1..5$$

$Ex_i :=$	$vxy_i :=$	$Gxy_i :=$	$Ey_i := Ex_i$	$\gamma_i :=$
$24.8 \cdot 10^9$	$.139$	$3.3 \cdot 10^9$		0
$22.3 \cdot 10^9$	$.2$	$7.0 \cdot 10^9$	$\Delta_i := 1 - (vxy_i)^2$	$.2$
$20.0 \cdot 10^9$	$.26$	$8.0 \cdot 10^9$		$.4$
$17 \cdot 10^9$	$.36$	$8.2 \cdot 10^9$		$.6$
$14.5 \cdot 10^9$	$.55$	$8.3 \cdot 10^9$		$.8$

$$D11_i := \frac{t^3 \cdot Ex_i}{12 \cdot \Delta_i} \qquad D22_i := \frac{t^3 \cdot Ey_i}{12 \cdot \Delta_i} \qquad D12_i := \frac{t^3 \cdot Ey_i \cdot vxy_i}{12 \cdot \Delta_i}$$

$$D66_i := \frac{t^3 \cdot Gxy_i}{12}$$

$$Nx_i := \frac{\pi^2 \cdot 2}{b^2} \cdot \left(\sqrt{D11_i \cdot D22_i} + D12_i + 2 \cdot D66_i \right)$$

Nx_i	$D11_i$	$D22_i$	$D12_i$	$D66_i$
$5.824 \cdot 10^4$	$2.107 \cdot 10^3$	$2.107 \cdot 10^3$	292.926	275
$6.888 \cdot 10^4$	$1.936 \cdot 10^3$	$1.936 \cdot 10^3$	387.153	583.333
$7.078 \cdot 10^4$	$1.788 \cdot 10^3$	$1.788 \cdot 10^3$	464.75	666.667
$7.067 \cdot 10^4$	$1.628 \cdot 10^3$	$1.628 \cdot 10^3$	585.938	683.333
$8.031 \cdot 10^4$	$1.732 \cdot 10^3$	$1.732 \cdot 10^3$	952.808	691.667

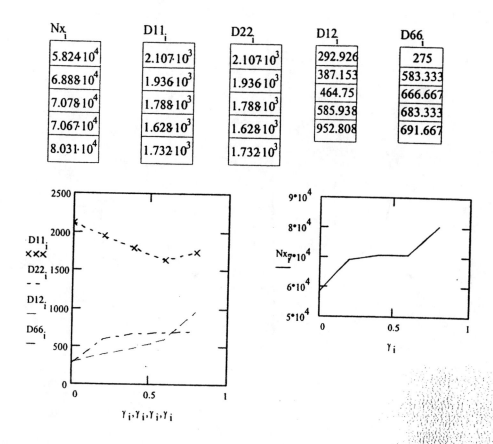

From the plot the values of N_x^{CR} appear to increase in the same manner as D_{12}. Check results against Fig. 9.5.

Exercise 9.5

Compute the critical edge load $N_x^{CR} = N_y^{CR}$ for the plate of Exercise 9.2 when it is loaded equally on all sides.

$$N_x^{CR} = \frac{\pi^2}{b^2} \frac{D_{11}m^2c^4 + 2Hm^2n^2c^2 + D_{22}n^4}{m^2c^2 + kn^2}$$

$k = 1$ because $N_x^{CR} = N_y^{CR}$
$c = b/a = 2$
$H = D_{12} + 2D_{66} = 2089$

$N_x^{CR} = 31.23 \; \frac{kN}{m}$

Exercise 9.6

Design a panel of length $a = 2$ m, width $b = 1.5$ m, stiffened along the length with 5 blade stiffeners spaced at 25 cm. The panel is simply supported all around and it must carry a uniform load $p = 24.8$ kPa transverse to the panel. Design the panel with a safety factor of 1.5 on the first ply failure strength. Use the carpet plots for E-glass-polyester. Assume a blade stiffener with a width to depth ratio $w/h = 1$, and a depth to thickness ratio $h/t = 5$.

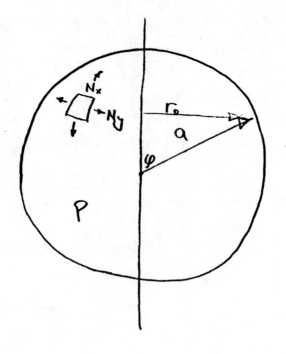

Exercise 10.1

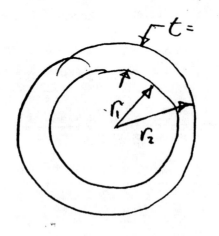

Exercise 10.2

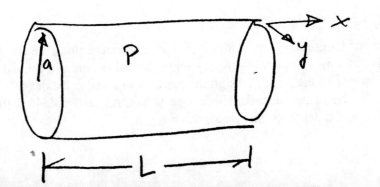

Exercise 10.3

$$L = 2.0 \text{ m}$$
$$a = 0.5 \text{ m}$$
$$P = 1.0 \text{ MPa}$$
$$\text{F.S.} = 1.5$$

Failure mode: FPF

$$\frac{N_x}{r_1} + \frac{N_y}{r_2} = P_z \ , \quad r_1 = \infty$$

$$\frac{N_y}{r_2} = P_z$$

$$N_y = P_z r_2 = (1.0 \text{ MPa})(0.5 \text{ m}) = 0.5 \text{ MN/m}$$

$$N_x = \frac{F}{2\pi r_0 \sin \varphi}$$

$$= \frac{\pi a^2 P}{2\pi a} = \frac{(0.5)^2 (1.0)}{2(0.5)} = 0.25 \text{MN/m}$$

Taking out the ±45 which are not needed:

	Ex 10.3	1st Try
α	0.3	0.33
β	0.6	0.66
γ	0.1	-

$$S_x = 78 \text{ MPa}$$
$$S_y = 115 \text{ MPa}$$

$$N_y/t < S_y/F.S. \Rightarrow t = \frac{(0.5)(1.5)}{115} = 6.52 \text{ mm}$$

$$N_x/t < S_x/F.S. \Rightarrow t = \frac{(0.25)(1.5)}{78} = 4.81 \text{ mm}$$

Use $t = 6.52$ mm

We see that the longitudinal direction (x) is overdesigned because $S_y < 2S_x$. The material can be further optimized to have $S_y = 2S_x$

	2nd Try
α	0.2
β	0.8
γ	-

$$S_x = 65 \text{ MPa}$$
$$S_y = 132.5 \text{MPa}$$

$$t = \frac{N_y(F.S.)}{S_y} = \frac{(0.5)(1.5)}{132.5} = 5.66 \text{ mm}$$

$$t = \frac{N_x(F.S.)}{S_x} = \frac{(0.25)(1.5)}{65} = 5.77 \text{ mm}$$

Exercise 10.4

Redesign the cylindrical tank of Ex. 10.3 using one or more layers of M1500 laminate. Use F.S. = 3 because only ultimate tensile strength values are listed in table 1.2.

Ex. 10.6

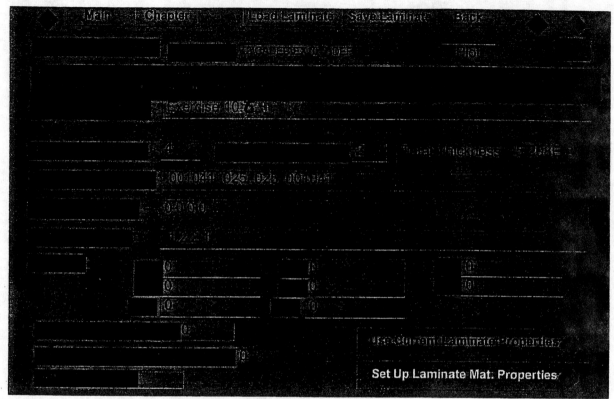

Ex 10.6

Ex. 10.6

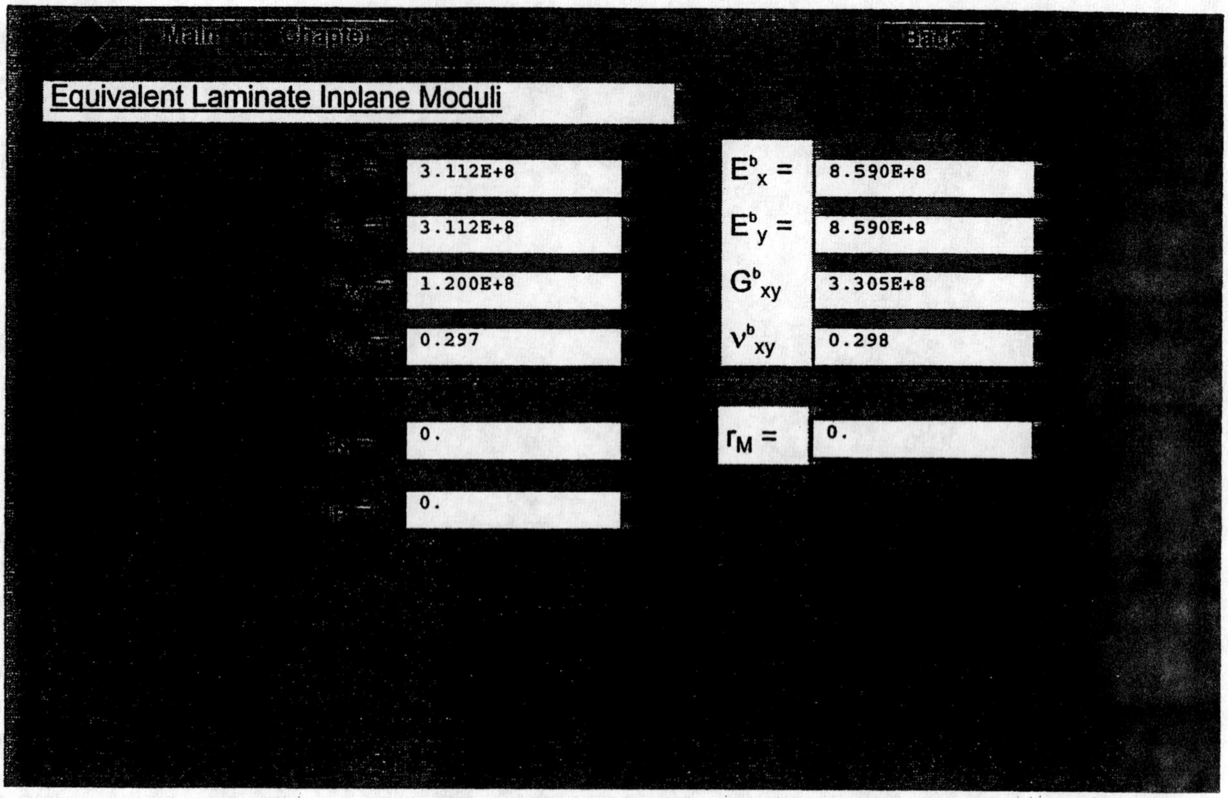

Equivalent Laminate Inplane Moduli

3.112E+8	$E^b_x =$	8.590E+8
3.112E+8	$E^b_y =$	8.590E+8
1.200E+8	G^b_{xy}	3.305E+8
0.297	ν^b_{xy}	0.298
0.	$r_M =$	0.
0.		

$t_{FOAM} = 0.050m$

Ex. 10.6

Ex. 10.6

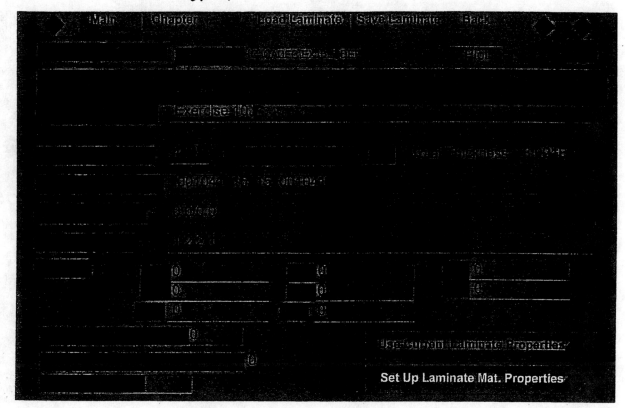

Set Up Laminate Mat. Properties

Ex. 10.6

Equivalent Laminate Inplane Moduli

7.028E+7	E^b_x = 1.694E+8
7.028E+7	E^b_y = 1.694E+8
2.736E+7	G^b_{xy} 6.565E+7
0.28	ν^b_{xy} 0.292
0.	r_M = 0.
0.000	

$t_{FOAM} = .300\,m$

Exercise 10.7 . See Figure next page.

$$a = \frac{h^2 + (s/2)^2}{2h} = 31.2$$

$$\varphi_{max} = 22.62°$$

P_D = Dead Load = 765 N/m²
P_L = Live Load = 955 N/m²
P_S = Snow Load = 1915 N/m²
P_P = Pres Load = 2000 N/m²

Live and Snow loads: See Figure next page.

Plan Area = $2\pi a^2 \sin\varphi \cos\varphi\, d\varphi$
$F_{L+S} = -\int (P_L + P_S) 2\pi a^2 \sin\varphi \cos\varphi\, d\varphi$
Dead load: See Figure next page.

Surface Area = $2\pi a^2 \sin\varphi\, d\varphi$
$F_D = -\int P_D (2\pi a^2 \sin\varphi)\, d\varphi$

Pres. Load: See Figure next page.

Plan Area = $2\pi a^2 \sin\varphi \cos\varphi\, d\varphi$
$F_P = +\int P_P \cos\varphi (2\pi a^2 \sin\varphi \cos\varphi)\, d\varphi$

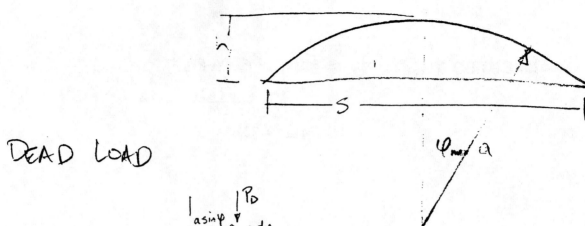

DEAD LOAD

$a\sin\psi$ P_D

$a\,d\psi$

ψ

$d\psi$

φ_{max} a

SURF AREA $= 2\pi a^2 \sin\psi\,d\psi$

$$F_D = -\int P_D (2\pi a^2 \sin\psi)\,d\psi$$

LIVE & SNOW LOAD

$P_L + P_S$

$a\sin\psi$

a

ψ

$d\psi$

PLAN AREA $= 2\pi a^2 \sin\psi\cos\psi\,d\psi$

$$F_{L+S} = -\int (P_L + P_S)\, 2\pi a^2 \sin\psi\cos\psi\,d\psi$$

PRES. LOAD

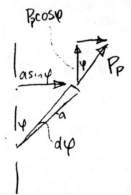

$a\,d\psi$

ψ

$a\,d\psi\cos\psi$

$P_P\cos\psi$

$a\sin\psi$

P_P

ψ

a

$d\psi$

PLAN AREA $= 2\pi a^2 \sin\psi\cos\psi\,d\psi$

$$F_P = +\int P_P\cos\psi (2\pi a^2 \sin\psi\cos\psi)\,d\psi$$

$N_x = 2.5(10^5) \frac{N}{m}$ $N_y = 5.0(10^5) \frac{N}{m}$
For M1500, $F_{xt} = F_{yt} = 128$ MPa

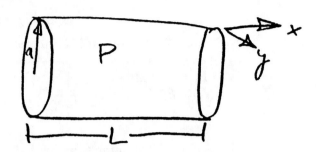

$t = N_y(F.S.)/F_{yt} = 11.72$ mm
Number of layers $= \frac{11.72}{1.041} = 11.26 = 12$ layers
$t = 12(1.041) = 12.492$ mm

Exercise 10.5

Optimize the design of Exercise 10.4 for minimum weight by using CM1808.
$F_{xt} = F_{yt} = 201$ MPa

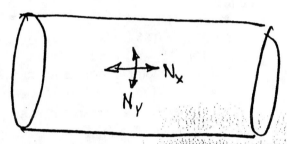

$t = N_y(F.S.)/F_{yt} = 7.46$ mm
Number of layers $= \frac{7.46}{1.219} = 6.12 = 7$ layers
$t = 7(1.219) = 8.533$ mm

Exercise 10.6

Replace the laminate in example 10.4 by a sandwich shell.
core material: $\rho = 50 \frac{kg}{m^3}$
                   $E = 20$ MPa
                   $v = 0.23$
dimensions: a = radius of hemisphere = 31.2 m
                   ϕ_{max} = half-center span = 22.62°
                   $P_{dead} = 654 \frac{N}{m^2}$
                   $P_{live} = 955 \frac{N}{m^2}$
                   $P_{snow} = 1915 \frac{N}{m^2}$
                   span $s = 24$ m
                   height $h = 2.4$ m

Fig.

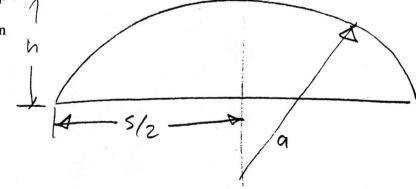

The maximum N_x value occurs at $\varphi = 22.62°$, and is found by Equation 10.3.

$$F = -2\pi a^2 \left\{ (P_L + P_S) \left[\frac{1}{2} \sin^2 \varphi \right] + P_D(1 + \cos\varphi) \right\}$$

$$N_x = \frac{F}{2\pi a \sin^2 \varphi} = -57.18 \text{ kN/m}$$

$$N_y = -(P_L + P_S + P_D)a(\cos\varphi) - N_x$$

The material properties of M1500 are:
$E_x = E_y = 7.303$ GPa
$F_x = F_y = 179$ MPa
$t = (N_x)(S.F.)/F_x = 1.917 \frac{mm}{layer}$

The sandwich shell allows for the finding of the laminate bending moduli, modelled by Eq. 6.33 or by using CADEC.

The shell thickness of each face of the sandwich panel will by set initially at 1.041 mm for each side, because that is the nominal thickness of M1500 (Table 1.2). The foam in the middle was set at 50 mm.

Using CADEC:
$E_x^b = E_y^b = 0.85$ GPa . *See CADEC output 10.6 (3/5)*
$t = 2(1.041) + 50 = 52.082$ mm

Using the empirical formula for buckling pressure on page 366:
$$q_{CR} = 0.3 E_x^b(t/a)^2 \left[1 - 0.00875(\varphi_{max} - 20°) \right]\left(1 - 0.000175 \, a/t \right) = 627.41 \, \frac{N}{m^2}$$

The value of q_{CR} is less than the total pressure acting on the shell: 3635 $\frac{N}{m^2}$, which means that the shell will fail in buckling.

For the second trial, we make the foam thickness 300 mm
From CADEC, $E_x^b = E_y^b = 0.1694$ GPa . *See CADEC output 10.6 (5/5)*
$q_{CR} = 4571 \frac{N}{m^2}$, which exceeds the value of the loading. See CADEC output.

See Ex. 9.1.

Skin design. Choose $\gamma = 0$, $\alpha = \beta = 0.5$. From Fig. 9.2-9.3
$F_x^b = 115$ MPa
$F_y^b = 44$ MPa
Stiffeners spaced at 0.25 m, using Eq. 9.21 for 1 cm wide strip of skin.

$$M = \frac{qL^2}{10} = \frac{(24.8)^3 10^3(0.01)(0.25)^2}{10} = 1.55 \text{ Nm}$$

Then the moment per unit width of skin is

$$M_x = \frac{M}{b} = \frac{1.55}{0.01} = 155 \text{ N}$$

From Eq. 9.1

$$t_{SK} = \sqrt{\frac{6M_x}{F_x^b/S.F.}} = \sqrt{\frac{6(155)}{115(10^6)/1.5}} = 0.0035 \text{ m}$$

Stiffener Design. $L = 2$ m. The load is applied over a strip wide as the stiffener spacing

$$q = 24.8(0.25) = 6.2 \text{ kN/m}$$

Using Eq. 9.18

$$M = \frac{6.2(10^3)(2)^2}{12} = 2,067 \text{ Nm}$$

From Fig. 7.10-7.12 for $\alpha = \beta = 0.5$
$F_{xt} = 96$ MPa
$F_{xc} = 96$ MPa
$F_{xy} = 44$ MPa
The required section modulus is

$$z = \frac{(F.S.)M}{F_{xc}} = \frac{(1.5)2,067}{96(10^6)} = 32.3(10^{-6}) \text{ m}^3$$

For a blade stiffener (Fig. 9.8) of uniform thickness $t = h/5$, and width $w = h$, h being the depth, assuming a thin-walled section; with respect to the neutral axis we have

$$I = 5h^4/120$$

$$c = (3/4)h$$

$$z = h^3/18$$

where c is the distance from the neutral axis to the farthest point in the cross section. Therefore

$$h = (32.3(10^{-6})18)^{1/3} = 0.083 \text{ m}$$

$$t = \frac{h}{5} = 0.0167 \text{ m}$$

Solutions Chapter 10

Exercise 10.1

Design a spherical pressure vessel with a radius $a = 0.5$ m and burst internal pressure $p = 10$ MPa.

Use one or more layers of M1500 laminate.
Use only the laminate strength and moduli given in Table 1.2.
Note that M1500 is a continuous strand mat fabric.
Use safety factor of 1.5. **See Figure on next page.**
Eq. 10.2
$$N_x + N_y = (p)(a) = 5.0(10^6) \ \tfrac{N}{m}$$
Eq. 10.3
$$N_x = F/2\pi r_o(\sin(\varphi))$$
The radial pressure acting on one-half of the sphere produces the same resultant as the pressure acting on the projected area πa^2, so $F = \pi a^2 p$.
$$N_x = \pi a^2 p/(2\pi a(\sin^2 \varphi)) = 2.5(10^6) \ \tfrac{N}{m}$$
$$N_y = pa - N_x = 2.5(10^6) \ \tfrac{N}{m}$$
From Table 1.2, $F_{xt} = 128$ MPa for M1500. Take F.S. = 1.5
Since the membrane force is the same in any direction on the spherical shell, $N_x = N_y = N$

$$\frac{N}{t} \le \frac{F_{xt}}{F.S.} \qquad\qquad t = \frac{N(S.F.)}{F_{xt}} = 29.3 \text{ mm}$$

Each layer of laminate is 1.041 mm thick, so $\frac{29.3}{1.041} = 28.15 = 29$ layers of laminate required.

Exercise 10.2

Optimize the design of Exercise 10.1 for minimum weight by selecting a laminate configuration in Table 1.2.

For M1500: $m = \rho V_{shell} = 1422.3\left[\frac{4}{3}\pi\left(r^3 - (r-t)^3\right)\right] = 1422.3(0.0890) = 126.6$ kg

For CM1808:
$F_{xt} = 201$ MPa
$N/t \le F_{xt}/(FS) \qquad t = 18.7$ mm
number of layers $= \frac{18.7}{1.219} = 15.34 = 16$ layers
$t = 16(1.219) = 19.50$ mm
$m = \rho V = 1546\left[\frac{4}{3}\pi\left(r^3 - (r-t)^3\right)\right] = 1546(0.0589) = 91.06$ kg

For XM2408: **See Figure on next page.**
$F_{xt} = 98$ MPa
$N/t \le F_{xt}/(FS) \qquad t = 38.3$ mm
number of layers $= \frac{38.3}{1.422} = 26.93 = 27.0$ layers
$t = 27(1.422) = 38.4$ mm
$m = \rho V = 1681\left[\frac{4}{3}\pi\left(r^3 - (r-t)^3\right)\right] = 1681(0.1116) = 187.6$ kg

CM1808 is the best design for minimum weight. Note that TVM3408 and UM1608 are not used because they do not have the same longitudinal and transverse strength. The later is not reported in Table 1.2 but we know it must be lower by looking at the fiber architecture in Table 2.3.

Exercise 10.3

Ex. 10.7

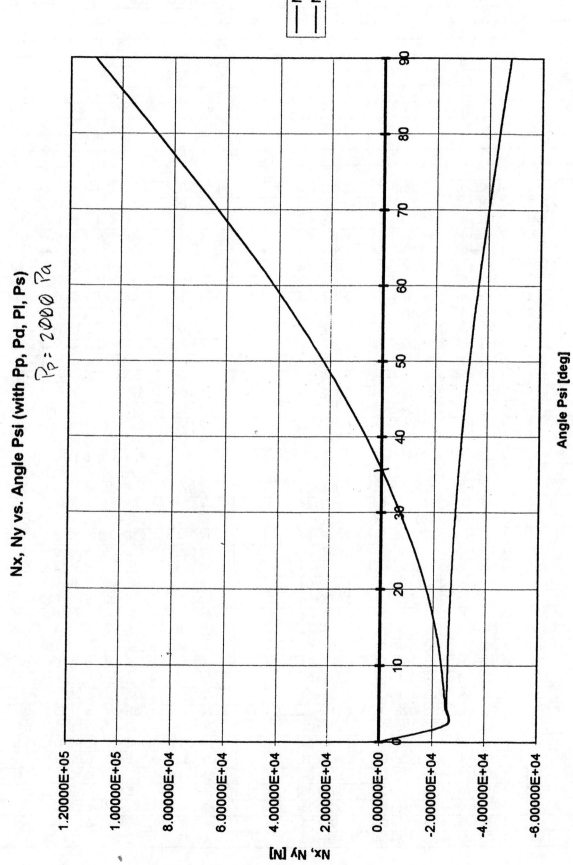

Nx, Ny vs. Angle Psi (with Pp, Pd, Pl, Ps)

Pp = 2000 Pa

Nx, Ny [N]

Angle Psi [deg]

Nx
Ny

C...t7

Page 1

Prob. 10.7

Spreadsheet for Prob. 10.7					Variation of Spreadsheet for Prob. 10.7			
a=	31.2				a=	31.2		
psimax=	22.62				psimax=	22.62	.	
Pd=	765				Pd=	765		
Pl=	955				Pl=	955		
Ps=	1915				Ps=	1915		
Pp=	2000				Pp=	500		
psi	F	Nx	Ny		psi	F	Nx	Ny
1.00E-05	-1.52E-07	-2.55323E+04	-2.55E+04		1.00E-05	-2.92E-07	-4.89004E+04	-4.89E+04
2.262	-7.79E+03	-2.55228E+04	-2.54E+04		2.262	-1.49E+04	-4.89137E+04	-4.88E+04
4.524	-3.12E+04	-2.55732E+04	-2.51E+04		4.524	-5.97E+04	-4.89368E+04	-4.85E+04
6.786	-7.02E+04	-2.56571E+04	-2.46E+04		6.786	-1.34E+05	-4.89752E+04	-4.80E+04
9.048	-1.25E+05	-2.57744E+04	-2.38E+04		9.048	-2.38E+05	-4.90291E+04	-4.74E+04
11.31	-1.95E+05	-2.59250E+04	-2.29E+04		11.31	-3.70E+05	-4.90985E+04	-4.65E+04
13.572	-2.82E+05	-2.61085E+04	-2.17E+04		13.572	-5.31E+05	-4.91834E+04	-4.55E+04
15.834	-3.84E+05	-2.63249E+04	-2.04E+04		15.834	-7.19E+05	-4.92838E+04	-4.42E+04
18.096	-5.03E+05	-2.65737E+04	-1.88E+04		18.096	-9.34E+05	-4.93999E+04	-4.28E+04
20.358	-6.37E+05	-2.68547E+04	-1.71E+04		20.358	-1.18E+06	-4.95317E+04	-4.12E+04
22.62	-7.88E+05	-2.71674E+04	-1.51E+04		22.62	-1.44E+06	-4.96794E+04	-3.94E+04
24.882	-9.55E+05	-2.75114E+04	-1.30E+04		24.882	-1.73E+06	-4.98430E+04	-3.74E+04
27.144	-1.14E+06	-2.78863E+04	-1.06E+04		27.144	-2.04E+06	-5.00227E+04	-3.53E+04
29.406	-1.34E+06	-2.82914E+04	-8.11E+03		29.406	-2.37E+06	-5.02186E+04	-3.30E+04
31.668	-1.55E+06	-2.87264E+04	-5.40E+03		31.668	-2.72E+06	-5.04310E+04	-3.05E+04
33.93	-1.78E+06	-2.91906E+04	-2.51E+03		33.93	-3.09E+06	-5.06601E+04	-2.78E+04
36.192	-2.03E+06	-2.96833E+04	5.55E+02		36.192	-3.48E+06	-5.09061E+04	-2.50E+04
38.454	-2.29E+06	-3.02040E+04	3.79E+03		38.454	-3.88E+06	-5.11692E+04	-2.20E+04
40.716	-2.57E+06	-3.07518E+04	7.19E+03		40.716	-4.29E+06	-5.14498E+04	-1.89E+04
42.978	-2.85E+06	-3.13262E+04	1.08E+04		42.978	-4.71E+06	-5.17483E+04	-1.56E+04
45.24	-3.16E+06	-3.19262E+04	1.45E+04		45.24	-5.15E+06	-5.20650E+04	-1.22E+04
47.502	-3.47E+06	-3.25513E+04	1.83E+04		47.502	-5.58E+06	-5.24004E+04	-8.62E+03
49.764	-3.79E+06	-3.32005E+04	2.23E+04		49.764	-6.03E+06	-5.27550E+04	-4.90E+03
52.026	-4.13E+06	-3.38730E+04	2.65E+04		52.026	-6.47E+06	-5.31294E+04	-1.05E+03
54.288	-4.47E+06	-3.45680E+04	3.08E+04		54.288	-6.92E+06	-5.35242E+04	2.92E+03
56.55	-4.82E+06	-3.52847E+04	3.52E+04		56.55	-7.36E+06	-5.39402E+04	7.03E+03
58.812	-5.17E+06	-3.60220E+04	3.97E+04		58.812	-7.80E+06	-5.43782E+04	1.12E+04
61.074	-5.52E+06	-3.67793E+04	4.43E+04		61.074	-8.24E+06	-5.48391E+04	1.56E+04
63.336	-5.88E+06	-3.75555E+04	4.91E+04		63.336	-8.66E+06	-5.53240E+04	2.00E+04
65.598	-6.23E+06	-3.83498E+04	5.39E+04		65.598	-9.08E+06	-5.58340E+04	2.46E+04
67.86	-6.59E+06	-3.91613E+04	5.88E+04		67.86	-9.48E+06	-5.63705E+04	2.92E+04
70.122	-6.93E+06	-3.99891E+04	6.38E+04		70.122	-9.87E+06	-5.69351E+04	3.40E+04
72.384	-7.27E+06	-4.08324E+04	6.89E+04		72.384	-1.02E+07	-5.75292E+04	3.88E+04
74.646	-7.60E+06	-4.16902E+04	7.41E+04		74.646	-1.06E+07	-5.81550E+04	4.37E+04
76.908	-7.92E+06	-4.25619E+04	7.93E+04		76.908	-1.09E+07	-5.88145E+04	4.87E+04
79.17	-8.22E+06	-4.34465E+04	8.45E+04		79.17	-1.13E+07	-5.95101E+04	5.38E+04
81.432	-8.50E+06	-4.43433E+04	8.98E+04		81.432	-1.15E+07	-6.02447E+04	5.89E+04
83.694	-8.76E+06	-4.52517E+04	9.52E+04		83.694	-1.18E+07	-6.10213E+04	6.42E+04
85.956	-9.01E+06	-4.61710E+04	1.01E+05		85.956	-1.21E+07	-6.18435E+04	6.94E+04
88.218	-9.22E+06	-4.71007E+04	1.06E+05		88.218	-1.23E+07	-6.27153E+04	7.48E+04
90	-9.38E+06	-4.78400E+04	1.10E+05		90	-1.24E+07	-6.34400E+04	7.90E+04
90.48	-9.42E+06	-4.80402E+04	1.11E+05		90.48	-1.25E+07	-6.36413E+04	8.02E+04

Buck ...4,6,7,8

	Example 10.4	Exercise 10.7	Exercise 10.6	Exercise 10.8
1st try:				
t comp=	0.0036	0.0017	0.002082	0.002844
t foam=	0	0	0.05	0.05
t total=	0.0036	0.0017	0.052082	0.052844
a=	31.2	31.2	31.2	31.2
Psimax=	22.62	22.62	22.62	22.62
E=	2.490E+10	2.490E+10	8.590E+08	2.337E+09
qcr [pa]=	-5.021E+01	-4.793E+01	6.281E+02	1.762E+03
2nd try:				
t comp=	0.0036	0.0017	0.002082	0.002844
t foam=	0.05	0.05	0.15	0.1
t total=	0.0536	0.0517	0.152082	0.102844
a=	31.2	31.2	31.2	31.2
Psimax=	22.62	22.62	22.62	22.62
E=	4.694E+09	2.392E+09	3.158E+08	1.243E+09
qcr [pa]=	3.647E+03	1.722E+03	2.120E+03	3.749E+03
3rd try:				
t comp=			0.002082	
t foam=			0.3	
t total=			0.302082	
a=			31.2	
Psimax=			22.62	
E=			1.694E+08	
qcr [pa]=			4.571E+03	

(10.7 continued)

$$F = F_{L+S} + F_D + F_P$$

$$= -2\pi a^2 \left[(P_L + P_S) \int_0^{\varphi_{max}} (\sin\varphi)(\cos\varphi)d\varphi + P_D \int_0^{\varphi_{max}} (\sin\varphi)d\varphi - P_P \int_0^{\varphi_{max}} \sin\varphi \cos^2\varphi d\varphi \right]$$

$$= -2\pi a^2 \left\{ (P_L + P_S)\left[\frac{1}{2}\sin^2\varphi \; |_0^{\varphi_{max}} \right] + P_D[-\cos\varphi \; |_0^{\varphi_{max}}] - P_P\left[-\frac{\cos^3\varphi}{3} \; |_0^{\varphi_{max}} \right] \right\}$$

$$= -2\pi a^2 \left\{ (P_L + P_S)\left[\frac{1}{2}\sin^2\varphi_{max} \right] + P_D[-\cos\varphi_{max} + 1] - P_P\left[-\frac{\cos^3\varphi_{max} + 1}{3} \right] \right\}$$

Substituting:

$$N_x = \frac{F}{2\pi a \sin^2\varphi}$$

$$N_y = P_z a - N_x = a[P_P - \cos\varphi(P_L + P_S + P_D)] - N_x$$

$N_x = -2.717 \; 10^4$ Pa
$N_y = -1.51 \; 10^4$ Pa
Use $\alpha = \beta = 0.5, \gamma = 0$
$S_{xt} = S_{yt} = S_{xc} = S_{yt} = 96$ MPa
$E_x = E_y = 24.9$ GPa
S.F. = 6.0

$$t = \frac{(2.717 \; 10^4 \text{ Pa})(6)}{96 \; 10^6 \text{ Pa}} \approx 1.70 \text{ mm}$$

Checking Buckling load:

$$q_{CR} = 0.3E\left(\frac{t}{a}\right)^2[1 - 0.00875(\varphi_{max} - 20°)]\left(1 - 0.000175\frac{a}{t}\right) \approx 47.9 \text{ Pa}$$

We must use sandwich shell
$\left\{ \begin{array}{l} E = 20 \; 10^6 \\ v = 0.23 \\ t = 0.050 \text{ m} \end{array} \right\}$ Polyurethane foam
$t_4 = 0.00085$ (top composite)
$t_3 = 0.025$ (foam)
$t_2 = 0.025$ (foam)
$t_1 = 0.00085$ (bottom composite)
$q_{CR} = 1722$ Pa

Exercise 10.8

$N_x = -5.72 \; 10^4$ Pa
$N_y = -4.75 \; 10^4$ Pa

$$t = \frac{(N_x)(S.F.)}{S_x} = \frac{(-5.72 \; 10^4)(6.0)}{228(10^6)} = 1.505 \text{ mm}$$

Material Properties
XM2408
$S_x = S_y = 228$ MPa ; $E_x = 15,158$ MPa, $E_y = 15,158$ MPa .
$E_x^b = E_y^b = 2.337 \; 10^9$ → See CADEC
First Try
$t_c = 1.422$ mm/layer $\quad q_{CR} = 1762$ Pa $\quad$ Too Low
$t_4 = 1.422$ (top composite face)
$t_3 = 25.0$ (foam)

$t_2 = 25.0$ (foam)
$t_1 = 1.422$ (bottom composite face)

Second Try
$t_c = 1.422$ mm/layer $q_{CR} = 3749$ Pa   O.K.
$t_4 = 1.422$ (top composite face)
$t_3 = 50.0$ (foam)
$t_2 = 50.0$ (foam)
$t_1 = 1.422$ (bottom composite face)

10.8 (1/5)

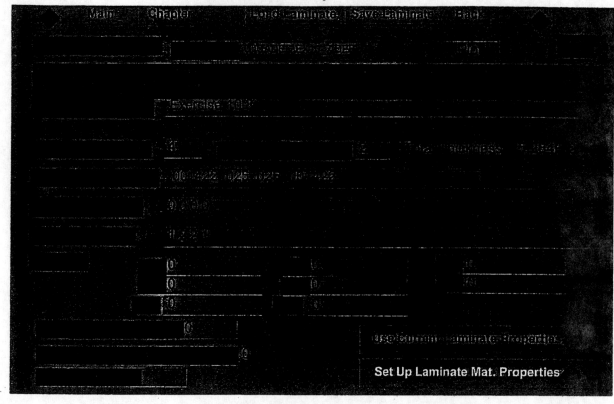

Set Up Laminate Mat. Properties

Ex 10.8 (2/5)

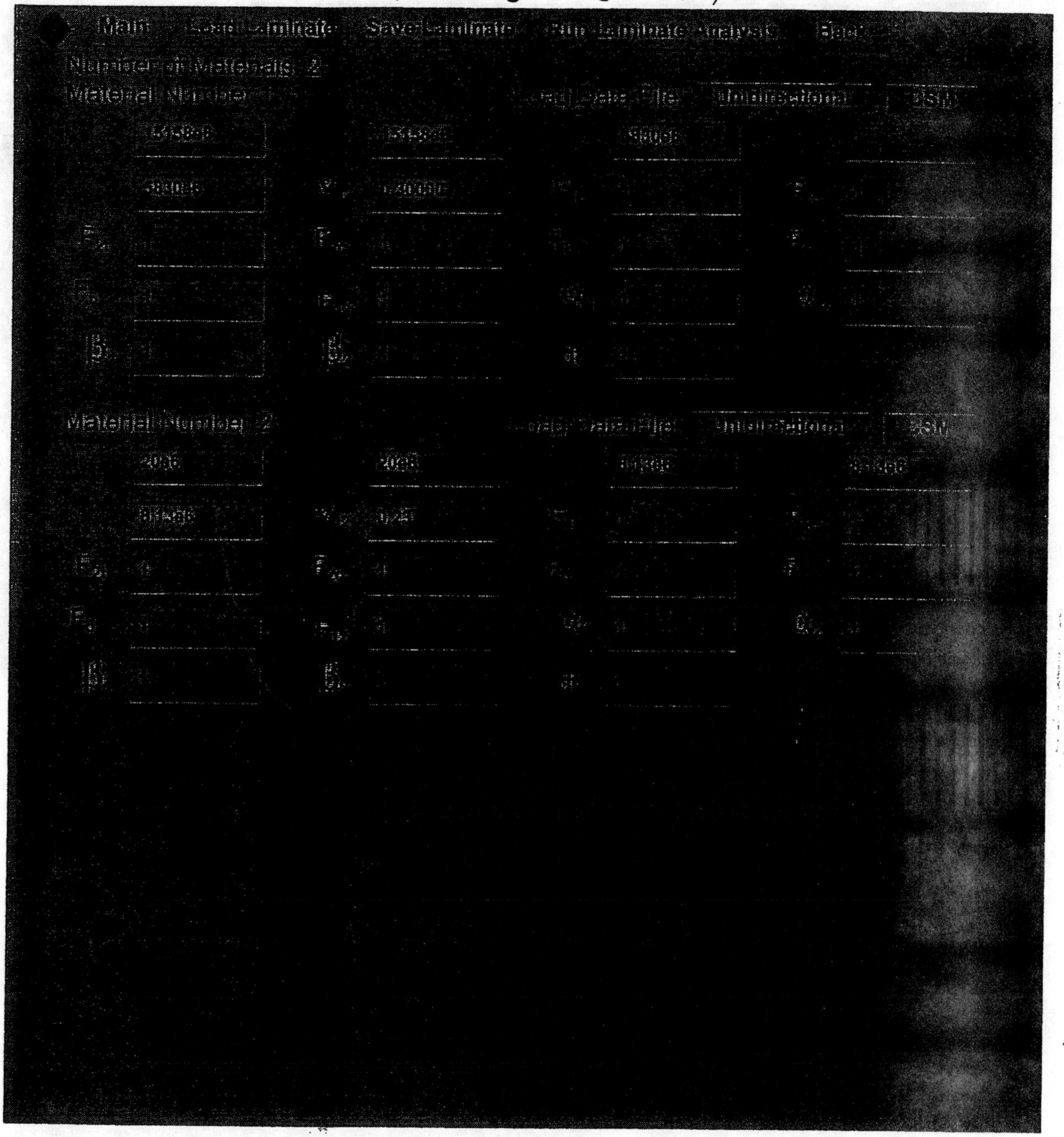
Main
Number of Materials: 2
Material Number 1
Unidirectional
Material Number 2

Fox 10.8 (3/5)

Equivalent Laminate Inplane Moduli

8.337E+8	$E^b_x =$ 2.337E+9
8.337E+8	$E^b_y =$ 2.337E+9
3.213E+8	G^b_{xy} 8.988E+8
0.298	ν^b_{xy} 0.299
0.	$r_M =$ 0.
0.	

$t_{foam} = 0.050m$

10.8 (4/5)

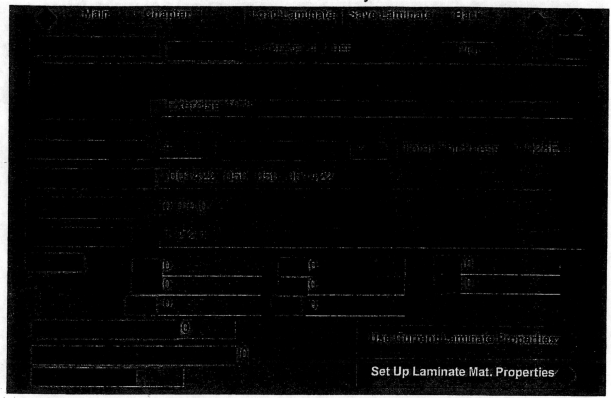

Set Up Laminate Mat. Properties

Ex 10.8 (5/5)

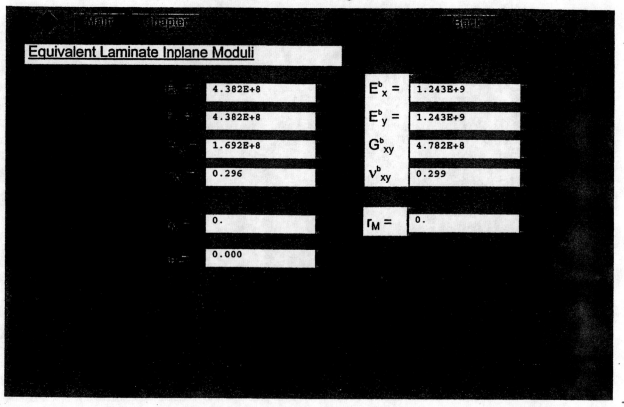

$t_{foam} = 0.100$ m

Taylor & Francis
Taylor & Francis Group
www.taylorandfrancisgroup.com

270 Madison Avenue
New York, NY 10016

2 Park Square, Milton Park
Abingdon, Oxon OX14 4RN, UK

GS940

ISBN 1-56032-702-2

90000

9 781560 327028

» to be returned on or before
        below.

New Materials by
Mechanical Alloying Techniques

E. Arzt and L. Schultz
Editors

INFORMATIONSGESELLSCHAFT · VERLAG

ISBN 3-88355-133-3

Papers presented at the DGM Conference
"New Materials by Mechanical Alloying Techniques",
Calw-Hirsau (FRG), October, 1988
under the direction of
Dr. E. Arzt, Stuttgart and Dr. L. Schultz, Erlangen.

D
671.3
NEW

HS

Hirsau, the historical location of the DGM Conference
„New Materials by Mechanical Alloying Techniques"

FOREWORD

The development of new engineering materials is increasingly considered a key technology, because it has the potential of stimulating innovation in all fields of engineering. In
particular, non-equilibrium processing methods, such as powder metallurgy and rapid
solidification, are receiving considerable attention in the research and development
community. Among the processes with the highest potential for tailoring advanced
materials, mechanical alloying techniques are of special interest because they offer
great flexibility in the choice of constituent materials to be combined. Invented in the
1970's as a method for producing unique dispersion-strengthened alloys for high-temperature applications, mechanical alloying has developed several side branches and
stimulated the development of related processes. Since the discovery in 1983 that even
amorphous alloys can be produced by mechanical alloying, a new wave of research
activity concerned with the possibilities and perspectives of this process has been
unleashed.

This book is the result of a conference on "New Materials by Mechanical Alloying Techniques", designed to present and discuss the state-of-the-art in this field and to provide
some "cross-fertilization" between the different research groups including materials
scientists, chemists, solid state physicists and producers. The conference was held on
October 3–5, 1988 in the Kloster Hotel at Calw-Hirsau in the picturesque Black Forest
region of West Germany. The group of participants consisted mainly of invited speakers
from eleven countries, including three overseas nations (USA, Japan, Israel). This
book, which contains the contributions presented orally and in poster form, will make
the results of the conference available to a larger community and serve as a future reference for the current state of development.

The editors gratefully acknowledge financial support for the conference by Deutsche
Forschungsgemeinschaft, Max-Planck-Gesellschaft and Siemens AG, and the patron-
age of Deutsche Gesellschaft für Metallkunde. We are indebted to Mrs. J. Heß and Mrs.
G. Poech for their invaluable assistance in running the conference and preparing this
book, and to Dr. R. Timmins for untiring proof reading.

Stuttgart and Erlangen, December 1988

E. Arzt
L. Schultz
(Editors)

Participants at the DGM Conference
"New Materials by Mechanical Alloying Techniques"
Calw–Hirsau, FRG, 3–5 October, 1988

CONTENTS

A:
MECHANICAL ALLOYING:
PROCESSES AND MECHANISMS

Mechanical Alloying -- A Perspective

John S. Benjamin
ALUMINUM COMPANY OF AMERICA
Alcoa Center, Pennsylvania, U.S.A.

Abstract

Mechanical alloying is a dry, high energy ball milling process for producing composite metal powders with a fine controlled microstructure. It was originally developed to make high temperature alloys combining oxide dispersion and intermetallic compound strengthening. The mechanism of the process is described using nickel-based superalloys as the vehicle. Early experiments leading to the invention of mechanical alloying are covered along with some peripheral problems that had to be solved in order for the process to be successfully applied. Use of the process with diverse systems including aluminum-based alloys, intermetallic compounds and amorphous materials is discussed.

Introduction and Background

Mechanical alloying (1) was developed around 1966 at INCO's Paul D. Merica Research Laboratory as part of a program to produce an alloy combining oxide dispersion strengthening with gamma prime precipitation hardening in a nickel-based superalloy intended for gas turbine applications. In the early 1960's INCO had developed a process for manufacturing graphitic aluminum alloys by injecting nickel-coated graphite particles into a molten bath by argon sparging. A modification of the same technique was tried to inoculate nickel-based alloys with a dispersion of nickel-coated, fine refractory oxide particles. The purpose of nickel coating was to render the normally unwetted oxide particles wettable by a nickel-chromium based alloy. Early experiments used metal-coated zirconium oxide purchased from an outside vendor. Chemical analysis, metallographic analysis and mechanical property measurements, however, revealed no differences between the inoculated materials and uninoculated alloys. Examination of the inoculants revealed that they were, in fact, zirconia-coated nickel rather than nickel-coated zirconia.

Two additional lines of experimentation were followed to overcome this problem. First, chemical coprecipitation and selective reduction processes, such as those employed by the DuPont Corporation to produce thoria dispersion strengthened nickel, were examined as a means of producing a composite nickel-refractory oxide particle with a high volume fraction of oxides (approximately 20-30%) for inoculation into Nichrome melts. None of these powders were any more successful in being retained in melts than the oxide coated nickel had been.

Second, a separate series of experiments examined the kinetics of the rejection of fine metal oxides from molten alloys in order to determine the difficulty of the problem of injecting and retaining the oxides in the melt. Because of the complexity of coprecipitation/selective reduction as a means of producing fine composite powders, attention was directed to ball milling as a means of coating oxide particles with nickel. Ball milling (2) had been applied to the coating of tungsten carbide with cobalt for well over 30 years. Small amounts of nickel-coated thoria and zirconia were successfully produced in a small high-speed shaker mill. An additional consideration came into play with the availability of this new tool for making composite powders. The role of alloying elements such as chromium and tungsten on the wetting of oxides in molten metals could now be examined relatively easily. The experimental procedure was to produce small amounts of composite metal powders (about 1 cubic centimeter of powder per experiment) in a high-speed shaker mill, compact these using an Arbor press and partially melt the compacts in a small arc melter. The melted compacts were sectioned to determine the severity of oxide rejection.

One theory regarding the success of the injection of nickel-coated graphite into aluminum was that the strong exothermic reaction between the nickel on the surface of the graphite and the molten aluminum cleansed the surface of the graphite and lowered the surface energy of the aluminum alloy/graphite interface. On this basis, it seemed that aluminum would be a better coating than nickel to apply to a refractory oxide for injection into a nickel alloy. A number of composite metal mixes were made with high contents of aluminum.

None of the experiments performed, with tungsten, chromium or aluminum additions, showed any significant effects of these alloying elements on retention of the oxide during arc melting of compacts of powder.

The kinetics of rejection were also examined by using a material in which an oxide had been successfully dispersed by another means. A small sample of TD nickel was placed in a button arc melter and partially melted by touching its edge briefly with an arc. The specimen was sectioned and its structure examined by stripping replicas from the surface in the vicinity of the melt/solid interface. The results of this single experiment were quite discouraging. Within microns of the solid/liquid interface, thoria particles began to agglomerate into long stringers which were then dragged by surface tension forces to the surface of the melt. It was estimated that the material in this vicinity could have been molten for no longer than a tenth of a second. Thus, it was unlikely that a fine, refractory oxide particle could be retained in a nickel alloy melt for sufficient time to cast and solidify an ingot even if it could be added successfully.

Sometime in mid-1966, out of a sense of desperation, attention was turned to the ball milling process that had been used to make metal powders for the wetting studies as a means of making the alloy itself by powder metallurgy. The reasoning was somewhat as follows: It was well known that ball milling could be used to coat hard phases such as tungsten carbide or zirconium oxide with a soft phase such as cobalt or nickel. It was also known that, under certain milling conditions, you could grind metal powders by plastically deforming them to the point where they began to fracture. If very fine particle sizes were required, liquids such as heptane or milling agents such as stearic acid had to be employed to prevent cold welding from occurring. This meant that, at some fine size, welding or coating became as rapid as grinding. It was also known that very fine metal powders, especially those containing reactive alloying elements such as chromium or aluminum, could be pyrophoric at worst or pick up large amounts of oxygen at best. Another factor to be considered was that the reactivity of an element in an alloy is directly related to its activity. Thus, the aluminum in a dilute nickel-aluminum alloy or nickel-aluminum intermetallic compound could be orders of magnitude less reactive than pure aluminum. All of these factors combined to suggest a method of producing a composite alloy powder. The idea was to use a high energy mill to favor the plastic deformation required for cold welding and shorten the process times; to use a mixture of elemental and master alloy powders; to put reactive alloying elements such as aluminum and titanium into a brittle, readily powderable form with low activities of the reactive elements; to eliminate the use of surface-active agents which would produce a fine pyrophoric powder as well as contaminate the powder and to rely on a constant interplay between welding and fracturing to yield a powder with a refined internal structure, typical of very fine powders normally produced, but having an overall particle size which was relatively coarse and, therefore, stable.

In order to test the hypothesis, the process was scaled up to a one-gallon attritor mill. Two sets of experiments were run in parallel. First a test on a mixture of pure nickel and thoria sought to duplicate the properties of TD nickel in a material produced by a completely different route. Second was a test to examine the feasibility of producing a simple nickel-chromium-aluminum-titanium alloy also containing a thoria dispersoid. This second test addressed another potential theoretical problem: that of destabilization of refractory oxides such as thoria, by the presence of reactive metals such as aluminum in the alloy matrix. Even before superior mechanical properties were demonstrated, a number of experiments involving aluminumizing TD nickel and extrusion consolidation of high energy milled aluminum containing powders indicated that the fine thoria dispersoid seemed to survive relatively long times at temperatures close to the melting point of the matrix.

As a historical note, it should be recorded that, at this stage, the process was referred to as "milling/mixing." The term "mechanical alloying" was actually coined in the late 1960's by Ewan C. MacQueen, a Patent Attorney for the International Nickel Company.

There were a number of problems other than the production of a superalloy powder with a fine oxide dispersoid that had to be solved in order to produce the first oxide dispersion strengthened gamma prime hardened superalloy. These problems included, for example:

1. the best method for consolidating the powders, whether they could be cold pressed and sintered, hot isostatically pressed, vacuum hot pressed or hot extruded;

2. if extruded whether the cans should be made of stainless steel to match the plastic deformation characteristics of the alloy within or a mild steel;

3. whether the powders had to be degassed prior to canning, evacuated within a can, heated when they were evacuated, or whether evacuation was required at all;

4. under what conditions hot extrusion should be carried out;

5. whether compaction was required prior to extrusion to insure complete consolidation of the material;

6. whether, like TD nickel, some post-extrusion working was required in order to develop optimum high temperature properties.

It is instructive, at this point, having described all of the uncertainties associated with the work in the late 1960's, to examine the powder manufacturing process itself with the perspective of today's understanding.

The Mechanical Alloying Process

Mechanical alloying is a dry, high energy milling process for the production of composite metallic powders with a fine controlled microstructure. It occurs by the repeated welding and fracturing of a mixture of metallic and nonmetallic powders in a highly activated ball charge. The raw materials used in mechanical alloying must include at least one fairly ductile metal to act as a host or binder to hold together the other ingredients. Ductile metals that have been used include nickel, copper, iron, aluminum and even such normally brittle metals as chromium. Under the extreme state of hydrostatic compression that exists when powder particles are trapped between colliding grinding balls, materials that are brittle under tensile loading will undergo fairly large amounts of plastic strain before fracture. Other ingredients can include brittle intermetallic compounds such as aluminides. Substantial amounts of nonmetallic ingredients including oxides, carbides, nitrides and elements such as carbon and boron can be added. Alloying additions can also be made by doping the milling atmosphere.

In the case of processing of very ductile metals such as aluminum (3), use of a suitable processing control agent (PCA) that decreases the clean metal-to-metal contact necessary for welding is required. Figure 1 gives the examples of representative constituents of starting powder that can be used in mechanical alloying and the effects that a single ball-powder-ball collision would have on these pure materials.

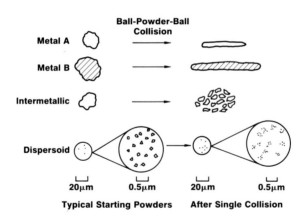

Fig. 1 Representative Constituents of Starting Powders Used in Mechanical Alloying, Showing Their Deformation Characteristics

Figure 2 shows a mixture of raw material powders such as is used in the production of a typical oxide dispersion strengthened superalloy. Visible in the photomicrograph are light particles of irregularly shaped chromium, somewhat smaller two-phased particles of a nickel-aluminum-titanium master alloy and small spiky particles of carbonyl nickel powder. Also present, but not visible at this magnification, are fine particles of Y_2O_3 having a crystallite size of about 35 nm.

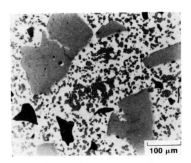

Fig. 2 Typical Raw Materials Mix for the Production of an Oxide Dispersion Strengthened Superalloy.

Figure 3 shows the structure of a mechanically alloyed superalloy powder after a short period of processing. Note that the initial ingredients can still be identified within the composite metal particles. This photograph also shows the characteristic lamellar structure of mechanically alloyed particles during early stages of processing.

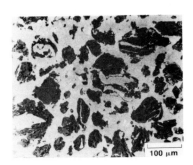

Fig. 3 Mechanically Alloyed Superalloy Powder after a Short Period of Processing. Note Original Ingredients Readily Identified.

As processing continues, the composites shown in Figure 3 are fractured and rewelded by the energetic grinding ball charge. This "kneading" action leads to a continual refinement of the internal structure of the metal powder. Because a balance is achieved between the amount of welding and the amount of fracturing, a steady state particle size distribution develops that is a function of the mechanical alloying process conditions and the composition of the alloy being processed. Figure 4 shows a well processed oxide dispersion strengthened superalloy powder. At this stage of processing, all of the lamellae within the powder particles have been reduced in thickness to below that of a wavelength of visible light. Thus, although the particle sizes of the coarsest starting ingredients are around one hundred microns, a degree of internal refinement has been achieved such that the inter-particle spacing of all the ingredients, including the fine oxide, is well below a half micron.

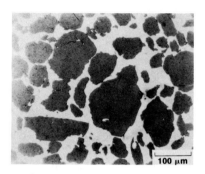

Fig. 4 Fully Processed Oxide Dispersion Strengthened Superalloy Powder. Structure Refined Below Optical Resolution.

Figure 5 shows a transmission electron photomicrograph of a mechanically alloyed superalloy powder particle (4). In this case, traces of the fine lamellar structure with an interlamellar spacing of well under .1 micron can readily be seen. The alloy processed here, INCONEL alloy MA 6000 (see Table 1), also contains the heavy element tungsten. The initial raw materials could be identified by the diffraction rings shown in the upper right hand corner of the illustration.

Nominal Compositions of High Temperature Alloys															
Alloy	Ni	Cr	Al	Ti	Ta	Mo	W	Co	V	Zr	Hf	B	C	Y_2O_3	ThO_2
INCONEL alloy MA 6000	BAL	15	4.5	2.5	2.0	2.0	4.0			.15		.01	.05	2.2	
A	BAL	19	1.2	2.4						.05		.002	.05		2.71
NIMONIC alloy 80A	BAL	19.5	1.4	2.4						.06		.003	.05	--	--
TD Nickel	BAL	--													2.0
B-alloy	BAL	12.5	4.7	2.0	2.4		11.7		.58	.07		.007	.085	1.2	
MAR-M-200+Hf	BAL	9.0	5.0	2.0			12.0	10.0		.08	1.0	.02	.14		

Table 1

Fig. 5 Transmission Electron Photomicrograph of Mechanically Alloyed Superalloy Particle.

Structure/Property Relationships in ODS Superalloys

Initially, it was thought that the major hurdle to be overcome in developing an ODS superalloy was simply to include oxide particles in a superalloy matrix. It was very quickly realized that the problem was far more complex than this. Figure 6 shows the 1093°C (2000°F) stress rupture properties of alloy A, (5) an extruded thorium oxide containing version of NIMONIC* alloy 80A (see Table 1 for composition). Also shown are the stress rupture properties for NIMONIC alloy 80A, representing the base metal, and TD nickel, then considered the most advanced oxide dispersion strengthened material but lacking intermediate temperature strength or high temperature corrosion resistance. Clearly, the results of this experiment were disappointing since the properties of the extruded, annealed oxide dispersion strengthened alloy were only comparable to those of the conventional alloy. Figure 7 shows the microstructure of the extruded and heat-treated oxide dispersion strengthened material. Aside from a fairly heavily twinned appearance and a somewhat irregular 100 to 200 μm grain structure, there were no obvious flaws which could explain the relatively low strength.

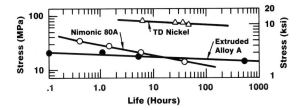

Fig. 6 1093°C (2000°F) Stress Rupture Properties of Early Thoriated Alloy A. Extruded and Annealed 1204°C (2200°F) 2 Hours

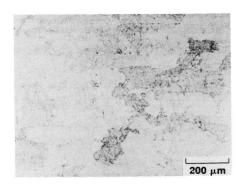

Fig. 7 Structure of Thoriated Alloy A. Extruded and Annealed 1204°C (2200°F)/4 Hours

On the assumption that some chemical heterogeniety, not visible optically, might be causing problems and relying on the general knowledge that alloys of

*NIMONIC is a trademark of the INCO family of companies.

this type are frequently improved by hot working, alloy A was hot bar rolled 50% reduction in area at 1175°C (2150°F) after a homogenization anneal at 1204°C (2200°F), the same homogenization anneal that had been given to the material tested in Figure 6. The material was then given standard heat treatments for NIMONIC alloy 80A. There was some improvement in the strength of the alloy and the slope of the log stress log rupture life plot as shown in Figure 8 was parallel to that of TD nickel. However, the strength of the alloy was only about half that of the oxide dispersion strengthened pure metal, again not the sort of result desired. Figure 9 shows the structure of the hot rolled material. Again, there is nothing unusual although there are some areas of much larger grains than the average in the field.

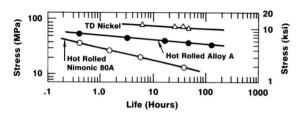

Fig. 8 1093°C (2000°F) Stress Rupture Properties of Hot Rolled Thoriated Alloy A. Extruded and Annealed and Hot Rolled and Solution Treat and Age

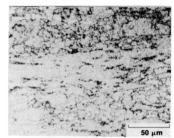

Fig. 9 Structure of Thoriated Alloy A. Extruded and Annealed, Hot Rolled 50% and Heat Treated

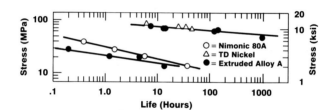

Fig. 10 Bimodal 1093°C (2000°F) Stress Rupture Behavior in Thoriated Alloy A

Figure 10 compares the properties of another extruded bar of the thoriated alloy which had been given an anneal at 1200°C (2200°F). Note that the stress rupture results are bimodal! From the same bar of material, some specimens gave properties inferior to those of cast and wrought NIMONIC alloy 80A, while other specimens gave properties approaching those of the TD nickel. It was with this set of experiments that the importance of grain structure was first realized. Recognize that (1) these graphs represent a collection of data from a number of bars, (2) we now know what was going on, and (3) these data points were coming in one at a time. In the actual experiments, a number of samples from the lower population loaded at stresses of 70 MPa (10 ksi) would fail on loading while other specimens loaded at lower stresses, thinking they were weak but in fact coming from this stronger population, would last for hundreds of hours. Examination of specimens from the stronger population showed that all of them had a very coarse and elongated grain structure such as shown in Figure 11. Here the grains average 3500 μm in length by about 600 μm in width. It was found that increasing the annealing temperature following extrusion from 1204°C (2200°F) to 1275°C (2325°F) eliminated the occurrence of the fine 100-200 μm grains seen in the weaker material. Specimens heat treated at the higher temperature gave results such as those shown in Figure 12.

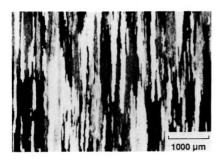

1000 μm

Fig. 11 Typical Coarse, Elongated Grain Structure

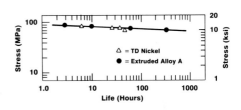

Fig. 12 1093°C (2000°F) Stress Rupture Properties of Thoriated Alloy A after Annealing at 1274°C (2325°F)

Having produced a few extruded bars of material which gave consistently high strengths following a 1274°C (2325°F) anneal, other extrusions, seemingly produced identically, would revert to the weaker properties or give mixtures of strong and weak specimens. It was now established that the latter correlated with the occurrence of fine grained areas. Attention was, therefore, directed at thirteen thermomechanical process variables. Even a minimum treatment of these variables at two or three levels apiece, depending on the specific variable, led to 70,000 combinations to be investigated if all were to be taken into consideration. Experiments were performed to identify which of those variables were important and some values of these variables which would consistently yield acceptable coarse elongated grain structures.

Forays into compositional modification were made. Some of these, such as small increases in dispersoid content, and additions of modest amounts of single alloying elements such as iron and cobalt, were reasonably successful. In other cases attempts to apply mechanical alloying to more complex compositions met with mixed success. It was apparent that thermomechanical processing was critical to achieving a desired grain structure and high temperature properties. It also appeared that certain compositions were far less responsive to the apparent correct thermomechanical processing procedures than others.

Because of the radioactivity of ThO_2, other stable refractory oxides were considered including, La_2O_3, Y_2O_3, CeO_2, Al_2O_3, and ZrO_2. Y_2O_3 was found to give the best combination of high strength and broad thermomechanical processing response. It was adopted for most subsequent alloy work.

Development of INCONEL Alloy MA 6000

The properties of the simple oxide dispersion strengthened gamma prime
hardened superalloys discussed above were insufficient for application as a
gas turbine blading material. Once extensive mapping of thermomechanical
processing conditions to produce coarse elongated grain structures had been
completed, a major alloy development program was undertaken in order to raise
intermediate temperature (760°C) properties to those required by centrifugally
loaded gas turbine blades.

Approximately 300 alloy compositions were prepared by mechanical alloying,
consolidated, recrystallized and evaluated against target properties (5).
Whenever possible, partial replicate, statistically designed experiments were
used to maximize the information obtained. Alloying elements examined were
divided according to their partitioning behavior in superalloys. Thus, strong
interactions were examined between gamma prime partitioning elements such as
aluminum, titanium, niobium, and tantalum. Many combinations of alloying
elements known to segregate to the gamma matrix such as chromium, tungsten,
molybdenum, and cobalt were examined and strong interactions between them
sought. Experiments were also performed where gamma prime partitioning and
gamma partitioning elements were both varied methodically. For example, in one
series of forty heats, thirteen compositional variables were evaluated at
three levels. This was an efficient approach since a full study of all
thirteen variables at three levels would have required nearly 1,600,000 heats.
The program had started with a material termed B-alloy derived, in turn, from
a cast alloy, IN-100. This series of compositional studies ultimately
resulted in an alloy designated as INCONEL alloy MA 6000. The composition
of the starting alloy and the final alloy MA 6000 are given in Table 1.

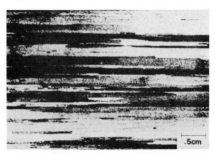

(a) **Optical micrograph of elongated grains**

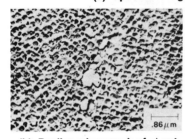

(b) **Replica micrograph of γ′ ppt**

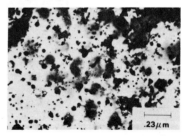

(c) **Thin foil micrograph showing
oxide dispersion and γ′ ppt**

Fig. 13 Microstructure of INCONEL Alloy MA 6000

The microstructure of fully heat treated INCONEL alloy MA 6000 is shown in Figure 13. Figure 13a shows the very coarse elongated grain structure which is formed by passing a hot zone through the material at a temperature below the solidus but above the gamma prime solvus. Figure 13b shows the high volume fraction of gamma prime while Figure 13c shows the fine dispersion of Y_2O_3 in both the gamma matrix and gamma prime precipitate. The properties of INCONEL alloy MA 6000 are compared to those of cast, directionally solidified alloy MAR-M-200+Hf* and the dispersion strengthened metal, TD nickel, in Figure 14. Note that MA 6000 combines the high strength of the cast alloy at lower temperatures, around 760°C, with the high strength and stability of the oxide dispersion strengthened metal at temperatures up to 1100°C. It is much stronger at high temperatures than TD nickel because of substantial solid solution hardening from heavy elements such as tungsten and molybdenum. The alloy also possesses an excellent combination of hot corrosion (sulphidation) and cyclic oxidation resistance.

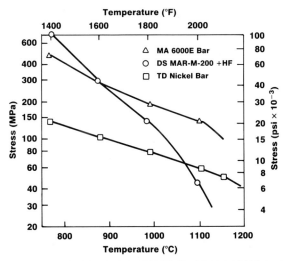

Fig. 14 Stress for 1000 Hour Life for INCONEL Alloy MA 6000, TD Nickel and MAR-M-200+Hf

Fig. 15 Mechanically Alloyed Aluminum Alloy Showing Fine Grain Structure and Uniform Dispersion of Equiaxed Particles.

*MAR-M is a trademark of Martin-Marietta

Aluminum Based Alloys

Mechanical alloying can be applied to aluminum based systems provided that
organic process control agents are used to control the extreme tendencies of
aluminum to weld to itself during high energy milling. Mechanical alloying
allows the development of extremely uniform dispersions of equiaxed Al_2O_3 and
carbide particles formed due to reaction of aluminum with oxygen and carbon
added in the process control agents (3).

Figure 15 shows the transmission electron microstructure of a typical
mechanically alloyed aluminum alloy. Note that the grain size is less than
0.5 microns. Also present are equiaxed particles of Al_2O_3 and carbides. The
major applications for mechanically alloyed aluminum alloys are at ambient
temperature rather than elevated temperature. One example is IN-9052 (Al-4%
Mg-.8% O-1.1% C). Although a solid solution alloy, IN-9052 gives combinations
of tensile strength, fatigue strength, and toughness normally only seen in
precipitation hardened 7000 series aluminum alloys (6). Because it is a solid
solution material, IN-9052 also possesses excellent corrosion and stress
corrosion cracking resistance. The high strengths of mechanically alloyed
aluminum based alloys are due to a combination of the extremely fine grain
size and direct dispersoid effects. The fatigue crack growth rates of IN-9052
and another mechanically alloyed aluminum alloy, IN-9021-T4, are compared with
that of alloy 7075 in Figure 16.

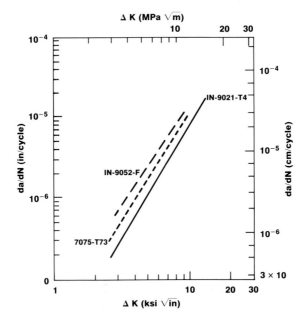

Fig. 16 Fatigue Crack Growth Characteristics of Mechanically Alloyed Aluminum Alloys

15

Exploring the Limits of Mechanical Alloying

With the successful proof, first, that mechanical alloying worked and, second, that it was possible to combine oxide dispersion strengthening and gamma prime precipitation hardening in the same alloy, a massive program was undertaken to explore just how broadly the process could be applied. Table 2 contains a partial listing of some of the systems and alloys to which mechanical alloying was applied by the INCO research organization in the late 1960's and early 1970's. Comments about a few of the specific systems are instructive, not only in light of recent developments, but perhaps as indications of developments yet to come.

Some Systems Studied in the INCO Mechanical Alloying Program C. 1970.

Chromium	Al_2O_3- NiO
Tool steel	Aluminum-nickel (Raney nickel)
Aluminum	Nb_3Al, Nb_3Sn
Copper	Mo_3Sn
Iron-copper alloys	Brass
Nickel-based superalloys	Titanium - Y_2O_3
Nickel	Magnesium-iron
Iron-chromium-aluminum alloys	Copper-lead
Stainless steels	Nickel-tungsten-zirconia

Table 2

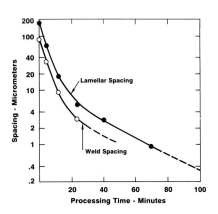

Fig. 17 Lamellar Spacing as a Function of Processing Time

That true alloying occurred in powders mechanically alloyed at ambient to slightly elevated (100-200°C) temperatures was demonstrated first by magnetic measurements on nickel-20% chromium powders (7) and, secondly, by observing color changes in copper-zinc alloys. Measurement of interlamellar spacing, as a function of processing time (see Figure 17) and the development of a theoretical equation to describe refinement of structure indicated that, for most practical applications, processing needed to be carried no further than the point at which the powder was optically featureless. It was understood that the process could be continued until homogeneity on an atomic order could be achieved mechanically. In the early 1970's there was no practical system or even one of academic interest where anything important could be achieved by doing so.

Work on Nb$_3$Sn (8) and Nb$_3$Al suggested that mechanical alloying was a very convenient method of exploring binary and high order systems for interesting intermetallic compounds. Reference to a fifteen year old edition of Hansen's Handbook of Binary Phase Diagrams indicated no such compound had been successfully prepared in the similar molybdenum-tin system. To make a point, a mixture of 75 atomic percent molybdenum and 25 atomic percent tin was mechanically alloyed for 90 minutes in a high-speed shaker mill. The powder was annealed at 1000^{0}C in an evacuated, fused quartz vial and an x-ray diffraction pattern run confirming the formation of an A15 structure. This was accomplished within a single eight-hour period of one work day. Unfortunately for us, reference to the latest edition of the Handbook of Binary Phase Diagrams indicated that such a compound had been prepared by other means. It had taken some 15 years between the two editions of the book for that to have been accomplished. The mechanical alloying process had allowed that to be done in less than one day.

A very simple consideration of what occurs during mechanical alloying allows one to form an idea about how powerful a process for homogenization of structure it is. Each time a powder particle is trapped between colliding grinding balls, it is plastically deformed sufficiently to reduce its thickness by a factor of two to three times. The manufacture of oxide dispersion strengthened superalloys, starting from particle sizes of around 100 microns and aiming to reduce the lamellar thickness within the composite powders to around .2 micron, requires that each composite powder particle take part in only six to eight mechanical alloying "events." To reduce the inter-lamellar spacing down to five angstroms, or an atomic diameter, would require between eleven and thirteen events.

Another interesting side effect was noted in experiments to examine whether Raney nickel catalyst could be produced more effectively from mechanically alloyed powders than from cast and crushed aluminum-nickel alloys. During one experiment to produce a powder containing about 25 weight percent aluminum, balance nickel, the entire powder charge became cold welded as a ring-shaped deposit in the lower, outer corners of the tank. The attritor was dismantled, the ball charge drained and the agitator removed from the tank. A technician used a long-handled cold chisel and hammer to chip the deposit of aluminum-nickel alloy from the tank. As soon as he had struck the first blow to the aluminum-nickel deposit, it began to glow dull red at the point where the chisel had made impact. Heating quickly spread to the entire deposit within the attritor tank raising it to dull red heat. The overall composition of the deposit of the attritor tank was relatively close to the Ni-Al intermetallic compound. The blow from the cold chisel had triggered the reaction of the intimately intermixed and contacted aluminum and nickel lamellae, liberating the heat of formation of the compound. This was enough to raise the temperature of the compacted ring in the attritor by some 800^{0}C. This heat of reaction is used in nickel-coated aluminum powders and nickel-aluminum powder agglomerates produced by other means for forming bond layers in plasma sprayed coatings. To my knowledge mechanical alloying has not, to date, been used for producing similar powders.

High energy milling of mixtures of alumina and nickel oxide powders led to greater ease of mixed oxide formation than by simple blending and calcining (9). This suggests that, as is the case deep within the earth, oxides subjected to high hydrostatic forces can deform plastically much like metals. Recent results by Koch (10) have shown that similar effects can be obtained in brittle semimetals such as silicon and germanium.

Some unusual variations of mechanical alloying can be shown by considering two systems: aluminum and titanium. Two approaches were taken during early work on applying mechanical alloying to aluminum. The one that has been given the most attention was the discovery that the extreme welding tendency of the aluminum could be overcome by the use of small amounts of process control agents such as stearic acid and methanol. Another approach that appeared to work equally well but was abandoned as impractical, was to process the aluminum at reduced temperatures. This was done by replacing the water in the cooling jacket of the stirred high energy ball mill by acetone that was cooled to -100°C by passing it through a bed of dry ice. This established the validity of using temperature as well as chemical agents to control the balance of welding and fracturing necessary to achieve mechanical alloying. In the case of titanium, the solution employed was the opposite. High temperature pressurized steam was passed through the "cooling" jacket of an attritor to raise the processing temperature several hundred degrees centigrade. Unfortunately, in this case, it meant that the stainless steel walls of the attritor tank were peened outward, effectively destroying the tank by changing the critical clearances between the agitator arm ends and the tank's walls. In theory, materials such as titanium, which forms a brittle hexagonal phase under plastic deformation at room temperature, could be processed in a specially constructed attritor by raising the process temperature to a regime where the metal would act more ductile. Of the systems listed in Table 2 the nickel-based superalloys, aluminum, and the iron-chromium aluminum system have been commercialized by INCO. Small amounts of iron-magnesium supercorroding alloys are being produced for military applications (11).

Recently, the application of mechanical alloying to relatively exotic systems has been renewed by the work of Koch, Schwarz and others (12) into the formation of amorphous materials. Table 3 adapted from Koch gives a listing of some of the systems that have been investigated. It has been found that mechanical alloying will yield amorphous compositions over a much broader range than other processes such as splat quenching. This exciting work, together with the earlier work of Hill et al on ceramics, suggests the possibility that mechanical alloying might be applied to the production of unique compounds, possibly including some of the new oxide superconducting materials.

Some Exotic Systems Prepared by Mechanical Alloying
(After Koch)

Ni-Nb	Cu-Zr	Fe-Zr
Ni-Ti	Cu-Er	
Ni-Zr		Co-Zr
Ni-Er	Ti-Cu	
	Ti-Co	Si-Ge
Nb-Ge	Ti-Pd	
Nb-Ge-Al	Ti-PD-Cu	
Nb-Sn		

Table 3

Summary

The mechanical alloying process was invented over 20 years ago as a method of producing oxide dispersion strengthened superalloys. This composite powder processing technique, together with controlled thermomechanical processing of powders, was successful in combining oxide dispersion strengthening and gamma prime precipitation hardening in the same material. Subsequent alloy development produced a number of systems possessing unique combinations of strength, oxidation resistance and sulfidation resistance at temperatures up to 1100°C. Following the initial success of mechanical alloying in superalloys, its application to a wide variety of systems was investigated. Other systems included aluminum based alloys, intermetallic compounds, ceramics, and unique composite materials. Recently, mechanical alloying has been expanded in its application to include amorphous systems. It is quite likely that many new discoveries involving the mechanical alloying process remain to be made.

References

(1) J. S. Benjamin: Met. Trans. (1970) Vol. 1, p.2943.
(2) S. L. Hoyt: AIME Trans. (1930) Vol. 89, p.9, see esp. p.25.
(3) J. S. Benjamin and M. J. Bomford: Met. Trans. A (1977) Vol. 8A, p.1301.
(4) S. K. Kang and R. C. Benn: Met. Trans. A (1987) Vol. 18A, p.747.
(5) R. C. Benn, J. S. Benjamin and C. M. Austin (1984) High Temperature Alloys: Theory and Design, p.419, Warrendale, PA TMS-AIME.
(6) D. L. Erich and S. J. Donachie: Met. Prog. (1982) Vol. 121, pp.22-25.
(7) J. S. Benjamin: Sci. Am. (1976) Vol. 40, p.234.
(8) H. F. Merrick, et al: unpublished research, INCO R&D Center (c. 1975).
(9) B. Hill: unpublished research, INCO R&D Center (c. 1974).
(10) R. M. Davis and C. C. Koch: Scripta Met. (1987) Vol. 21, p.305.
(11) S. A. Black: (1979) Development of Supercorroding Alloys for Use as Timed Releases for Ocean Engineering Applications, Civil Eng. Lab. (Navy), Port Hueneme, CA, 40 pp.
(12) R. B. Schwarz and C. C. Koch: Appl. Phys. Lett. (1986) Vol. 49, No. 3, p.146.

Dispersion Strengthening by Mechanical Alloying

R.C. Benn and P.K. Mirchandani, Inco Alloys International,
Inc., Huntington, WV, USA

Abstract

The structural and physical metallurgy requirements for
dispersion strengthening materials can be conveniently, if not
uniquely, met by Mechanical Alloying. The requirements for
dispersion strengthening in a wide range of materials ranging
from superalloys to light metals are reviewed. The design of
optimum microstructures to maximize dispersion strengthening
using MA is then discussed with particular emphasis on tailoring
properties to meet given aerospace and industrial applications.

Introduction

The development of dispersion strengthened (DS) alloys by
internal oxidation circa 1930(1) and the invention of dispersion
strengthened aluminum in the form of sintered aluminum powder
(SAP) in 1949(2) promoted interest in using the potential of
dispersion strengthening for extending performance in many other
alloy systems. The emergence of experimental facts about DS
metals stimulated a number of theoretical investigations of
dispersion strengthening at high homologous temperatures invoking
Orowan(3) bowing of dislocations. In its simplest form, the
critical stress for a dislocation to bypass a dispersoid leaving
behind a dislocation loop around the dispersoid is given(4) by

$$\sigma_{Or} = \frac{0.8 \ GbM}{L} \qquad [1]$$

where G is the shear modulus, b is the Burgers vector, M is the
Taylor factor and L is the interdispersoid distance. The appropriate interdispersoid distance is the mean square lattice
spacing (4).

$$L = \left[(\pi/f)^{0.5} - 2 \right] (2/3)^{0.5} r \qquad [2]$$

where f is the dispersoid volume fraction and r is the dispersoid
radius. Consideration of equations (1) and (2) suggests that it
is essential to distribute the dispersoids finely and evenly to
produce a strong alloy. Significant increases in, say, room
temperature strength due to the addition of dispersoids may be
obtained in alloys which exhibit interdispersoid spacings of
approximately 100 nm (e.g. 50-300 nm). Dispersions with an
interparticle distance of much more than 100 nm will not significantly increase the yield strength. Moreover, for optimum
dispersion strengthening, distributing a given volume fraction
(e.g. 0.002-0.10) of dispersoids (e.g. 10-50 nm diameter) more
finely is more effective than increasing the volume fraction per
se because of the square root dependence of equation [2].

Early attempts to produce dispersion strengthened nickel by ball milling were not successful owing to the inability to incorporate the dispersoid on a sufficiently fine and intimate scale. In general, mechanical mixing of powders produces a material whose internal homogeneity is dependent on starting powder particle size. Internal oxidation or selective reduction techniques are inefficient and limited,from thermodynamic considerations, to simple alloy systems. Chemical co-precipitation of fine dispersoid and matrix powders did meet the structural requirements, described above, for dispersoid strengthening and culminated in the commercial introduction, by DuPont, of thoria dispersed nickel, TD-Ni(5) followed by TD-NiCr(6,7). Although the theoretical geometric requirements of fine dispersoid and interparticle spacing was achieved via chemical co-precipitation methods, it was at a high production cost and subject to the limitation that only a few alloying elements could be added. The technical breakthrough for dispersion strengthened materials came in the form of the Mechanical Alloying (MA) process discovered by Benjamin(8). The MA process immediately overcame the metallurgical problems associated with the incorporation of fine particles by allowing the production of fairly coarse composite powders but with a finely distributed dispersoid. The process also permits use of reactive elements such as Al and Ti to produce the full range of complex oxide dispersion strengthened (ODS) superalloys. Adaptations of the MA process have led to the production of dispersion strengthened light alloys (eg Al base) and essentially the unique synthesis of any alloy (including intermetallics per se) or composite material in general with a controlled microstructure.

The theme of this paper will be to illustrate how MA technology can be applied to produce advanced materials in which the microstructure is designed to optimize dispersion strengthening effects, inter alia, while retaining the compositional and macrostructural control to meet other performance requirements for any given application. Depending on the physical metallurgy of the specific alloy system, the optimum microstructural design will be inherently different in, for example, Ni-base ODS superalloys compared with Al-based DS alloys. We shall, for convenience, discuss DS effects ostensibly as a function of equilibrium grain size ranging from nominally large grained superalloys to fine grained Al alloys with particular emphasis on recent alloy developments and some future prospects.

ODS Superalloys by MA

The Mechanical Alloying Process

Mechanical alloying is a dry-milling method for processing composite powders with a controlled microstructure. The process was originally developed to produce Ni-based ODS superalloys(8). By virtue of being a solid-state process, MA allows the materials scientist to circumvent materials limitations and, in particular, make alloys from normally incompatible components such as oxide-metal systems (e.g. ODS alloys), cermets in general, immiscible liquids, incongruent-melting components, intermetallics, metastable phases (e.g. amorphous materials), ceramic and/or polymer "alloys" and overcoming the practical size limit for metal powders as related to contamination and pyrophoricity. Further, the scale of microstructure is relatively independent of alloy composition.

General descriptions of the MA process and the mechanisms by which the process operates have been provided by Benjamin and co-workers (8-11) with a perspective view in this symposium(12) to which the reader is directed for details. It is sufficient here to emphasize that the repetitive cold welding, deformation and fracturing of the powder particles during mechanical alloying produces an equilibrium composite powder structure in which the average dispersoid (e.g. added oxide and/or naturally occurring oxide films) interparticle spacing is approximately the same as the welding interspace as shown schematically in Fig. 1(a). This "microstructural spacing" observed in various systems studied has shown a maximum value of 0.7 μm but in general is seen to be far less as observed, e.g. Fig. 1 (b), in heat treated INCONEL* alloy MA 6000 (Table I composition). This spacing also coincides with the random interparticle spacing of oxide dispersoids calculated on the basis of average oxide particle size and volume fraction added to the original powder mix.

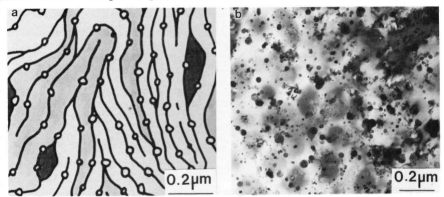

Fig. 1(a) Schematic showing microstructure of a powder particle at the final state of mechanical alloying in which the interdispersoid spacing is approximately the same as the welding interspace. (b) Microstructure of INCONEL alloy MA 6000 showing fine oxice dispersoids distributed in the γ/γ"-phase matrix.

Strengthening Mechanisms in ODS Alloys

There are three structural features which make ODS alloys different from conventional cast alloys(13): dispersoids, grain structure and texture.

Dispersoids

Conventional Ni-based superalloys are limited in their application at high temperature by the stability of the γ'-phase

22

Table 1. Nominal Compositions (Wt%) of MA Dispersion Strengthened Alloys

	Ni	Fe	Al	Cr	Mg	Li	C	O	Y_2O_3	Ti	W	Mo	Ta	B	Zr
SUPERALLOYS															
INCOLOY* alloy alloy MA 956		Bal	4.5	20					0.5	0.5					
INCONEL* alloy alloy MA 754	Bal	1	0.3	20			0.05		0.6	0.5					
INCONEL alloy alloy MA 6000	Bal		4.5	15			0.05		1.1	2.5	4	2	2	0.01	0.15
INCONEL alloy alloy MA 760	Bal		6	20			0.05		0.95		3.5	2		0.01	0.15
Alloy 51	Bal		8.5	9.5			0.05		1.1		6.6	3.4		0.01	0.15
Alloy 92**	Bal		6.5	8			0.05		1.0	1	6	1.5	3	0.01	0.15
MA Ni$_3$Al	Bal		12				0.05		1.0	0.5				0.05	
ALUMINUM ALLOYS															
IncoMAP* alloy AL-9052			Bal		4		1.15	0.5							
IncoMAP alloy AL-905XL			Bal		4	1.3	1.15	0.5							

**Includes 3% Re + 5% Co
*INCONEL, INCOLOY and IncoMAP are trademarks of the Inco family of companies.

and refractory carbides upon which they depend for strength. The ability of these dispersed phases to impede dislocation motion at elevated temperature is related to their free energy of formation. Refractory oxides such as Y_2O_3 (ΔG_f = 920 kJ/Mole at 1000°C), HfO_2, ZrO_2 and Al_2O_3, with ΔG_f values several-fold higher than the γ' or carbides, offer more stable alternatives. Negligible coarsening of the Y_2O_3-based dispersoid occurs, for example, in INCONEL alloy MA 6000 during exposure to temperatures below 1000°C for even more than 10,000h as shown in Figs. 2 and 3(13). Above 1100°C, coarsening has been observed in INCONEL alloy MA 6000 and INCOLOY alloy MA 956(14) but it is still controversial as to whether the increase in dispersoid diameter is due to the dissolution of small dispersoids ("Ostwald ripen-

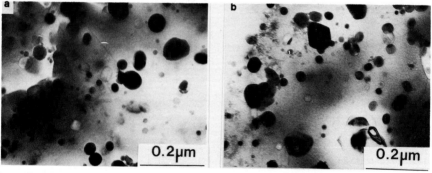

Fig. 2 TEM showing dispersoid morphology after stress rupture testing at 982°C: (a) 193 MPa/303 h, (b) 166 MPa/12,554 h. Note: larger particles that tend to be faceted are carbides (27).

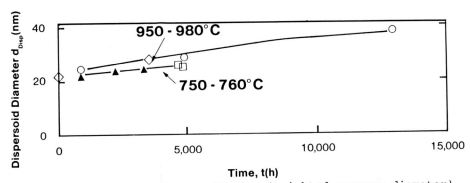

Fig. 3 Variation of dispersoid size (weighted average diameter)

ing") and/or the interaction of the Y_2O_3 particles with Al dissolved in the matrix to form mixed Y-Al oxides during high temperature exposure such as $3Y_2O_3 \cdot 5Al_2O_3$. In any event it may be deduced from Fig. 3 that there is no serious loss in the load-bearing capability of INCONEL alloy MA 6000 for practical applications. Earlier work on IN-853, which is similar to INCONEL alloy MA 754, exposed for 50h at 1315°C found a rupture stress drop of ~15% for a 100h life at 1038°C(15). Similarly, reductions in the yield strength of INCOLOY alloy MA 956 are negligible for exposures up to 1100°C and only become significant for extended (i.e. ≥2000h) exposures at or above 1150°C where reductions in the 1093°C/100h rupture strength of approximately 25% have been measured(16).

The effect of dispersoids on creep strength has been invest-igated by numerous authors. Work by Lund and Nix(17) compared creep rates of an ODS single crystal with the same alloy with no dispersoids. By normalizing the creep rates, $\dot{\varepsilon}$ and stress, σ,

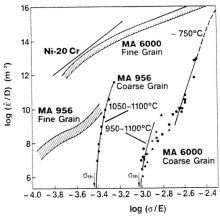

Fig. 4. Diffusion-compensated creep rates as a function of modulus-compensated applied stress for INCONEL alloy MA 6000 and INCOLOY alloy MA 956 (18).

with respect to the diffusivity, D and Young's modulus, E, respectively they were able to plot all the data onto a single master curve. This implies that the "power-law creep" semi-empirical equation that holds for many high temperature materials can, in principle, be applied to ODS alloys as well:

$$\dot{\varepsilon} = A. \; D. \; \left[\frac{\sigma}{E}\right]^n \qquad\qquad [3]$$

where A is a material constant and n is the stress exponent. Minimum creep rate data for INCONEL alloy MA 6000 and INCOLOY alloy MA 956 are accordingly plotted in Fig. 4(18). For comparison, the data for fine-grained (unrecrystallized) material and for Ni-20Cr are also included. The effect of dispersoids in the coarse grained materials is to lower the creep rates and to increase the stress exponent, n, to 20-40, which is far beyond the values typical of conventional superalloys. Such a high stress sensitivity is best described by invoking a "threshold stress" below which the creep rates are considered to be negligible. The data then follows a modified constitutive equation:

$$\dot{\varepsilon} = A. \; D. \; \left[\frac{\sigma-\sigma_{th}}{E}\right]^{n'} \qquad\qquad [4]$$

where σ_{th} is the threshold stress and n' the new stress exponent. Equation(4) has been verified for INCONEL alloy MA 6000 using n'= 3.5 and fitting the numerical value of σ_{th}(19). It is important to note that in the fine-grained condition, neither alloy exhibits a threshold stress of similar magnitude. This fact, which is due to the occurrence of grain boundary sliding and interface-controlled diffusion creep(20) is exploited in forming operations and will be discussed further in the context of fine-grained MA Al alloys.

Although the threshold stress concept itself has become widely accepted, the physical interpretation depends on what is thought to be the deformation mechanism. In many models σ_{th} turns out to be some fraction of σ_{Or} (21-23). More recent studies reviewed by Arzt et al.(24) with additional work reported in this symposium(25) have revealed an attractive interaction between dispersoids and dislocations at high temperatures. Theoretically it has been shown that only a small attractive interaction is required in order to bring a new strengthening mechanism into the foreground; the detachment of dislocations from the departure side of dispersoids over which the dislocations have climbed as shown schematically in Fig. 5 (26). The threshold stress for this process depends on a parameter, k, describing the relaxation of the dislocation and is proportional to the Orowan stress:

$$\sigma_{th} = \sigma_{Or} \sqrt{1-k^2} \qquad\qquad [5]$$

This approach may eventually lead to a better understanding of threshold stresses and allow a more rational exploitation of dispersion strengthening in high temperature alloys.

Detailed analysis of the stress rupture behavior of INCONEL alloy MA 6000 has revealed the existence of two relatively dis-

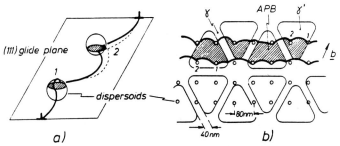

Fig. 5: Idealized representation of the dislocation-particle interaction in ODS alloys. (a) perspective view illustrating climb over dispersoid particles, 1, and subsequent "interfacial pinning," 2. (b) (111)-glide plane view showing pairwise γ ' shearing in INCONEL alloy MA 6000. (26)

tinct regions in the data plot as shown in Fig. 6(27). The locus of the inflection points in Fig. 6 is a complex function defined by the intersection of two property surfaces that characterize the behavior of the alloy in Regions I and II respectively. Each surface is a function, inter alia, of T, σ, γ '-phase volume fraction and dispersoid parameters. The rupture life data, which was shown to be compatible with the strain rate data in Figure 6, indicates that in region I (high stress, short lives) the stress sensitivity is much lower than in region II (low stress, long lives). This is of great importance for the extrapolation to long lives which will be over-conservative when based on data in

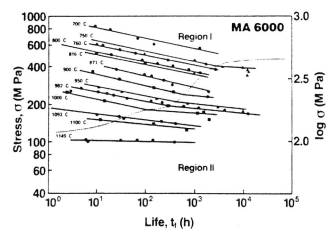

Fig. 6. Stress rupture behavior of INCONEL alloy MA 6000 (27).

region I. It appears likely that γ '-phase and dispersoid parameters predominantly influence the behavior in regions I and II respectively. These observations have an important influence on alloy design as discussed ahead.

Grain Structure and Texture

In the as-consolidated condition ODS alloys exhibit very fine grain sizes ($\leq$1µm diameter) that produce, at least, the yield strength increase expected from decreasing grain size (Hall-Petch) effects(28). However, ODS alloys are generally given a high temperature recrystallization heat treatment and used in the coarse grained condition ($\geq$0.5mm) where grain boundary hardening is minimal. The recrystallization heat treatment produces grain structures that are highly elongated, having grain aspect ratios (GAR) of $\geq$10, with serrated grain boundaries that enable neighboring grains to interlock with each other. High GAR's minimize the detrimental effect of transverse grain boundaries and are a prerequisite for high temperature creep strength(29).

Recrystallization textures developed in some ODS superalloys are favorable such as <100> (INCONEL alloy MA 754) or less favorable such as <110> (INCONEL alloy MA 6000) or higher indices (INCOLOY alloy MA 956) for good thermal fatigue resistance. More suitable textures for fatigue resistance and intermediate temperature strength improvement may be attainable through controlled thermomechanical processing(30,31).

Alloy Design Options

The transition between Region I and II behavior in Figure 6 is somewhat arbitrary but it indicates various alloy design options based on our understanding of dispersion strengthening in MA alloys such as INCONEL alloy MA 6000. As a first generation commercial alloy combining γ'-phase precipitation hardening at intermediate temperatures and oxide dispersion strengthening at high temperatures, INCONEL alloy MA 6000 has been of paramount importance as a learning vehicle for production routes to market and in focusing the design of second generation alloys to meet increasing technological demands on materials.

While INCONEL alloy MA 6000 has excellent corrosion resistance in applications for which it was designed, such as advanced aerospace gas turbines, a new class of ODS superalloys is being developed (31,32), such as INCONEL alloy MA 760 (Table I) for industrial applications that require extremely high corrosion resistance typical of IN-939. INCONEL alloy MA 760 not only meets these corrosion resistance requirements but also has exceptional oxidation resistance as shown by the performance under Burner Rig testing at 1093°C (Table II). The Region II characteristics of good, long-term 10^4-10^5h rupture strength capability have been retained as shown in Fig. 7, resulting in a new alloy with an unprecedented combination of environmental resistance and strength.

Region I stress rupture performance has also been improved in experimental ODS superalloys such as Alloy 51 (34,35). Recent developments(36) such as experimental Alloy 92 (Table I and Fig. 8) have improved the intermediate temperature strength to levels similar to certain cast single crystal alloys with high tempera-

Table II: Burner Rig Corrosion Results[a]

Alloy	Metal Loss (μm)		Depth of Attack (μm)	
	324(h)	504(h)	324(h)	504(h)
INCONEL alloy MA 760[b]	+7.6	+10.2	+10.2	172.7
INCONEL alloy MA 6000	20.3	91.4	81.3	358.1
INCONEL alloy MA 754	40.6	322.6	40.6	469.9
INCONEL alloy 617	(c)	551.2	(c)	736.6
IN-738	726.4	(c)	(c)	(c)
IN-939	1485.9	(c)	1546.9	(c)

Notes: (a) 1093°C (2000°F) 1 cycle/hour (58 min. in flame 2 min
            out in air) 30:1 air to fuel (0.3% S in JP-5 fuel +
            5 ppm sea salt) ratio. Mass flow 5.5 ft^3/min.
        (b) Gain in diameter due to very adherent scale
        (c) Not determined

ture strength and oxidation resistance levels beyond INCONEL
alloy MA 6000. These alloys were designed to meet some of the
requirements of, for example, more advanced aerospace engines.
Similar alloy developments are being researched elsewhere(37,38).

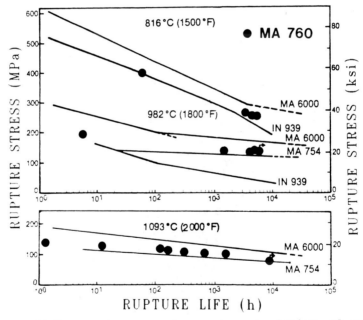

Fig. 7. Preliminary stress rupture characteristics of INCONEL
alloy MA 760 (33).

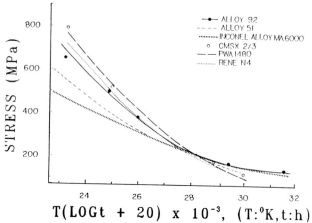

Fig. 8 Comparison of stress rupture characteristics of advanced superalloys (36)

From research on ODS alloys with majority γ' (Ni$_3$Al)-phase, it has been a logical development to examine the compositional ranges of the Ni-Al system stoichiometric intermetallic compounds and other intermetallics through mechanical alloying(39). MA Ni$_3$Al-based intermetallics such as that given in Table 1 present interesting micro-structures that tend to resist grain coarsening by standard secondary recrystallization heat treatments. However, such stable, fine-grained MA Ni$_3$Al alloys appear to show significant yield strength increases over similar rapidly solidified or conventionally cast Ni$_3$Al-based alloys while retaining useful ductility(39). Grain size refinement by the dispersoids appears to be the primary strengthening method with secondary contributions from Orowan strengthening and other structural features in these alloys. Moreover, similar dispersion strengthening characteristics may well control the behavior of other MA intermetallics with fine grain sizes such as NiAl, Ti- and Nb-aluminides and Al$_3$Ti. In general, dispersion strengthening in light metals such as Al alloys has many important differences from ODS superalloys that will be discussed next.

Dispersion Strengthened Aluminum Alloys By MA

Dispersion strengthened Al alloys have, in recent years, increased significantly in their technological importance [40,41]. In contrast to conventional precipitation hardening Al alloys, dispersion strengthened alloys rely upon the presence of fine, stable, hard, high melting point dispersoids for improvements in their elevated temperature strength, stiffness, corrosion resistance, and thermal stability. Improvements in elevated temperature strength arise as a result of the ability of the dispersoids to interact with mobile dislocations over a much wider range of temperature compared to the precipitates present in conventional alloys. Dispersoids can also indirectly contribute to strength as well as corrosion resistance by stabilizing a fine grain/subgrain structure within the alloy. Further, dispersoids can provide constraint to elastic deformation of the alloy, and in this manner, significantly increase the stiffness of Al alloys. For these reasons, advanced Al alloy development has focused on alloys in which dispersion strengthening, rather than precipitation hardening, is the dominant strengthening mechanism. Moreover, as described earlier, MA offers the greatest flexibility in tailoring the structure and hence the properties of dispersion strengthened Al alloys to suit a wide variety of applications.

Microstructure Evolution and Strengthening Mechanisms in MA-processed dispersion strengthened Al alloys

MA processing of Al alloys has been discussed in detail elsewhere [42-44]. A process control agent (PCA) is usually added to maintain a balance between fracturing and welding of the Al powder particles during the milling operation and to supply C and O for subsequent conversion into inert dispersoids. The PCA is intimately incorporated into the particles during the repeated welding, fracturing, and rewelding of the Al particles similar to the process shown in Fig. 1. During a subsequent degassing step, the PCA is converted to ultrafine (30-40 nm average diameter) stable carbides and oxides, very uniformly distributed within the powder particles. Incorporation of such dispersoids by any technique which involves melting and subsequent solidification steps would, in all likelihood, prove futile because of segregation problems. MA thus offers a unique method of incorporating very fine inert disperoids within Al alloys.

MA can also be utilized to introduce very fine Al_nX (where X is a transition element(s)) type intermetallic phases within Al alloys [45]. Such phases often have melting points significantly higher than that of aluminum. Further, transition elements typically exhibit low rates of diffusion within Al. Because of their stability and inherent hardness, Al_nX type intermetallics can contribute to the elevated temperature strength, stiffness, and thermal stability of dispersion strengthened Al alloys.

In contrast to ODS Ni-based alloys in which grain growth can be enhanced by directional recrystallization (DR), MA dispersion strengthened Al alloys are usually characterized by a fine grain structure which is highly resistant to coarsening. Fine grain sizes in MA Al alloys arise due to the following considerations:

(a) Al, being a high stacking fault energy metal, exhibits cross slip during deformation. In general, cross slip promotes the formation of a well defined dislocation cell structure during deformation rather than a high density of dislocation tangles as in the case of metals with low stacking fault energies. The formation of the cell structure, often referred to as dynamic recovery, leads to a reduction in the stored energy, and hence the driving force, available for recrystallization.

(b) MA Al alloys typically contain at least 4-5 v/o of fine inert dispersoids. The presence of such dispersoids provides a drag pressure on moving boundaries, thus retarding their motion. The drag pressure, P, is approximately given by [46]:

$$P = \frac{3f}{2r} \, \gamma \qquad\qquad [6]$$

where f = volume fraction of dispersoids, r = their average radius, and γ = the grain boundary free energy. The disper- soids are responsible for retarding the recrystallization and grain growth kinetics, and hence the excellent microstructural stability of dispersion strengthened MA Al alloys.

The microstructural stability of MA Al alloys can be illustrated by comparing grain sizes before and after long-term annealing of such alloys. For example, the microstructure of MA-processed unalloyed Al in the as-extruded condition typically consists of Al grains in 200-300 nm range with fine (30-40 nm) oxide and carbide dispersoids present mainly along grain bound- aries. Fig. 9 shows the microstructure of MA unalloyed Al after exposure to 500°C for 100 hr (essentially a fully annealed condition). The grain size in this condition increases only slightly to ~400 nm. The alloy shown in Fig. 9 had a C concen- tration of ~1.2 w/o and an O concentration of ~ 0.8 w/o. Assuming that all of the C and O are present as Al_4C_3 and Al_2O_3, the total volume fraction of dispersoids can be estimated to be ~6.7 v/o. The maximum grain size in an alloy containing stable dispersoids is usually given by [46]:

$$D_{max} = \frac{4r}{3f} \qquad\qquad [7]$$

where r and f denote the same quantitites as in equation (1). Taking r = 20 nm gives D_{max} ~400 nm. Comparison with the actually observed grain size shows that equation (2) provides a reasonable upper bound estimate of the grain size in dispersion strengthened MA Al alloys.

Role of the dispersoids in controlling the properties of dispersion strengthened MA Al alloys

As stated earlier, at least two types of dispersoids can be introduced within Al alloys by MA, namely (1) inert dispersoids such as carbides and oxides, and (2) Al_nX type intermetallic phases. We shall consider the role of these two types of dispersoids individually:

Carbide and oxide dispersoids present in MA Al alloys are extremely small in size, usually 30-40 nm in average diameter, and present mainly along grain boundaries. Theoretically such dispersoids can contribute to the strength of MA Al alloys in at least two distinct ways: (a) they can provide resistance to dislocation motion and cause the formation of dislocation loops (viz. Orowan [3]), and/or (b) they can stabilize a fine grain/subgrain structure (as discussed earlier) and hence indirectly contribute to grain boundary strengthening (Hall- Petch effects). In a recent study [47] on the deformation

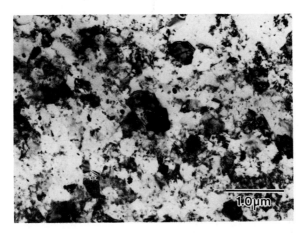

Fig. 9. Microstructure of MA unalloyed Al specimen annealed for 100 h at 500°C

behavior of MA unalloyed Al, it was shown that Orowan looping was not observed even after an MA Al alloy was deformed 11% in tension. Instead, dislocations were found to be pinned at grain boundaries or at individual particles located at grain boundaries. It is possible, therefore, that carbide and oxide dispersoids contribute to strength indirectly through Hall-Petch effects rather than through the Orowan mechanism in MA Al alloys. In practice, it is difficult to separate the contributions of the individual mechanisms on alloy strength since the achievement of a fine grain structure relies upon the presence of fine dispersoids. However, regardless of which mechanism is operative, it can be expected that an increase in the volume fraction of carbide and oxide dispersoids will cause a concurrent increase in strength. Indeed this is observed in practice (see Fig. 10a). Control of the volume fraction of the inert dispersoids thus provides an effective technique for designing alloys to meet specific strength targets.

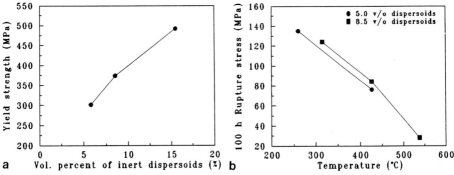

Fig. 10(a) Variation in the yield stress of MA Al with inert dispersoid volume fraction. (b) Variation in the 100h rupture strength of MA Al with temperature.

MA dispersion strengthened unalloyed Al has also been found to exhibit remarkable elevated temperature strength as well as creep resistance [43]. The 100 h rupture strength of MA un-alloyed Al is shown in Fig. 10(b) as a function of temperature. unalloyed Al exhibits far superior elevated temperature strength and resistance to creep deformation compared to any conventional Al alloy. This is due in part to the fact that MA Al alloys are characterized by extremely fine grain sizes. While dispersoid and grain/subgrain strengthening mechanisms can be used to explain the resistance to dislocation glide and dislocation creep, such mechanisms are usually not considered to be effective against diffusional creep since the rate of diffusional creep is generally considered to increase with decreasing grain size. An interesting hypothesis has been proposed by Arzt et al. (20) to explain this apparent anomalous behavior. They have proposed that when diffusion distances are small, as in fine grained MA Al alloys, elevated temperature deformation processes are controlled by motion of dislocations along grain boundaries rather than the lattice diffusion of vacancies. The creep rate is then determined by the resistance offered to the motion of boundary dislocations by dispersoids situated along grain boundaries in much the same way as conventional Orowan strengthening. The process is illustrated schematically in Fig. 11. Since the dispersoids in MA Al alloys are situated mainly

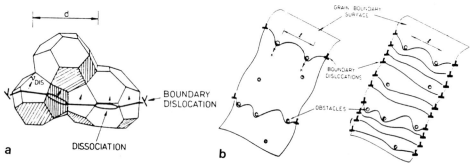

Fig. 11. (a) Schematic representation of a boundary dislocation on a grain boundary surface and (b) boundary dislocations interacting with discrete obstacles in a grain boundary, single dislocation (left) and dislocation pile-ups (right). (20)

along grain boundaries, the hypothesis of Arzt et al. provides a plausible explanation for the excellent resistance to creep deformation exhibited by MA Al alloys in spite of their very fine grain size.

In some ways analogous to the ODS superalloys, there appears a transition in the log-log plot of secondary creep rate against stress for MA Al alloys such as IncoMAP alloy AL-9052. At this transition stress, σ_c (48), the creep rate increases steeply with increase in stress (48), and the profile of the creep curve changes from strain hardening type in the low stress regime to strain softening type in the high stress regime above σ_c. It has been suggested (48), that the strain could be

ascribed to such a microstructural change as subgrain growth
during creep which can only occur above σ_c. Further work is
required to elucidate these hypotheses.

Besides contributing to the room and elevated temperature
strength of MA Al alloys, carbide and oxide dispersoids can also
indirectly provide improvements in the corrosion resistance of
such alloys (44). The excellent corrosion resistance exhibited
by MA Al alloys is due to the ability of the dispersoids to
stabilize an ultrafine scale of microstructure, combined with
the inherent process-related improvements in the microstructural
homogeneity, of such alloys.

As noted earlier, fine submicron Al_nX type intermetallic
dispersoids can also be introduced during the MA processing of
Al-X type alloys. The microstructure of an MA Al-Ti based alloy
containing about 20 v/o of Al_3Ti is illustrated in Fig. 12 as an
example of such alloys. The Al_3Ti dispersoids are of about the
same size as the Al grains present in the alloy. Because of
their relatively large size, such dispersoids do not contribute
as strongly to strength as carbide and oxide dispersoids.
However, large volume fractions of such dispersoids can be
introduced without substantial losses in ductility. As will be
shown in the next section, alloys containing both inert and

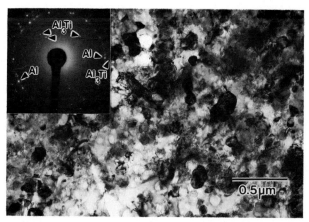

Fig. 12. Microstructure of MA Al-9 Ti alloy.

intermetallic phases, and having very attractive combinations of
room and elevated temperature strength, stiffness, and ductility, can be fabricated by judicious alloy design.

Alloy Design Options

The foregoing discussion provided an overview of the manner
in which different strengthening mechanisms can be invoked by
the presence of stable dispersoids within MA Al alloys. In
addition, solid solution strengthening can be obtained in such
alloys by the addition of suitable alloying elements. The MA

process has been applied by Inco Alloys International, Inc. to produce different classes of alloys suitable for a variety of applications. Examples of such alloys, some of which are commercially available and some of which are in the experimental stages, are provided below:

(a) Light-weight high strength corrosion resistant alloys:

IncoMAP alloy AL-9052 combines solid solution strengthening (through the presence of 4 w/o Mg) and dispersion strengthening (through the presence of 5-6 v/o of carbide and oxide disper-soids). Because of its fine grain size, this alloy combines good ambient temperature strength with excellent corrosion resistance. The typical properties of IncoMAP alloy AL-9052 have been previously reported (49) and are summarized in Table III. Besides improved corrosion resistance, an important advantage that this alloy offers over conventional high strength Al alloys is that it does not require heat treatment to achieve its high strength level. In this manner, close tolerance parts can be fabricated from this alloy directly in the worked condition.

Table III. Measured Properties of IncoMAP alloy AL-9052

Density: 2.68 Mg/m^3	Melting Point: 608°C
Mechanical Properties	
	Longitudinal
Ultimate Tensile Strength	450 MPa
Yield Strength 0.2%	380 MPa
Elongation	13%
Fracture Toughness K_{IC}	44 MPa $\sqrt{m}$
Elastic Modulus	76 GPa

IncoMAP alloy AL-905XL is a Li-containing version of the above alloy. The Li level in IncoMAP alloy AL-905XL is delibe-rately maintained at a particular low level (1.3 w/o) to avoid complications arising from the precipitation of embrittling phases while maintaining a balance between density, corrosion resistance and mechanical properties. Besides having a low density (2.58 g/cm^3), IncoMAP alloy AL-905XL offers good tensile properties (equivalent to 7XXX aerospace alloys), excellent corrosion resistance, and importantly, near-isotropic properties in the forged condition (difficult to achieve with conventional Al-Li based alloys). IncoMAP alloy AL-905XL is finding appli-cations as a low density aircraft forging alloy. The typical properties of IncoMAP alloy AL-905XL in the forged condition have been previously reported (50) and are summarized in Table IV.

(b) High temperature alloys:

Reference was made earlier to the excellent elevated temperature strength and thermal stability of MA unalloyed Al. The properties of such alloys have been reviewed by Benjamin and

Bomford (43). The properties of variants of MA Al (containing deliberate additions of C or Y_2O_3) have also recently been reported (51). MA unalloyed Al represents the earliest high temperature dispersion strengthened Al alloy made using MA techniques and has served as a model alloy for the development of the next generation of high temperature MA Al-based alloys.

Table IV. Measured Properties of IncoMAP alloy AL-905XL Forgings

Density: 2.58 Mg/m^3	Melting Point: 593°C	
Mechanical Properties		
	Longitudinal	Short Transverse
Ultimate Tensile Strength (MPa)	520	480
Yield Strength 0.2% (MPa)	450	420
Elongation (%)	9	6
Fracture Toughness K_{IC} (MPa $\sqrt{m}$)	30	30
Elastic Modulus (GPa)	80	
Corrosion		
General	0.003 mm/yr	
Stress Corrosion Cracking Threshold	345 MPa	
(Alternate immersion in 3.5% NaCl ASTM G44-75 and G38-75, 90 Day Exposure)		

More recently, developmental activity on high temperature alloys has focused on alloys containing Al_nX type intermetallic phases besides the inert oxide and carbide dispersoids present in MA unalloyed Al. The incorporation of intermetallic phases provides a means of enhancing some of the properties of MA unalloyed Al. For example, large volume fractions of inter-metallic phases can be incorporated without significant losses in ductility. Such phases can make substantial contributions to the stiffness of MA unalloyed Al, besides making modest contri-butions to strength. In this manner, combining intermetallic and inert dispersoid strengthening provides a means of fabri-cating alloys with excellent combinations of room and elevated temperature strength, stiffness, corrosion resistance, thermal stability, and ductility. The incorporation of intermetallic phases has also been found to enhance the elevated temperature ductility of MA unalloyed Al, thus providing a means of im-proving the high temperature fabricability of MA Al materials.

The properties of MA Al-Ti based alloys containing large volume fractions (up to ~35 v/o) of Al_3Ti have been reported recently (52). Ti is considered a preferred alloying element because Al_3Ti, the phase in equilibrium with Al, has the highest density-compensated melting point (~405 $cm^3 \cdot °C/g$) among the common Al_nX type phases. Further, Ti has one of the lowest rates of diffusion within Al (1.69 x 10^{-14} cm^2/s) among the common transition elements.

The modulus of MA Al-Ti alloys increases almost linearly with increasing Ti content (see Fig. 13a). At the same modulus

36

level (say 105 GPa), MA Al-Ti alloys exhibit much better ductility compared to currently available Al-SiC composites. Further, the tensile strength of an MA Al-Ti based alloy is compared with that of cast and wrought Al alloy 2014-T6 as well as that of a rapid solidification (RS) processed high temperature Al-8Fe-4Ce alloy in Figure 13(b). The MA Al-Ti alloy is substantially stronger than 2014-T6 at temperatures exceeding about 200°C. Also, the MA Al-Ti alloy exhibits superior strength retention at temperatures greater than about 300°C, as compared to Al-8Fe-4Ce. MA Al-Ti based alloys have also been found to exhibit excellent microstructural stability when exposed to elevated temperatures for extended periods of time (53).

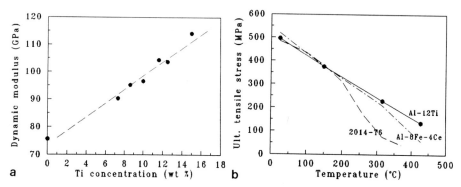

Fig. 13(a) Influence of Ti concentration on the dynamic modulus of MA Al-Ti alloys, (b) Comparison between the tensile strength of MA Al-12Ti, cast and wrought 2014-T6, and Al-8Fe-4Ce over a range of temperatures.

Concluding Remarks

In the evolution of new materials to meet ever-increasing performance requirements, there is a growing need to tailor and control microstructures and properties at the most fundamental levels of material composition possible. This rationale has led to better performance of existing materials through improved compositional control, microstructural refinement and the design of "composite" materials utilizing equilibrium and/or non-equilibrium processing. The mechanical alloying process is uniquely suited to synthesizing such composite materials in general and specifically to combining the benefits of dispersion strengthening with matrix structures and compositions designed to meet wide-ranging performance goals.

37

References

1. C. S. Smith, Mining and Metallurgy, (1930), 11, 213.
2. R. Irman, Techn. Rundschan (Bern), (1949), 36, 9.
3. E. Orowan, Discussion Symposium on Internal Stresses in Metals and Alloys, Monograph and Rept. Series No. 5, Institute of Metals, London, (1948), p. 451.
4. L. M. Brown and R. K. Ham., in "Strengthening Methods in Crystals" (Eds: A. Kelly and R. B. Nicholson), Elsevier Amsterdam, (1971), p. 9.
5. G. B. Alexander, U.S. Patent No. 2,972,529, Feb. 21 (1961).
6. G. B. Alexander, U.S. Patent No. 3,019,103, Jan. 30, (1962).
7. G. B. Alexander, French Patent No. 1,424,902, (1965).
8. J. S. Benjamin, Metall. Trans., (1970), 1, 2943.
9. J. S. Benjamin, Scientific American, 234, (1976), 40-48.
10. J. S. Benjamin and T. E. Volin, Metall. Trans., 5, (1974), 1929.
11. P. S. Gilman and J. S. Benjamin, Ann. Rev. Mater. Sci, 13, (1983), 279.
12. J. S. Benjamin in Proc. DGM Conf. on "New Materials by Mechanical Alloying Techniques" (Eds: E. Arzt and L. Schultz), Calw-Hirsau, Oct. 1988 DGM Informationsgesellschaft, Oberursel, 1989.
13. R. F. Singer, R. C. Benn and S. K. Kang in Proc. Conf. "Frontiers of High Temperature Materials II" (Eds: J. S. Benjamin and R. C. Benn), London, 1983, IncoMAP, Inco Alloys Int'l., p. 336.
14. J. K. Tien ibid. ref. 13 Discussion Section p. 119.
15. J. S. Benjamin, T. E. Volin and J. H. Weber, High Temperatures/ High Pressures, 16, (1974), 443.
16. Unpublished research, Inco Alloys Int'l., Inc.
17. R. W. Lund and W. D. Nix, Acta. Met., 24 (1976), 469.
18. R. F. Singer and E. Arzt in Proc. Conf., "High Temperature Alloys for Gas Turbines and Other Applications", (Eds; W. Betz et al.), Vol. 1, (1986) Reidel, Dordrecht, 97-126.
19. J. D. Whittenberger, Metall. Trans., 15A, (1984), 1753.
20. E. Arzt, M. F. Ashby and R. A. Verall, Acta. Met., 31, (1983), 1977.
21. R. S. Shewfelt and L. M. Brown, Phil. Mag., 35., (1977), 945.
22. C. M. Sellars and R. A. Petkovic - Luton, Mats. Sci. and Eng'g., 46, (1980), 75.
23. E. Arzt and M. F. Ashby, Scripta Met., 16, (1982), 1285.
24. E. Arzt, J. Rösler and J. H. Schröder ibid ref. 18 p. 217.
25. E. Arzt, ibid, ref. 12.
26. B. Reppich, W. Listl and T. Meyer, ibid. ref. 18, 1023.
27. R. C. Benn and S. K. Kang in Proc. Conf. "Superalloys 1984", (Eds: M. Gell et al.), TMS-AIME, (1984), 319.
28. R. F. Singer and G. H. Gessinger, "Powder Metallurgy of Superalloys" by G. H. Gessinger, Butterworth & Co., London, (1984), 213-292.
29. B. A. Wilcox and A. H. Clauer, Acta. Met., 20, (1972), 743.
30. R. C. Benn, L. R. Curwick and G. A. J. Hack, Powder Met., 4, (1981), 191.

31. E. Grundy and M. J. Bomford in Proc. Conf "Advanced Materials and Processing Techniques for Structural Applications" (Eds: T. Kahn and A. Lasalmonie) Sept. 1987, pp. 357 - 366, ASM-Europe, Paris, France.
32. R. C. Benn, Proc. "Third Int'l. Conf. on Creep and Fracture of Engineering Materials and Structures", (Eds: B. Wilshire and R. W. Evans), The Inst. of Metals, 1987, 319.
33. R. C. Benn and G. M. McColvin in Proc. Conf. "Superalloys 1988" (Eds: D. N Duhl et al.) TMS-AIME, (1988),
34 R. C Benn in "Superalloys 1980", (Eds: J. K. Tien et al.), ASM Metals Park, Ohio, (1980), 541-550.
35. S. K. Kang and R. C. Benn, Metall. Trans., (1985), 16A, 1285.
36. R. C. Benn unpublished research, Inco Alloys Int'l. (1986).
37. K. Mino, in Proc. Int'l. Conf. on "PM Aerospace Materials '87", Paper No. 13, Metal Powder Report - published conf., Luzern Nov. 2-4 (1987).
38. S. Ochi, N. Kuroishi and Y. Doi, Progr. in Powder Metall. 41 (1986) 439.
39. R. C. Benn, P. K. Mirchandani and A. S. Watwe in "P/M 88" Int'l. Powder Metallurgy Conf. June '88, to be published in "Modern Developments in Powder Metallurgy, APMI.
40. W. M. Griffith, R.E. Sanders, and G.J. Hildeman, in High Strength Powder Metallurgy Aluminum Alloys, (Eds. M.J. Koczak and G.J. Hildeman), TMS-AIME, Warrendale, PA, 1982, pp 209-224.
41. S.L. Lagenbeck et al, Final Report, AFWAL TR-86-4027, May 1986.
42. U.S. Patent No. 3,740,210 (1973).
43. J.S. Benjamin and M.J. Bomford, Met. Trans., 8A, 1977, pp 1301-1305.
44. J.S. Benjamin and R.D. Schelleng, Met. Trans., 12A, 1981, pp 1827-1832.
45. D.L. Erich, USAF Report, AFML-TR-79-4210, (1980).
46. Zener, as quoted by C.S. Smith, Trans. AIME, 175, p 15, (1948).
47. H.G.F. Wilsdorf, in Dispersion Strengthened Aluminum Alloys, (Eds: Y.W. Kim and W.M. Griffith), in print, TMS-AIME, Warrendale, PA.
48. M. Otsuka, Y. Abe and R. Horiuchi, ibid. ref. 32, p. 307.
49. J.H. Weber and R.D. Schelleng, ibid. ref. 47.
50. R.D. Schelleng, A.I. Kempinnen, and J.H. Weber, in Space Age Metals Technology, (Eds: F.H. Froes and R.A. Cull), SAMPE, Covina, CA, (1988), pp 177-187.
51. J.A. Hawk, P.K. Mirchandani, R.C. Benn, and H.G.F. Wilsdorf, in Dispersion Strengthened Aluminum Alloys, (Eds: Y.W. Kim and W.M. Griffith), in print, TMS-AIME, Warrendale, PA.
52. P.K. Mirchandani and R.C. Benn, in Space Age Metals Technology, (Eds: F.H. Froes and R.A. Cull), SAMPE, Covina, CA, (1988), pp 188- 201.
53. J.A. Hawk, P.K. Mirchandani, R.C. Benn, and H.G.F. Wilsdorf, in Dispersion Strengthened Aluminum Alloys, (Eds: Y.W. Kim and W.M. Griffith), in print, TMS-AIME, Warrendale, PA.

Reaction Milling of Aluminium Alloys

G. Jangg, Technische Universität Wien, Vienna, Austria

1. Introduction

Reaction milling, i.e. intense milling of metal powder mixtures (occasionally with additives) is a variant of mechanical alloying, which has been established for many years, mainly for production of homogeneous alloys and alloys with fine subgrain structure, for the production of alloys with an amorphous structure, but also for the manufacture of dispersion hardened materials.

For producing dispersion hardened materials, suitable dispersoid powders are homogeneously introduced into a matrix by milling. In order to obtain satisfactory strengthening the powder size should be very small (approx. 0.02-0.1 μm) and the particles should be very hard and of high melting point, insoluble in and not reacting with the matrix. Frequently, suitable materials such as Al_2O_3, ThO_2, Y_2O_3 are unavailable as sufficiently fine powders or are too expensive.

Alternatively, suitable dispersoids can be obtained in a milling process by introducing materials that react with the matrix either during milling or during subsequent heat treatment. Based on this idea, the process of reaction milling was developed in Vienna (1-5).

Pre-conditions for obtaining high quality materials - which require homogeneous distribution of sufficiently fine, dispersion strengthening particles - are

- The milling behaviour - which in the case of Al is complicated by fast particle coarsening - must not be affected adversely but in contrast improved by the additive.

- It must be possible to introduce the additives homogeneously into the milling granules.

- They must react to form hard dispersoids that are inert with respect to the matrix also at elevated temperatures; formation of agglomerates is also undesirable.

Numerous experiments showed that from the many materials tested only a few fulfil these requirements. Most successful was the introduction of carbon as lampblack or graphite. The carbon introduced by milling reacts with Al to form fine and well distributed Al_4C_3 particles. Al_4C_3 exhibits high hardness and shear strength, is practically insoluble in Al even close to the melting point and therefore does not tend to coarsen during thermal exposure of the final dispersion hardened products.

The experience gained hitherto and the problems encountered in the production of dispersion hardened Al by reaction milling are described below in detail.

2. Milling

2.1 General milling behaviour

As is known from mechanical alloying in general, the particle size of the metal powders decreases drastically at first, when metal powders or powder mixtures are milled. During further milling the particle size increases and finally very coarse granulates are formed (Fig. 1). The changes in particle size during milling are determined mainly by the type of material milled and by the milling energy. With the same material – e.g. Al – the particle size changes will proceed faster the higher the milling energy.

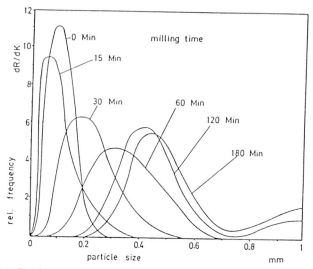

<u>Fig. 1:</u> Particle size distribution of Al+3% carbon black mixtures milled for different times

In contrast, the type of mill used is only of secondary importance. With mills of very high milling effect, such as attritors, the stages of refining and coarsening are attained faster than e.g. in vibration ball mills. Mechanical alloying is possible also with high enery ball mills, the time necessary for obtaining satisfactory granulates is however markedly longer. For a rotating ball mill (laboratory attritor, 3 l vessel, 200 g Al powder + 3 g lampblack, 6 kg milling balls), Fig. 2 depicts the pronounced influence of the milling energy (rotor revolution number) on the average particle size. Experiments with different types of mills under varying conditions have shown that by introduction of a given milling energy roughly comparable granulates are attained. Sufficient cooling of the materials is of decisive importance in order to remain below the recovery temperature of the material; otherwise, very fast welding of the particles will occur. Al alloy powders with higher hardness and recovery temperature are milled more easily than pure Al powders.

Milling of Al powder mixtures offers various problems. Frequently, welding of the material to the balls, the vessel walls and the rotor are encountered. On the other hand, loose agglomerates of superfines can be formed in less agitated areas of the mill (which are encountered in all mills of similar shape), mainly in the first stage of milling in which very fine granulates are present. During further milling these deposits are mostly destroyed. This is clearly visible from the "output", that is the amount of granulate obtained after emptying the mill and screening off the granulate from the balls compared to the amount of starting powder (Fig. 3).

As mentioned above, during further milling coarser particles are formed and welding of the material to the mill occurs. This may disrupt the milling operation, because the welded material is difficult to remove from the mill. Welding however is rather typical for long time milling, during which the granulates coarsen substantially, and it can be avoided by stopping the milling process in time. The welded material is virtually dense ($\approx 100\%$) while the deposits are very porous (approx. 60-70% theoretical density) and are removed more easily.

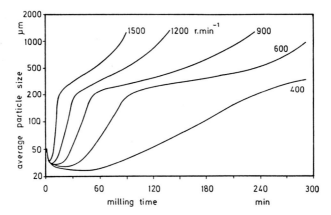

Fig. 2: Particle size of granulates of Al+3%C mixtures milled for different times with different rotor revolutions

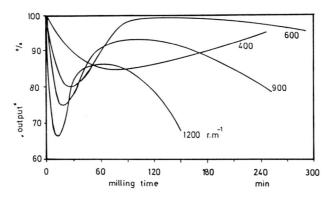

Fig. 3: "Output" of milling Al+1%C mixtures at different times and different rotor revolutions

2.2   Influence of additives

Addition of very fine, non-metallic powders strongly affects the milling behaviour, as these materials may deposit at the particle surfaces and inhibit welding, thus preventing particle coarsening and formation of deposited agglomerates.

The results obtained with a large number of experiments indicate that the milling behaviour is not significantly changed by inert materials, while it may be decisively improved by materials that react with Al (6). The milling behaviour of Al powder with very fine Al_2O_3, ThO_2 or Y_2O_3 powders is virtually identical to that of plain Al. Addition of lampblack however results in markedly slower particle coarsening and lower tendency to welding (Fig. 4). Easily grindable Acheson graphite may also be used. Improvements may also be attained by addition of water, and especially using surface active organic materials such as oils, waxes, alcohols etc. Stearic acid is used as a "milling aid", "process control agent" in the NOVAMET process (7,8). Not only the type but also the amount of additive affects the milling behaviour.

Because it must be assumed that a de-activation of the new surfaces continuously generated during milling inhibits welding, it is understandable that also the milling atmosphere affects the milling behaviour (Figs. 5 and 6).

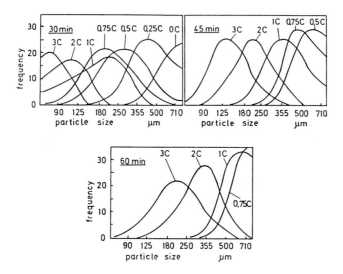

Fig. 4: Particle size distribution of granulates with different C content after milling for different times (3 l attritor; 600 r.min⁻¹)

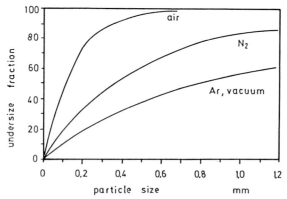

Fig. 5: Particle size distribution of Al+1%C granulates after 60 min milling with 600 r.min⁻¹ in different atmospheres

Milling of Al in vacuum results in particularly fast particle coarsening and pronounced welding. Argon is comparable to vacuum, while N_2 offers some improvement. Much better milling behaviour is encountered when milling is performed in air. This however results in oxidation and the milled granulates contain larger amounts of Al_2O_3. Since the oxide particles are fine and act as hardening dispersoids they are not particularly undesirable.

Excellent milling behaviour is encountered with flowing protective gas loaded with gaseous hydrocarbons or methanol. It must be kept in mind however that these materials may be milled into the granulates and react with Al to form H_2 which remains in the granulates; this is of importance for further processing.

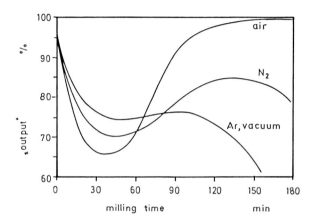

Fig. 6: "Output" after 60 min milling of Al+1%C with 600 r.min[-1] in different atmospheres

2.3   Distribution of the additives during milling

As mentioned above, satisfactory mechanical properties are attained of the additives are introduced homogeneously into the matrix and not as agglomerates. Understandably, this is easier the longer the granulates remain fine during milling. Coarse particles are not further crushed during milling, and the additives can no more be introduced into these particles.

Numerous experiments have shown that a sufficiently homogeneous distribution of the additives is attained if the granulate remains finer than approx. 200 μm for a sufficiently long time. When milling Al with graphite in the attritor, homogeneous distribution is attained after 60-90 min. Appraisal of the C particle distribution is however difficult. Statistical evaluation of a large number of TEM photographs (obtained with extruded material because thin foils could not be produced directly from the granulates) allows reliable conclusions to be drawn. Experiments showed that the irregularities found in metallographic sections do not depict the dispersoid distribution but are apparently caused by local differences in the dislocation density.

Slightly less reliable information about the distribution is obtained by measuring the C content in different particle fractions after different milling times (Figs. 7 and 8). After short milling marked differences in the C content are still found which however disappear after prolonged milling.

At given parameters for further processing, granulates containing inhomogeneously distributed dispersoids result in extruded products with lower strength and especially lower ductility. As shown in Fig. 9, the mechanical properties also confirm that sufficient homogeneity is attained as soon as the particles start to coarsen after being refined. Further prolongation of the milling does not yield any improvement owing to the welding problems described above. A useful tool, also for estimating the dispersoid distribution, is the "quality factor" of the extruded products, which is described below in detail.

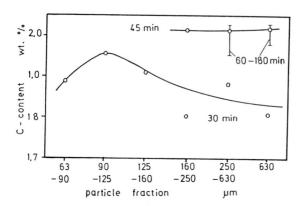

Fig. 7: C content in different particle fractions of Al+2%C granulates after different milling times

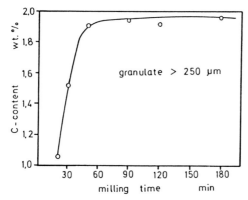

Fig. 8: C content in the >250 μm fraction of Al+2%C granulates after different milling times

3.   Further processing of the granulates

During reaction milling, dispersoids themselves are not introduced but materials that form dispersoids by reaction with Al are. The reaction takes place in part during milling, it must however be completed by heat treating the granulates. Granulates directly extruded without heat treatment result in products of poor quality as the unreacted materials, such as lampblack, H_2O or hydrocarbons, are not suitable dispersoids.

The reaction of lampblack and graphite, respectively, introduced into Al was investigated, in detail, as a function of the heat treatment parameters. Analytical determination of the Al_4C_3 was carried out by gas chromatography. The samples are dissolved in dilute NaOH, and Al_4C_3 forms an equivalent amount of methane, while the remaining elemental carbon remains unaffected.

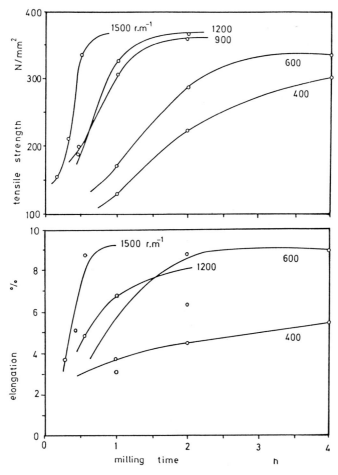

Fig. 9: Tensile strength and elongation of bars extruded from differently milled Al+2%C granulates

3.1 Reaction rate of Al_4C_3 formation

Thorough investigations showed that the reaction rate depends on the type of carbon used (Fig. 10). It is clear that the reaction rate depends on the grindability and crushability of the carbon types during milling with Al (9,10). Loose lampblack types produced at lower temperatures (type A1) are more suitable than furnace types (type A2) produced at higher temperatures. The so-called "coke-soot" (type C) consists of coarse, hard particles which are only insufficiently crushed during milling and are therefore unsuitable for reaction milling. Graphite (type B), especially Acheson graphite, has good grindability and is thus suitable, due to its lamellar structure. Graphite can be added also as coarse powder.

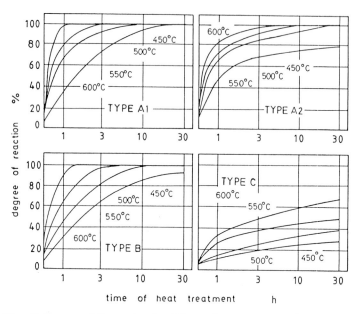

Fig. 10: Degree of C reaction by different heat treatment of granulates prepared from different carbon types

As shown in Fig. 10, granulates prepared using suitable lampblack or graphite require at least 3 h heat treatment at 550°C for complete reaction. During technical scale production, such heat treatment is usually attained during heating to the extrusion temperature of the compacts pressed from the granulates. Up to now, no adverse effect of more intense heat treatment (longer time, higher temperature) on the properties of the extruded products has been found. As shown in more detail in (11-14), the microstructures are very stable at elevated temperatures. Up to 600°C no recovery or recrystallization was found. The thermal stability of the Al_4C_3 particles has been confirmed by TEM investigations. Thorough statistical measurement of the Al_4C_3 particle size showed only very slight correlation with the heat treatment parameters. Even long time heat treatment does not result in noticeable coarsening. There is however a correlation between the amount of carbon introduced and the Al_4C_3 particle size, with finer matrix microstructure and somewhat coarser dispersoid particles being found at higher dispersoid content (13).

Heat treatment can be carried out in air. At temperatures up to 600°C only slight surface oxidation of the granulate particles is observed. The increase of the oxide content correlates with the average granulate particle size. Only with very fine (< 45 μm) granulates does a significant increase of the oxide content (> 0.2 wt.%) occur.

Figs. 10 furthermore show that practically no Al_4C_3 is generated during milling when using graphite, while "humid" lampblack types containing some amounts of hydrocarbons react more strongly. C compounds such as oils, waxes, stearic acid etc. react still more with Al during milling; complete reaction is however not attained even after long-time milling.

The hydrocarbon content of "humid" lampblack types, as well as "milling aids" such as alcohol or stearic acid, milling in humid atmosphere or using Al powders with increased hydroxide content, results in the formation of hydrogen which remains in the granulates and may be present in fairly high concentrations. Increased H_2 content and the presence of additives that have not reacted are undesirable because materials extruded from such granulates possess poor quality.

However, a heat treatment carried out to complete the formation of Al_4C_3 removes all hydroxides and almost all the H_2.

3.2   De-gassing

Detailed investigations were carried out together with DFVLR (15). During slow heating of the granulates in vacuum (Figs. 11 and 12) H_2O is at first removed above approx. 100°C.

At approximately 350°C all adsorbed H_2O is evaporated and all hydroxides have been dissociated. At 450°C, H_2O starts to evaporate, diffusing slowly from the interior of the granulate particles. With granulates heat treated in air and then stored in air for some time, H_2O re-emerges as the granulates have adsorbed H_2O during storage in air and form hydroxide skins. H_2, in contrast, is not re-adsorbed, and during de-gassing of heat treated granulates markedly lower amounts of H_2 are set free. Even the high gas content of granulates milled with stearic acid as milling aid can be completely removed by a suitable heat treatment.

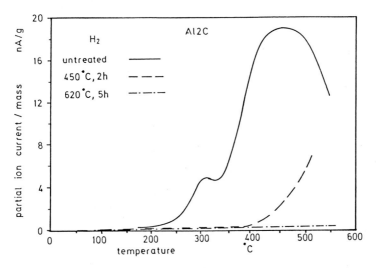

Fig. 11: De-gassing curve for H_2O of differently heat treated Al+2%C granulates

48

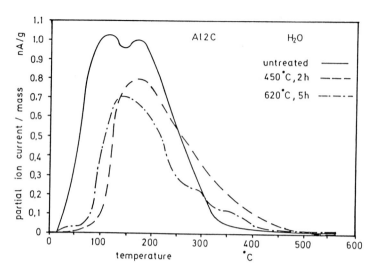

Fig. 12: De-gassing curve for H_2O of differently heat treated Al+2%C granulates

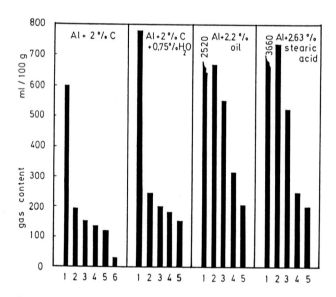

Fig. 13: Gas content of samples after different heat treatment
      granulates: 1: as-milled   3: h.t. 3h 500°C
                                 2: h.t. 3h 450°C   4: h.t. 3h 550°C
      extrusions: 5: as-extruded from granulate, h.t. 3h 550°C
                                 6: as 5, bars additionally h.t. 500h 550°C

Investigation of the total gas content of granulates – by measuring the gas volumes set free during melting of the samples – shows that by heat treating the granulates in air the gas content can be decreased to very low levels (Fig. 13). Vacuum treatment offers no significant advantages compared to that in air (Fig. 14). H_2 is removed fairly slowly. It was found that the residual hydrogen content in the extruded products can be further decreased by long time annealing (see Fig. 13).

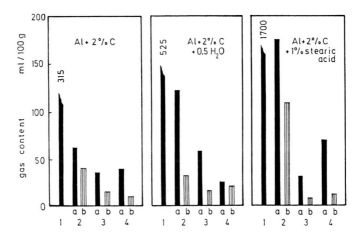

Fig. 14: Gas content of different processed granulates after 3h h.t. at 530°C
1: granulate without h.t.   3: h.t. in N2
2: h.t. in air   4: h.t in vacuum
a: granulate
b: bar extruded from the heat treated granulate

3.3   Other dispersoids

With special materials, e.g. when precipitation hardening is to be combined with dispersion strengthening, the heat treatment necessary for formation of the Al_4C_3 dispersoids is too intense, resulting in coarse precipitates that lower the mechanical properties.

Therefore, materials were required that react with Al and form dispersoids at lower temperatures but do not form H_2, which can only be removed at higher temperatures. In this respect BN, which can form AlN and Al_2B, looked promising. Hexagonal BN was easily milled with Al, such as graphite, and homogeneous granulates were obtained. X-ray investigations showed that BN partly reacts during milling and this reaction is completed at low temperatures (Fig. 15). Al_2B formed together with AlN however ages very fast, and coarse, metallographically visible particles are formed. As discussed below, the presence of such particles results in inferior mechanical properties, especially ductility and fatigue strength.

Direct introduction of the dispersoids by milling could also render heat treatment unnnecessary. However, suitably fine Al_2O_3 with low hydroxide content proved to be difficult to mill. In contrast to lampblack or graphite, alumina does not act as a "milling aid". The granulates coarsen quite fast, and a homogeneous distribution of Al_2O_3 can hardly be attained.

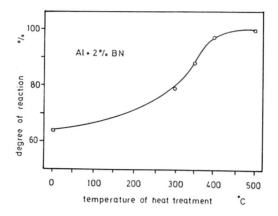

Fig. 15: Degree of reaction of BN in milled Al+2%BN granulates by 3h h.t. at different temperatures

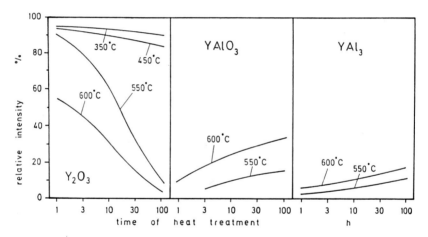

Fig. 16: Degree of reaction (measured by X-ray intensities) in milled Al+5% Y_2O_3 granulates at different heat treatments

At the moment, experiments are being carried out to produce Al_2O_3 dispersoids by "reaction milling" of Al in O_2 containing atmospheres, but there are several problems being encountered concerning the milling behaviour.

Y_2O_3, which is being used for ODS superalloys, would be sufficiently fine and hydroxidefree. However, it is difficult to mill into Al. Furthermore, it reacts at moderate heat treatment temperatures to form YAl_3 and Y-spinel (Fig. 16). While the spinel particles, which are finely distributed, would be suitable as dispersoids, YAl_3 tends to age and coarsen rapidly.

Best results were obtained therefore with graphite as additive. The DISPAL materials, which are technically produced, are obtained by milling graphite into plain Al, heat treating the granulates in air and hot extruding.

4.   Production of compact materials from the granulates

Due to the oxide and hydroxide layers covering the Al surfaces, which are further thickened by heat treatment of the granulates in air, compaction offers special problems. In order to obtain sufficiently strong contacts between the granulate particles, shearing–off of the oxide layers is necessary, which is achieved only at sufficiently high deformation ratios. Powder forging was successful only in the laboratory, under closely controlled conditions. Technically, forging of parts directly from the granulates does not seem feasible, nor does HIPing. With hot extrusion, sufficient shearing is attained at an extrusion ratio in excess of 1:16. Experimentally, satisfactory compaction was attained by the CONFORM process.

5.   Quality control

With correctly produced materials, high tensile strength R_m and, at the same time, elongation values A are measured. As observed with several thousand samples so far, R_m and A are specially correlated. When plotting R_m vs. A, for satisfactory materials, the data are found to lie in a narrow band (Fig. 17). Materials with higher dispersoid content exhibit higher tensile strength and lower elongation. More intense heat treatment of the granulates – in excess of that necessary for complete C reaction – results in materials with lower tensile strength and higher elongation. However, the values are found within the band.

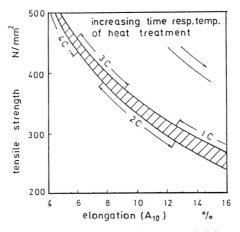

Fig. 17: Tensile strength and elongation of bars extruded from differently heat treated granulates with different contents of various carbon types

From the experimental correlation between R_m and A an empirical equation can be derived (16):

$$QF = \frac{(R_m + 500)}{520} \cdot A^{0.219}$$

QF = "Quality factor"

For the case when the value couples are found within the band, QF = 1. According to the experience gained so far, QF < 1 can be attributed to avoidable mistakes during production:

- Inhomogeneous dispersoid distribution by too short a milling time or too rapid coarsening of the granulate during milling

- Presence of agglomerates of dispersoids

- Presence of coarse inclusions (> approx. 1.5 μm), which lower the ductility (16). Such inclusions can be introduced by slag particles from the atomized Al powder or by dust from handling of the powders and granulates

- Presence of aged intermetallic particles, which emerge when using too impure Al powder (e.g. too high Fe content) or unsuitable alloy powders

- Residues of carbon that have not reacted due to insufficient heat treatment or when using unsuitable carbon types.

- Too low shear effect during extruding

- Extrusion defects, such as internal shear cracks when extruding with too high deformation energy

Generally, the work has proceeded sufficiently far to enable production of Al-C materials by reaction milling (trade mark DISPAL) with consistently high quality.

Acknowledgement

This work was partly sponsored by BMFT and was carried out in close cooperation with Fa. KREBSÖGE, ERBSLÖH and ECKART.

References

(1)   G. Jangg, G. Korb, F. Kutner: Proc. of the 6th Int. Leichtmetalltagung, Al-Verlag, Düsseldorf (1975), p.61
(2)   G. Jangg, F. Kutner, G. Korb: Aluminium 51 (1975) 641
(3)   G. Jangg, F. Kutner, G. Korb: Powder Met. Int. 9 (1977) 24
(4)   G. Korb, G. Jangg, F. Kutner: Wire 28 (1979) 6
(5)   G. Jangg: Radex-Rundschau 1986, Vol. 2/3, p. 169
(6)   H. Oppenheim, G. Jangg: Z. F. Werkstofftechnik 14 (1983) 170
(7)   J.S. Benjamin, R. Schelling: Met. Trans. A 12A (1981) 1827
(8)   R.F. Singer, W.C. Oliver, W.D. Nix: Met. Trans. A 11A (1980) 1895
(9)   J. Zbiral, K. Schröder, G. Jangg, M. Slesar, M. Besterci: Proc. 7th Int. PM Conf., Pardubice 1987, Vol. 1, p.151
(10)   G. Jangg, M. Slesar, M. Besterci, J. Zbiral: to be published
(11)   M. Slesar, G. Jangg, M. Besterci, J. Durisin: Proc. 7th Int. Leichtmetalltagung, Leoben, Aluminium-Verlag, Düsseldorf (1981) p. 238
(12)   G. Jangg, M. Slesar, M. Besterci, H. Oppenheim: Z. F. Werkstofftechnik 18 (1987) 36
(13)   M. Slesar, G. Jangg, M. Besterci, J. Durisin, M. Drolinova: to be published
(14)   J. Schalunov, M. Slesar, M. Besterci, H. Oppenheim, G. Jangg: Metall 40 (1986) 601
(15)   H. Schlich, M. Thumann, G. Wirth: unpublished reports
(16)   G. Jangg, V. Arnhold, K. Hummert, G. Brockmann, H.C. Neubing: DGM-Symposium Verbundwerkstoffe, Konstanz 1988, Proc. in press

Basic Phenomena of Glass Formation by Mechanical Alloying

Ludwig Schultz, Siemens AG, Research Laboratories, Erlangen, FRG

Abstract

Since mechanical alloying is a non-equilibrium processing tool, it can be used to form metastable alloys. In this contribution, the basic phenomena of the formation of amorphous metals by mechanical alloying, as the process of glass formation, the characterization of the atomic structure, the glass-forming ranges and the glass-forming systems, are described.

Introduction

Amorphous metals were first investigated by Buckel and Hilsch [1] over 30 years ago. They succeeded in depositing thin amorphous films by quench condensation of metals on a substrate cooled to cryogenic temperatures. In 1960, Duwez et al. [2] reported on the formation of amorphous metals by rapid quenching from the melt. Contrary to this, in 1983, Johnson and coworkers [3,4] showed that amorphous metals can also be formed by an interdiffusion reaction within the solid state, as by the diffusion of hydrogen gas into the metastable crystalline Zr_3Rh phase [3] or the interdiffusion in ultrafine layered evaporated Au-La sandwich films [4]. Since no quenching processes are involved, amorphous metals of basically unlimited thickness can be formed, which has been demonstrated for macroscopic composites of two finely layered elemental metals produced by mechanical co-deformation for the case of Ni-Zr [5]. At about the same time, it was demonstrated that amorphous metals can also be formed within the solid state by mechanical alloying [6].

The basic principles of glass formation by a solid-state reaction [7] are described in Fig. 1. The free enthalpy of the equilibrium crystalline state (G_x) is always lower than that of the amorphous state (G_a) for metallic systems below the melting temperature. The amorphous state is a metastable state, i.e. an energy barrier exists which prevents amorphous metals from spontaneous crystallization. In order to form an amorphous metal by a solid-state reaction it is necessary to create an initial crystalline state (G_0) with a high free enthalpy (see Fig. 1). Depending on the formation process this initial state can be provided, for example, by:

1) the system intermetallic phase (Zr_3Rh) plus hydrogen gas [3],
2) an ultrafine layered system of two crystalline elemental metals [4,5], if the alloy system has a negative free enthalpy of mixing,
3) an intermetallic phase that is disordered as a result of the introduction of lattice defects [8-10],
4) a metastable intermetallic phase far from equilibrium [11].

Starting from this initial state G_0, the free enthalpy of the sy-

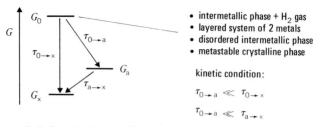

energetically favored: intermetallic phase

kinetically favored: amorphous phase

<u>Fig. 1:</u> Basic principles of glass formation by a solid-state reaction.

stem can be lowered either by the formation of the metastable amorphous phase or by the formation of the crystalline intermetallic phase (or phase mixture). Energetically favored is, of course, the crystalline equilibrium phase, but the kinetics of the phase formation decides which phase is in fact formed. To evaluate this, the time scales of the possible reactions must be considered. The formation of the amorphous phase is possible and likely if the formation reaction for the amorphous phase is much faster than that for the crystalline phase, i.e.:

$$\tau_{o->a} \ll \tau_{o->x}$$

($\tau_{i->j}$: characteristic time scale of the reaction). During this reaction the amorphous phase should not crystallize:

$$\tau_{o->a} \ll \tau_{a->x}$$

i.e. the reaction temperature T_r must be well below the crystallization temperature T_x.

In the following sections this principle of glass formation by a solid-state reaction (for a more detailed discussion see [12]) will be discussed with respect to glass formation by mechanical alloying. Mechanical alloying produces an ultrafine layered microstructure of two crystalline elemental metals. The free enthalpy of this initial state can be lowered by the transition to the amorphous state. Contrary to this, milling of powder of an intermetallic phase can create a highly disordered crystalline state from which a transition to the amorphous state becomes possible [9].

<u>The process of glass formation</u>

Mechanical alloying is performed as dry ball milling of elemental powders in a high-energy ball mill under an inert atmosphere. The powder particles are repeatedly cold-welded by the colliding balls

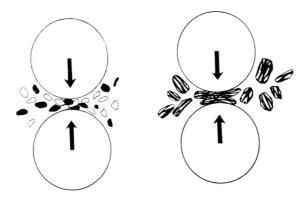

Fig. 2: Formation of layered powder particles during ball milling.

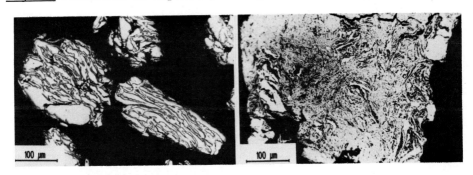

a) after 0.5 h   b) after 4 h

Fig. 3: Microstructure of mechanically alloyed Fe-Zr powder
particles after different milling periods [13].

c) after 16 h of milling.

(Fig. 2). Thus, composite powder particles with a characteristically layered microstructure are formed. With regard to glass formation, the development of the microstructure has been investigated by Hellstern et al. [13] in detail for an $Fe_{50}Zr_{50}$ composi-

tion. Light microscopy reveals that the layered microstructure with a preferred orientation has developed after a short period of milling (Fig. 3a). Further milling lets the particles grow as a result of cold welding and reduces the individual layer thickness as a result of deformation. After 4 h of milling the individual layers are at the resolution limit of light microscopy (Fig. 3b). After 16 h of milling (Fig. 3c) the powder particles look completely different from those of Figs. 3a and b. With the exception of only a few particles they are amorphous. X-ray diffraction patterns (Fig. 4) of these powders, taken after different milling times, confirm this. The intensity of the diffraction peaks of crystalline Fe and Zr is gradually reduced with increasing processing time. A diffuse maximum which is attributed to the amorphous phase grows separately at a position between the three crystalline low-angle Zr lines and the Fe(110) line. After 16 h of milling only small residual crystalline peaks are still superimposed on the broad amorphous maximum. After 60 h of milling no crystalline material can be detected any more and the X-ray pattern reveals the broad diffuse maximum of the amorphous phase.

Whereas a qualitative description of glass formation is given by X-ray diffraction, the amount of amorphous phase formed during milling can be determined quantitatively from crystallization enthalpy measurements by differential scanning calorimetry (DSC). Figure 5 shows the crystallization enthalpy of $Fe_{20}Zr_{80}$ vs. milling time. The amorphization reaction takes place mainly between 4 and 16 h.

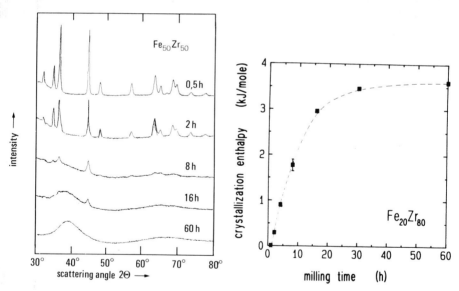

Fig. 4: X-ray diffraction patterns of mechanically alloyed $Fe_{50}Zr_{50}$ powder after increasing milling time [13].

Fig. 5: Crystallization enthalpy of $Fe_{20}Zr_{80}$ vs. milling time.

For compositions away from the central part of the phase diagram, such as $Fe_{80}Zr_{20}$ or $Fe_{20}Zr_{80}$, mixtures of the amorphous phase and the major element are often formed (see the discussion of the glass-forming ranges below). In this case the crystalline lines of the major element do not vanish. This allows to follow the micro-crystallite size of the elemental Fe and Zr, which is obtained from the peak width $\Delta\Theta$ of the Fe(110) and Zr(101) peaks using the Scherrer formula, during the milling process (Fig. 6). The micro-crystallite size decreases from approximately 80 nm for the un-milled elemental powders to less than 40 nm at the very beginning of the milling process, and is only slightly further reduced from 2 to 16 h (to about 20 nm). Recently, Hellstern et al. [14] inve-stigated this problem in more detail by ball milling of elemental Ru and of the intermetallic AlRu phase which both do not become amorphous during this treatment. The crystallite size saturates at about 12 nm for Ru and 7 nm for AlRu. This behavior is explained by considering the Hall-Petch relation which describes the depen-dence of the yield stress σ_y on the grain size d of the material:

$$\sigma_y = \sigma_o + k \cdot d^{-1/2}.$$

(σ_o and k are constants.) Using typical values for k and σ_o, the yield stress σ_y at d = 10 nm reaches about 20 % of the theoretical shear stress of Ru. Therefore, a deformation by dislocation gli-ding is prohibited at a very small crystallite size. A further de-formation is then only possible by a grain boundary gliding mecha-nism which does not further reduce the crystallite size. This is also confirmed by the dependence of the lattice strain, obtained from the shift of the X-ray diffraction peaks, on the crystallite size. It increases during the initial milling but decreases for extensive milling when the crystallite size has been reduced to less than 15 nm for Ru or 10 nm for AlRu [14]. A similar observa-tion was made by Frommeyer and Wassermann [15] about 15 years ago, when they investigated the properties of heavily deformed filamen-tary composites. When the filament diameter was reduced to about

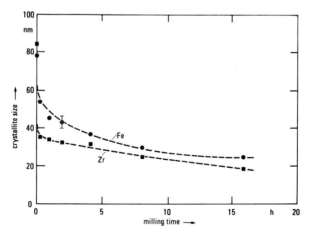

Fig. 6: Size of the Fe and Zr crystallites vs. milling time [13].

20 to 30 nm, the yield stress strongly increased and the dislocation density dropped, because the dislocations were annihilated in
the grain boundaries due to image forces.

This shows that by mechanical alloying the crystallite size can
not be indefinitely reduced until atomic mixing occurs. Since
nearly 80 % of the amorphization occurs within the time interval
from 2 to 16 h (see Fig. 5), glass formation cannot be ascribed to
a continuous breaking into smaller crystallites, as has been assumed before for Ni-Nb [6] and Cu-Ti [16]. This process would result
in a gradual increase of the crystalline line width until it became impossible to distinguish the X-ray scans of the amorphous
and the microcrystalline phase. Of course, in some cases it can
happen that the amorphous diffraction peak is located at the same
angle 2Θ as a crystalline peak of one of the involved elements.
Then the formation of the amorphous phase as a newly formed phase
cannot be recognized as clearly as in the case of Fe-Zr.

So far, it has been shown that mechanical alloying creates an ultrafine layered microstructure of the two crystalline elemental
metals as an intermediate product. The free enthalpy of this initial state is higher than that of the amorphous state (compare
Fig. 1) if the alloy system has a negative heat of mixing term. It
can, therefore, be lowered by an interdiffusion reaction if the
temperature during processing becomes high enough to allow the
diffusional process. To check this, Eckert et al. [17] investigated the influence of the milling intensity on glass formation for
Ni-Zr. Figure 7 shows the X-ray diffraction patterns of mechanically alloyed Ni-Zr powders for different compositions milled at
intensities 3, 5, or 7 which corresponds to velocities of 2.5, 3.6
or 4.7 m/s of the balls during the collisions. (In these experiments a Fritsch Pulverisette 5 planetary ball mill was used.) For

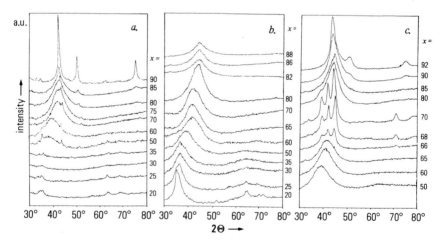

Fig. 7: X-ray diffraction patterns of mechanically alloyed
Ni_xZr_{100-x} powders milled at different intensities [17]:
(a) intensity 3; (b) intensity 5; (c) intensity 7.

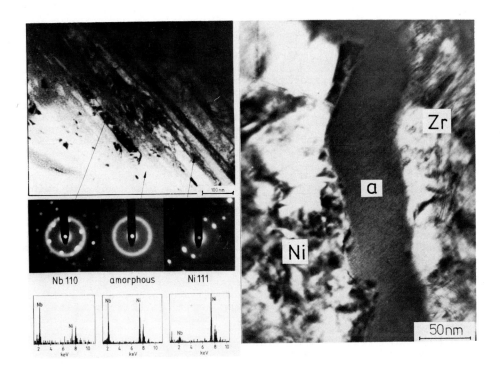

Fig. 8: TEM picture and EDX spectra of mechanically alloyed and partially amorphized $Ni_{60}Nb_{40}$ (Petzoldt et al. [18]).

Fig. 9: TEM picture of a mechanically co-deformed and partially reacted Ni-Zr composite [19].

the low milling intensity 3 a complete amorphization could not be achieved within 60 h of milling as revealed by the residual crystalline lines. For intensity 5 complete amorphization occurs between about 30 and 82 at% Ni whereas for intensity 7 the crystalline intermetallic Ni_3Zr phase appears from 66 to 75 at% Ni. It must be concluded that, at intensity 7, the individual powder particle is heated up considerably during the collision event, so that even crystallization occurs. An estimate of the maximum temperature of the powder particles resulting from shear deformation during collision of the balls, using a model proposed by Schwarz and Koch [9], gives temperatures of 130 °C, 250 °C and 407 °C for milling intensities 3, 5 and 7, respectively [17]. Since amorphization by interdiffusion is always the faster process than crystallization for these layered composites of early and late transition metals, the temperature reached in the powder particles during milling is sufficient for the interdiffusion reaction.

Also transmission electron microscopy (TEM) investigations by Petzoldt et al. [18] of partially amorphized Ni-Nb powders prove

that the amorphous phase is formed in between layers of elemental
Ni and Nb. Figure 8 shows the sequence "crystalline Nb/amorphous
Ni-Nb/crystalline Ni", which is confirmed by electron diffraction
patterns and energy-dispersive X-ray (EDX) analysis (lower part of
Fig. 8). For comparison, Fig. 9 shows a TEM picture of mechanically co-deformed and partially reacted Ni-Zr composites obtained by
Eckert et al. [19], which clearly demonstrates that for amorphization by an interdiffusion reaction in these layered composites the
amorphous phase is formed at the interfaces between the two elements. The similarity of the two TEM pictures (Figs. 8 and 9)
gives further evidence that the glass-forming processes are closely related.

These results, i.e. the formation of an ultrafine layered microstructure, the lower limit of the crystallite size, the TEM observations and the fact that considerably high temperatures are reached in the powder particles, indicate that glass formation by mechanical alloying is due to an interdiffusion reaction in a similar way as for amorphization by solid-state reaction in layered
composites which were artificially prepared by evaporation [4] or
mechanical co-deformation [5]. Further evidence is given by the
investigation of the glass-forming ranges which will prove that
the reaction takes place under a metastable equilibrium.

The very long milling times mentioned above are characteristic for
the alloying process in a planetary ball mill where the material
sticks to the container wall most of the time. Otherwise, the use
of a planetary ball mill is quite convenient for laboratory experiments because of its versatility. An attritor-type ball mill reduces the milling time by an order of magnitude.

As well as the mechanical alloying of elemental powders, ball milling of an intermetallic compound can also lead to amorphization,
which has been demonstrated for Y-Co [8], Ni-Nb, Ni-Ti [9], Ni-Zr
[20] and Fe-Zr [21]. This case cannot be explained by the above
statements. The amorphization by milling, starting from powders of
crystalline intermetallics, is instead attributed to the accumulation of lattice defects which raise the free enthalpy of the faul-
ted intermetallic above that of the amorphous alloy [9]. Therefore, some similarity to irradiation-induced amorphization exists
[10]. A more detailed description is given by Koch et al. [22] in
these proceedings.

Characterization of the atomic structure

In most investigations, the formation of the amorphous phase by
mechanical alloying is monitored by X-ray diffraction and by DSC
studies by which the crystallization of the initially formed amorphous phase is observed. In a few cases, TEM studies have also
been performed. These investigations can at best only prove the
amorphous nature of the newly formed phase, but they do not reveal
any detailed information on the actual atomic structure. A more
sophisticated structural analysis has been performed recently by
Biegel et al. [23] and Burblies and Petzoldt [24] who compared the
radial distribution functions G(r) of mechanically alloyed amor-

phous Fe-Zr and Ni-Zr, respectively, with those of the corresponding rapidly quenched material. No significant differences were
detected. Mößbauer spectroscopy of mechanically alloyed, rapidly
quenched and sputtered Fe-Zr samples by Michaelsen and Hellstern
[25] also could not reveal any significant differences, demonstrating that the local atomic structures of the differently prepared
samples are similar.

A further possibility for characterizing the atomic structure is
to investigate structure-sensitive physical properties such as superconductivity or hydrogen absorption or to study the structural
relaxation by calorimetry. Fig. 10 shows the superconducting transition temperature T_c of mechanically alloyed Ni_xZr_{100-x} powder
vs. Ni content in comparison with results for melt-spun amorphous
ribbons [26]. The almost constant $T_c(x)$ dependence for x < 25 is
due to a two-phase material (see next section), whereas for x > 30
both materials show the same linear dependence on composition. The
difference between both curves is attributed to the different measuring techniques (ac susceptibility vs. resistivity) and a small
oxygen content (less than 2 at%) of the mechanically alloyed samples [27]. These results definitely prove the amorphous structure
of these samples since superconductivity in Ni-Zr sensitively depends on the atomic structure. Thus, for amorphous Ni-Zr samples
which are structurally relaxed by annealing below the crystallization temperature, the transition temperature is considerably reduced. Crystallized samples do not show superconductivity above
1.3 K.

Structural relaxation of these samples during annealing has been
studied in detail by Brüning et al. [27]. Figure 11 shows the DSC
traces of the irreversible structural relaxation in melt-spun and
mechanically alloyed $Ni_{30}Zr_{70}$ samples. In principle, the same behavior is observed for both samples, but the effect of thermal relaxation is about twice as large for the mechanically alloyed sam-

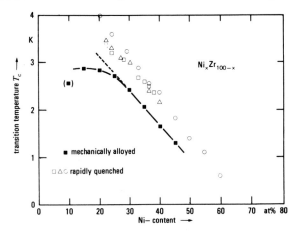

Fig. 10: Superconducting transition temperature of mechanically
alloyed and melt-spun amorphous Ni_xZr_{100-x} samples [26].

ples. The relaxation of these samples starts at a somewhat lower temperature, which increases slightly when the samples are stored at room temperature for a long period. Therefore, it is concluded that the mechanically alloyed samples are in a very unrelaxed state and that thermal relaxation can take place even at room temperature.

The hydrogen absorption of amorphous metals strongly depends on the atomic order, expecially on the nature and number of interstitial sites present, on the composition and on a possible chemical

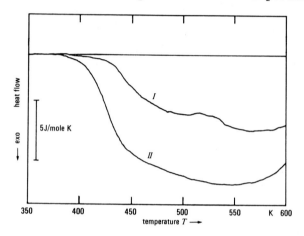

Fig. 11: DSC-traces of the irreversible structural relaxation in melt-spun (curve I) and mechanically alloyed (curve II) $Ni_{30}Zr_{70}$ (Brüning et al. [27]).

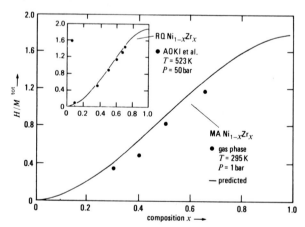

Fig. 12: Total hydrogen-to-metal ratio vs. alloy composition for gas-phase absorption of mechanically alloyed and melt-spun amorphous Ni_xZr_{100-x} samples (Harris et al. [28]).

short range order. Harris et al. [28] compared the hydrogen occupation statistics and thermodynamics of hydrogen absorption of mechanically alloyed and rapidly quenched amorphous Ni-Zr samples.
Figure 12 shows the total hydrogen-to-metal ratio vs. alloy composition for gas phase absorption as predicted theoretically and
as obtained experimentally for both types of material. From the
similar experimental results and the good fit with the theoretically predicted curve it is concluded that the maximum possible
amounts of hydrogen absorption, i.e. the number of available sites
and the type of sites occupied by hydrogen, are identical. Otherwise, electrochemical absorption, which allows to study the chemical potential range where the Ni_2Zr_2 tetrahedral sites are occupied, reveals serious differences. Within the available electrochemical window only less than half of the hydrogen is absorbed by
the mechanically alloyed samples compared with the rapidly quenched samples. Harris et al. [28] conclude that the peak of the
density of states for hydrogen occupation of the Ni_2Zr_2 sites is
significantly wider and energetically shifted. Annealing of the
samples did not change the peak width and position. Since hydrogen
energetics are extremely sensitive to the size of the interstitial
sites, the differences in energetics and width of the density of
states peak could be due to intrinsic strain in the mechanically
alloyed material or they may reflect slightly different atomic
packing characteristics for the two types of alloys.

These results show that mechanically alloyed and rapidly quenched
amorphous metals are indeed structurally very similar, both topologically and chemically. Minor differences as revealed by hydrogen absorption have still to be studied in detail.

Glass-forming ranges

As shown in the previous sections, the process of glass formation
by mechanical alloying differs completely from rapid quenching.
Therefore it is likely that the composition range where glass formation is possible will also be different. In this section, the
glass-forming ranges which are theoretically predicted, the appearance of the resulting metastable phase diagram and how the glass-
forming ranges can be determined accurately are described. Figure
13 shows the thermodynamically stable phase diagram of a hypothetical A-B alloy and the free enthalpy curves at a reaction temperature T_r [29,30]. The equilibrium phases are the two primary phases α and β and the intermetallic phase γ. The phase boundaries
are obtained from the double-tangent construction at the free enthalpy curves. At T_r the free enthalpy curve for the liquid or
(assuming that T_r is below the glass transition temperature T_g)
the amorphous phase (broken line) is always above the composed
curve for the stable state. Therefore, the amorphous phase is not
present in the equilibrium phase diagram (full lines). If we consider the experimental fact that during amorphization by mechanical alloying the intermetallic phase γ does not form, we can ig-
nore the free enthalpy curve in Fig. 13. The free-enthalpy curve
of the amorphous phase is then lowest in the central part making
the amorphous phase the most stable of the possible phases. The
corresponding metastable phase diagram is then obtained from the

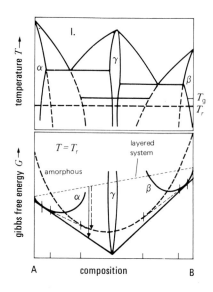

Fig. 13: Free enthalpy curves and the resulting equilibrium (full lines) and metastable equilibrium phase diagram (neglecting the intermetallic phase; broken lines) of a hypothetical A-B alloy (after [29] and [30]).

free enthalpy curves for α, amorphous and β phases by applying the double-tangent construction to these curves (broken lines in Fig. 13). As a result, we find a wide glass-forming range in the central part of the phase diagram, two two-phase regions (α plus amorphous and β plus amorphous) and extended solubilities in the primary phases. To check these predictions the glass-forming ranges of mechanically alloyed samples were experimentally determined in detail.

Amorphization by mechanical alloying has been studied in a wide composition range for Ni-Ti [31], Cu-Ti [32], Fe-Zr [33], Ni-Zr [17,26,34,35], Co-Zr [36,37], Ni-Nb [38], and Nb-Al [39] (for a review see [30,40]). In most cases, X-ray diffraction patterns are used to show the formation of an amorphous phase and to determine the glass-forming range, but these patterns mostly give only a qualitative description. Again, as for the structural characterization of the amorphous state, it is much more reliable to measure physical properties as a function of composition [30]. Especially useful are intensive physical properties such as the superconducting transition temperature (Fig. 10) and the crystallization temperature (Fig. 14) which are qualitative properties depending on composition within the homogeneity range of the amorphous phase and being constant in the two-phase region. Fig. 14 gives an excellent example of such a composition dependence [26]. Whereas the crystallization temperature of rapidly quenched amorphous Ni_xZr_{100-x} samples decreases monotonically with decreasing Ni content (at $x < 45$), it stays almost constant for mechanically alloyed samples with less than 27 at% Ni. The phase boundary between the amorphous phase and the two-phase region is therefore determined to be 27 at% Ni. The plot of the superconducting transition temperature vs. composition (Fig. 10) gives the same result. In a

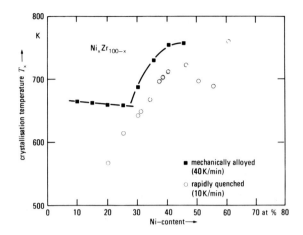

Fig. 14: Crystallization temperature vs. composition of mechanically alloyed and melt-spun Ni_xZr_{100-x} samples [26].

similar way, extensive (quantitative) physical properties which depend on composition within the homogeneity range and vary linearly with the amount of phase present in the two-phase region according to the lever rule can be used.

Following these considerations we determined the glass-forming ranges of Fe-Zr [33], Ni-Zr [26,34], and Co-Zr [37] (Table I). Amorphous metals can be formed by mechanical alloying in the central part of the phase diagrams. To both sides of the amorphous phase two-phase regions exist. The solubility of the primary phases is largely extended (5 at% Zr in Fe [33], 7 at% Ni in Zr [26]). This demonstrates that the qualitative predictions based on the above thermodynamic considerations are, in fact, fulfilled, i.e. the results can be at least qualitatively described by a metastable phase diagram which is constructed from the free enthalpy

Table I: Experimentally and theoretically obtained glass-forming ranges:

	experimental x [at%]	theoretical Miedema model [40] x [at%]	theoretical CALPHAD [41] x [at%]
Ni_xZr_{100-x}	27 - 83	24 - 83	33 - 83
Fe_xZr_{100-x}	30 - 78	27 - 79	
Co_xZr_{100-x}	27 - 92		

curves of the two primary phases and of the amorphous phase neglecting the intermetallic phases. For a quantitative evaluation
the free enthalpy curves must be calculated. Bormann et al. [41]
used the CALPHAD technique for the Ni-Zr system, predicting the
glass-forming range for mechanically alloyed Ni-Zr to extend from
33 to 83 at% Ni (Table I). Miedema calculations [41] performed by
Weeber and Bakker [40] predict 24 to 83 at% Ni (Table I). On the
Ni-rich side, the value determined experimentally by Eckert et al.
[17] exactly hits the value predicted by both theoretical techniques, whereas on the Zr-rich side, the experimental value lies
between the values of the two theoretical procedures, but here
also the results are very close. Table I also compares the experimentally obtained glass-forming range for Fe-Zr (30 to 78 at% Fe)
[33] with the Miedema results (27 to 79 at% Fe) [40]. Again, the
results agree quite well. These results show that an excellent
agreement of the experimentally and theoretically obtained glass-
forming ranges is achieved if they are carefully determined. The
glass-forming ranges by mechanical alloying can be derived from
the free enthalpy curves of the primary phases and the amorphous
phase simply by neglecting the existence of intermetallic phases
which are prevented from forming by the kinetic restrictions of
the mechanical alloying process.

Finally, the glass-forming range of mechanically alloyed samples
should be compared with that of melt-spun samples. The most interesting example for this is the Fe-Zr system (Fig. 15; [33]), where
the glass-forming ranges overlap only in a very small region, indicating again that the two processes are completely different.
Whereas rapid quenching is a truely non-equilibrium process allowing glass formation preferentially close to deep eutectics, glass
formation by mechanical alloying takes place under a metastable
equilibrium preferentially in the central part of the phase diagram, i.e. also in the range of high-melting intermetallic phases.
Eutectic compositions of the equilibrium phase diagram do not play
a role.

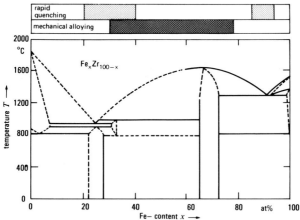

Fig. 15: Fe-Zr phase diagram and glass-forming ranges for mechanical alloying and rapid quenching [33].

Glass-forming systems

Glass formation by mechanical alloying has been found in a large number of transition metal-transition metal (TM_1-TM_2) alloys, where TM_1 is Cr, Mn, Fe, Co, Ni, Cu, Pd or Ru, and TM_2 is Zr, Ti, V, Nb or Hf [6,16,31-33,43-47]. A systematic study of the glass-forming ability has been performed for 3d transition metal alloys with Zr [44] and Ti [46]. Figure 16 shows the X-ray diffraction patterns of $TM_{60}Ti_{40}$ alloys (where TM varies from Cu to V) after mechanical alloying. A complete amorphization is achieved only in the Ni-Ti, Cu-Ti and Co-Ti systems. The Fe-Ti, Mn-Ti and Cr-Ti samples show a remaining crystalline portion increasing in this sequence. In the V-Ti system, the crystalline solid solution forms which is stable over the whole composition range at higher temperatures and which, therefore, has no nucleation barrier. Within the TM-Zr alloy series with the same TM elements, only Cr-Zr and V-Zr do not become amorphous [44].

The different alloying behavior can be explained by considering the free enthalpy of mixing as estimated using the Miedema model [42]. It is remarkable that those alloys which have a large nega-

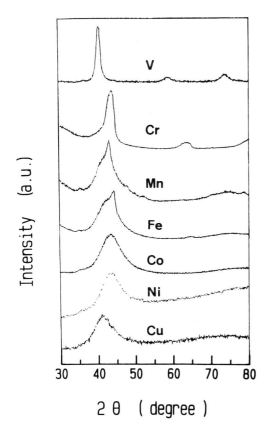

Fig. 16: X-ray diffraction patterns of several $TM_{60}Ti_{40}$ samples after mechanical alloying [46].

tive heat of mixing show easy amorphization and that the V-Ti, V-Zr and Cr-Zr alloys possess the lowest mixing tendency [46]. A similar example is Fe-Nd where the difference of the free enthalpies between the amorphous phase and the layered composite is positive. Therefore, neither alloying nor amorphization occurs during milling [48].

Most of the TM-TM alloys mentioned here can also be amorphized by melt spinning, although the glass-forming ranges are different from those of the mechanically alloyed samples. Ti-Al is an example where an amorphous phase cannot be obtained by melt spinning. Figure 17 shows the X-ray diffraction patterns of various Ti_xAl_{100-x} samples after mechanical alloying [30]. Only the $Ti_{52}Al_{58}$ and the $Ti_{40}Al_{60}$ samples become amorphous. $Ti_{60}Al_{40}$ and other Ti-rich samples show mainly crystalline material, as the $Ti_{25}Al_{75}$ sample does. The glass-forming range, therefore, extends from about 45 to 65 at% Al, i.e. it is in the central part of the phase diagram but shifted slightly to the Al-rich side (see also [49] in this volume). The Ti-Al system also has a sufficiently high negative heat of mixing.

Further binary alloy systems in which amorphization by mechanical alloying has been found are Nb-Al [50], Nb-Ge [51,52], Nb-Sn [51], Co-Sn [53] and Co-Gd [54]. Therefore, amorphization by mechanical alloying is possible for a wide variety of alloy systems.

However, the commercially most important group of amorphous metals, the metal-metalloid systems such as Fe-Ni-B, has been omitted so far. Our first experiments with Fe-Ni-B were unsuccessful. We therefore studied the Fe-Zr-B system [30,55]. Milling of elemental Fe, Zr and submicron amorphous B powder first produces a layered microstructure of Fe and Zr. The undeformed B particles are caught by the colliding Fe and Zr particles and are embedded

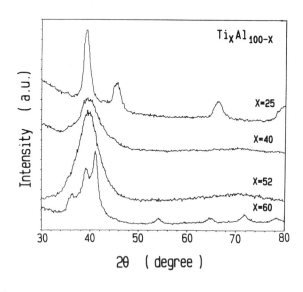

Fig. 17: X-ray diffraction patterns of Ti_xAl_{100-x} samples prepared by mechanical alloying [30].

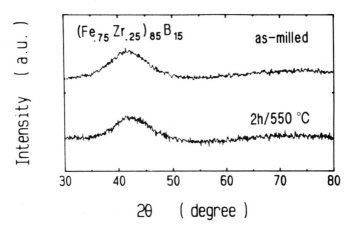

<u>Fig. 18:</u> X-ray diffraction patterns of $(Fe_{0.75}Zr_{0.25})_{85}B_{15}$ samples in the as-milled state and after annealing of 2 h at 550 °C.

in the Fe/Zr interfaces. Further milling leads to a refinement of this layered microstructure until finally the Fe and the Zr layers react to form amorphous Fe-Zr as in the case of binary Fe-Zr (Fig. 18). The B particles are still finely dispersed within the amorphous Fe-Zr. An additional solid-state reaction process (2 h at 550 °C) enables the boron to diffuse into the Fe-Zr, forming amorphous Fe-Zr-B (Fig. 18). For a series of $(Fe_{0.75}Zr_{0.25})_{100-x}B_x$ compositions, we succeeded in forming perfectly amorphous material up to x = 15. For 20 at% B the samples crystallized, possibly because of structural changes when the TM-TM amorphous structure transforms to the TM-metalloid amorphous structure. Using the same process we also succeeded in forming amorphous Ni-Nb-B.

<u>Conclusions</u>

In this contribution, the basic phenomena of glass formation by mechanical alloying were discussed. Starting from the elemental, crystalline powders, ball milling first produces powder particles with a characteristically layered microstructure. Further milling leads to an ultrafine composite in which amorphization by solid-state reaction takes place. A study of the influence of the milling intensity on glass formation shows that the effective temperature during mechanical alloying can rise significantly. Necessary conditions for glass formation are: (1) a sufficiently negative heat of mixing, (2) a sufficient diffusivity at temperatures below the crystallization temperature and (3) the suppression of the formation of the crystalline intermetallic phases.

Besides X-ray diffraction and differential scanning calorimetry to study crystallization, the atomic arrangement is characterized by the measurement of structure-sensitive physical properties as superconductivity, Mößbauer effect, structural relaxation, or

hydrogen absorption. As a result, mechanically alloyed and rapidly
quenched amorphous metals are structurally very similar, both
topologically and chemically. Minor differences have been revealed
by the hydrogen absorption studies. Therefore, the amorphous state
should not be considered a frozen liquid but a metastable phase
which can be equally approached by quenching from the liquid state
and by a crystal-to-glass transition via interdiffusion. The
glass-forming ranges for mechanical alloying are exactly determined also by measuring physical properties as a function of composition. These experimentally obtained glass-forming ranges agree
quite well with those derived theoretically from the free enthalpy
curves of the primary phases and the amorphous phase by simply
neglecting the existence of intermetallic phases which are prevented from forming by the kinetic restrictions of the mechanical alloying process. This indicates that glass formation by mechanical
alloying can be described as a metastable equilibrium process.
Amorphous metallic powders can be formed by mechanical alloying in
a large number of alloy systems. Also, compositions that are not
accessible by melt spinning can be made amorphous. Mechanical alloying therefore offers an attractive technique for forming amorphous metals with possible technological applications.

The author thanks R. Brüning, R. Bormann, J. Eckert, J.H. Harris,
E. Hellstern, C. Michaelsen, F. Petzoldt and A. Thomä for effective cooperation and stimulating discussions.

References

[1] W. Buckel and R. Hilsch, Z. Phys. 138 (1954) 109.
[2] P. Duwez, R.H. Willens and W. Klement, Jr., J. Appl. Phys. 31
    (1960) 1136.
[3] X.L. Yeh, K. Samwer and W.L. Johnson, Appl. Phys. Lett. 42
    (1983) 242.
[4] R.B. Schwarz and W.L. Johnson, Phys. Rev. Lett. 51 (1983)
    415.
[5] L. Schultz, in M. von Allmen (ed.): Proc. MRS Europe Meeting
    on Amorphous Metals and Non-Equilibrium Processing,
    Strasbourg, Les Editions de Physique, Les Ulis 1984, p. 135;
    L. Schultz, in S. Steeb and H. Warlimont (eds.): Proc. 5th
    Int. Conf. on Rapidly Quenched Metals, Würzburg, Elsevier,
    Amsterdam 1985, p. 1585.
[6] C.C. Koch, O.B. Cavin, C.G. McKamey and J.O. Scarbrough,
    Appl. Phys. Lett. 43 (1983) 1017.
[7] W. L. Johnson, Mater. Sci. Eng. 97 (1988) 1.
[8] A.E. Ermakov, E.E. Yurchikov and V.A. Barinov, Fiz. metal.
    metalloved. 52 (1981) 1184.
[9] R.B. Schwarz and C.C. Koch, Appl. Phys. Lett. 49 (1986) 146.
[10] L.E. Rehn, P.R. Okamoto, J. Pearson, R. Bhadra and M. Grims-
    ditch, Phys. Rev. Lett. 59 (1987) 2987.
[11] A. Blatter and M. von Allmen, Phys. Rev. Lett. 54 (1985)
    2103.
[12] W. L. Johnson, Prog. Mater. Sci. 30 (1986) 81.
[13] E. Hellstern and L. Schultz, J. Appl. Phys. 63 (1988) 1408.

[14] E. Hellstern, H.J. Fecht, Z. Fu and W.L. Johnson, Appl. Phys. Lett. 54 (January 1989).
[15] G. Frommeyer and G. Wassermann, Phys. Status Solidi A 27 (1975) 99.
[16] C. Politis, Z. Phys. Chem. 157 (1988) 209.
[17] J. Eckert, L. Schultz, E. Hellstern and K. Urban, J. Appl. Phys. 64 (1988) 3224.
[18] F. Petzoldt, B. Scholz and H.D. Kunze, Mater. Lett. 5 (1987) 280.
[19] J. Eckert, 1988, unpublished.
[20] A.W. Weeber, H. Bakker and F.R. Boer, Europhys. Lett. 2 (1986) 445.
[21] E. Hellstern and L. Schultz, 1986, unpublished.
[22] C.C. Koch, J.S.C. Jang and P.Y. Lee, in E. Arzt and L. Schultz (eds.), Proc. DGM Conference on New Materials by Mechanical Alloying Techniques, Calw/Hirsau, October 1988, DGM Informationsgesellschaft, Oberursel, these proceedings.
[23] W. Biegel, H.U. Krebs, C. Michaelsen, H.C. Freyhardt and E. Hellstern, Mater. Sci. Eng. 97 (1988) 59.
[24] A. Burblies and F. Petzoldt, J. Non-Cryst. Solids (in print).
[25] C.Michaelsen and E. Hellstern, J. Appl. Phys., 62 (1987) 117.
[26] L. Schultz, E. Hellstern and A. Thomä, Europhys. Lett. 3 (1987) 921.
[27] R. Brüning, Z. Altounian, J.O. Strom-Olsen and L. Schultz, Mater. Sci. Eng. 97 (1988) 317.
[28] J.H. Harris, W.A. Curtin and L. Schultz, J. Mater. Research 3 (1988) 872.
[29] R.B. Schwarz, MRS Bull. 1986, p. 55.
[30] L. Schultz, Mater. Sci. Eng. 97 (1988) 15.
[31] R.B. Schwarz, R.R. Petrich and C.K. Saw, J. Non-Cryst. Solids 76 (1985) 281.
[32] C. Politis and W.L. Johnson, J. Appl. Phys. 60 (1986) 1147.
[33] E. Hellstern and L. Schultz, Appl. Phys. Lett. 49 (1986) 1163.
[34] J. Eckert, L. Schultz and K. Urban, J. Less-Common Met. 145 (1988).
[35] F. Petzoldt, B. Scholz and H.D. Kunze, Mater. Sci. Eng. 97 (1988) 25.
[36] A. Thomä, G. Saemann-Ischenko, L. Schultz and E. Hellstern, Jap. J. Appl. Phys. 26 (1987) 977.
[37] E. Hellstern, L. Schultz and J. Eckert, J. Less-Common Met. 140 (1988) 93.
[38] P.Y. Lee and C.C. Koch, J. Non-Cryst. Solids 94 (1988) 88.
[39] E. Hellstern, L. Schultz, R. Bormann and D. Lee, Appl. Phys. Lett. 53 (1988) 1399.
[40] A.W. Weeber and H. Bakker, Physica B 153 (1988) 93.
[41] R. Bormann, F. Gärtner, K. Zöltzer and R. Busch, J. Less-Common Met. 145 (1988).
[42] A.K. Niessen, F.R. de Boer, P.F. de Chatel, W.C.M. Mattens and A.R. Miedema, CALPHAD 7 (1983) 51.
[43] E. Hellstern and L. Schultz, Appl. Phys. Lett. 48 (1986) 124.
[44] E. Hellstern and L. Schultz, Philos. Mag. B 56 (1987) 443.
[45] B.P. Dolgin, M.A. Vanek, T. McGory and D.J. Ham, J. Non-Cryst. Solids 87 (1986) 281.
[46] E. Hellstern and L. Schultz, Mater. Sci. Eng. 93 (1987) 213.

[47] E. Hellstern, L. Schultz, R. Bormann and D. Lee, Appl. Phys. Lett. 53 (1988) 1399.
[48] L. Schultz and J. Wecker, Mater. Sci. Eng. 99 (1988) 127.
[49] G. Cocco, S.Enzo, L. Schiffini and L. Battezzati, in E. Arzt and L. Schultz (eds.), Proc. DGM Conference on New Materials by Mechanical Alloying Techniques, Calw/Hirsau, October 1988, DGM Informationsgesellschaft, Oberursel, these proceedings.
[50] J. Eckert, L. Schultz and K. Urban, in E. Arzt and L. Schultz (eds.), Proc. DGM Conference on New Materials by Mechanical Alloying Techniques, Calw/Hirsau, October 1988, DGM Informationsgesellschaft, Oberursel, these proceedings.
[51] C.C. Koch and M.S. Kim, J. Phys. (Paris) Colloque 46 (1985) 573.
[52] C. Politis, Physica 135B (1985) 286.
[53] A. Hikata, M.J. McKenna and C. Elbaum, Appl. Phys. Lett. 50 (1987) 478.
[54] D. Girardin and M. Maurer, in E. Arzt and L. Schultz (eds.), Proc. DGM Conference on New Materials by Mechanical Alloying Techniques, Calw/Hirsau, October 1988, DGM Informationsgesellschaft, Oberursel, these proceedings.
[55] L. Schultz, E. Hellstern and G. Zorn, Z. Phys. Chem. 157 (1988) 203.

Amorphous Phase Formation by Mechanical Alloying: Thermodynamic and Kinetic Requirements

R. Bormann and R. Busch
Institut für Metallphysik, Universität Göttingen, D-3400 Göttingen and Sonderforschungsbereich 126 Göttingen/Clausthal, FRG

Introduction

Besides melt-quenching techniques mechanical alloying of powders has become a major route for preparing metastable crystalline as well as amorphous phases. The formation of an amorphous alloy by this technique has first been demonstrated by Koch et al. (1), who produced amorphous $Nb_{40}Ni_{60}$ by mechanical alloying Ni and Nb powders. Since then many amorphous alloys have been prepared by this method usually from the elemental components of the alloy (2-5). Recently, it has been shown by Schwarz and Koch (6) that also intermetallic compounds can be transformed into an amorphous alloy by a ball milling process.

The investigated alloy systems often contain at least one transition metal component. In particular, the combination of a late and an early transition metal usually results in a large compositional range of amorphous alloy formation, as has been demonstrated for a variety of Zr, Nb, and Hf systems.

The process of amorphous phase formation in the case of mechanical alloying of the elemental powders is often described as a diffusion reaction, which occurs in the layered microstructure developed during the milling procedure (5) (7). The thermodynamic and kinetic conditions under which amorphous phases are formed are therefore assumed to be similar to the one required for amorphous phase formation during a thermal anneal of layered composites, e.g. a large (negative) driving force of the reaction and a kinetic suppression of the formation of intermetallic compounds. The compositional range of amorphous phase formation is approximately determined by the assumption of a metastable equilibrium between the terminal solid solutions and the amorphous phase (2) (8).

It is the purpose of this paper to demonstrate that this assumption is not sufficient to explain the experimentally observed concentration range of amorphous phase formation during mechanical alloying. In particular, the formation of intermetallic compounds during the milling procedure has to be considered, which can result in an **enhanced** formation range of amorphous alloys.

Thermodynamic considerations

In order to calculate the (chemical) driving force for the reactions occuring during mechanical alloying the (Gibbs) free energy functions of the competing phases have to be determined. This can be most favorably performed by utilizing the CALPHAD (*Cal*culation of *Pha*se *Di*agrams) method (8) (9). Thereby the thermodynamic functions of the stable and the metastable phases are calculated by using the experimental thermodynamic data of the alloy system. In particular, the amorphous phase is treated as an undercooled liquid at the glass transition temperature. An exact thermodynamic determination of the free energy of the amorphous phase requires the

consideration of the excess specific heat of the undercooled alloy melt and experimental data on the thermodynamic properties of the amorphous phase, such as the
heat of crystallization and the chemical potentials measured by the electromotive
force in a galvanic cell (10-12).

The calculations of thermodynamic functions have been performed on a variety of
transition metal alloys being of interest for amorphous phase formation (13). In Fig.
1 the free energy curves in the Zr-Ni system are given at a temperature of T = 350 °C
(14). It demonstrates the large driving force for the amorphous phase formation from
the elemental components. Compared to the amorphous phase the terminal solid
solutions exhibit a smaller stability in the middle concentration range - a necessary
requirement for amorphous phase formation (8). By considering a metastable equilibrium
between the amorphous phase and terminal solid solutions (2) (a reasonable assumption

in the case of a **diffusional reaction** between the elemental precursors) the homogeneity
of the amorphous phase can be predicted to extend from 32 to 83 at.% Ni. This is in
good agreement with the concentration range of amorphous phase formation by
mechanical alloying Ni and Zr powder which is observed to be from 27 to 83 at.% Ni
(15) (16). (The determination of the concentration range is accurate within a few percent
for both methods with respect to the Zr-Ni system.)

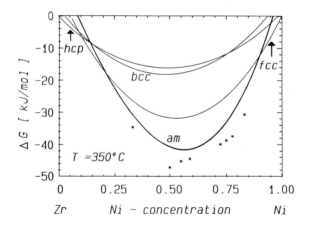

Fig. 1: Diagram of the free energies vs. concentration for the Zr-Ni system at
T= 350 °C (14). * represent the free energy of the intermetallic line compounds.
The free energy of h.c.p. Zr and f.c.c. Ni are chosen as reference states. By
assuming metastable equilibrium (common tangent rule to the free energy
curves) between the amorphous phase and the terminal solutions the concentration range of amorphous phase formation can be determined to be between
32 and 83 at.% Ni.

In Fig. 2 the free energy curves in the Nb-Ni system are displayed (17). The concentration range of amorphous phase formation is determined to be between 45 and 80 at.% Ni - again under the assumption of a metastable equilibrium between the solid solutions and the amorphous alloy. However, the experimental results indicate that amorphous phases can be achieved over a much wider concentration from 20 to 80 at.% Ni (5). In addition to X-ray analyses, the amorphous structure has been confirmed on the Nb-rich side by TEM investigations down to 30 at.% Ni (18). The discrepancy between the calculated formation range and the experimentally observed one exceeds by far the uncertainty of the calculations, which is (as in the Zr-Ni system) within a few at.%.

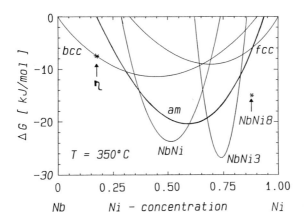

Fig. 2: Diagram of the free energies vs. concentration for the Nb-Ni system at T = 350 °C (17). The free energy of b.c.c. Nb and f.c.c. Ni are chosen as reference state. The metastable amorphous phase region with respect to the terminal solid solutions extend from 45 to 80 at.% Ni.

Kinetic considerations

Metastable phases will only be formed in a diffusion reaction if they are kinetically favored with respect to the stable phases. In the view of the classical nucleation theory this can be due to differences in the nucleation rate and in the growth velocity. It is reasonable to assume that any new phase - whether amorphous or crystalline - will precipitate heterogenously at the interface boundary between the layered components. However, contrary to usual boundaries the interface energy between elements forming amorphous phases can be quite small or even negative due to the strongly attractive interaction energy between unlike atoms. The nucleation barrier should therefore be comparable to the homogeneous nucleation (13) as there is no pronounced saving of the interface energy by precipitating the phase in the boundary.

The nucleation barrier of the competing phases can be estimated by using the calculated free energies. For an incoherent interface between an intermetallic compound and the solid solution phases the interfacial energy is typically of the order of $1 J/m^2$. This results in a nucleation barrier of (at least) 4eV for the intermetallic compound formation in the Ni-Zr system. The interface energy of the amorphous phase with respect to the terminal solutions can be assumed to be much lower, typically by (at least) a factor of 2. From Figs. 1 and 2 it is obvious that the (chemical) driving force for the precipitation of the amorphous phase is quite comparable to the one of the intermetallic compounds. Therefore, the lower interface energy in the case of the amorphous phase yields a rather small nucleation barrier of about 0.5eV. However, these barriers can be lowered further if precipitation at a high-angle grain boundary is considered (19).

From these calculations it can be concluded that a nucleation barrier of the intermetallic compounds cannot be neglected. As it exceeds greatly the barrier of the amorphous phase a kinetic phase selection can occur already in the nucleation regime of the phase reaction. In addition, a further phase selection can be achieved during the growth of the phases which has been described by Gösele and Tu, for example (20).The model takes into account differences in the particle flux through the competing phases. The phase with the highest interface velocity and diffusivity will be favored and suppresses the formation of the other phases.

The formation of an amorphous phase in the early stages of a phase reaction will therefore only be possible in a certain temperature range, in which the intermetallic compound exhibits a pronounced difference in the nucleation rate and/or growth velocity with respect to the amorphous alloy.

If the precipitation of intermetallic compounds can be kinetically by-passed in the phase reaction the concentration range of amorphous phase formation can be estimated by assuming a metastable equilibrium between the amorphous phase and the terminal solutions. However, the formation of intermetallic compounds can often not be neglected during ball-milling, even if the final product is completely amorphous. In the Zr-Ni system, for example, Petzoldt et al. (21) observed the formation of a (metastable) intermetallic compound which amorphized by further processing. This is in agreement with results from Lee et al. (22) who demonstrated that Zr-Ni intermetallic compounds can be amorphized by mechanical milling. The transformation into the amorphous phase can be understood by the destruction of the (chemical) order in the intermetallic compound resulting in an increase of its free energy (23). The amount of stored energy required for a polymorphous transformation into the amorphous phase can be determined from the free energy curves. From Fig. 1, for example, it can be concluded that an increase of the free energy typically by a few kJ/mol is required in order to amorphize intermetallic compounds.

The formation of intermetallic compounds during the mechanical alloying is assumed to be caused by an increase of the material's temperature above the crystallization temperature. This indicates that the primary crystallization products formed at low temperatures from the amorphous phase have to be considered in particular. In the case of the Zr-Ni system these are the Zr_2Ni, $ZrNi$, Zr_7Ni_{10} and the (probably metastable) $ZrNi_2$ compound. As seen from Fig. 1, the concentrations of these compounds lay **inside** the amorphous composition range determined under the assumption of metastable equilibrium between the amorphous phase and the solid solutions. There-fore, a formation and subsequent amorphization of these compounds during the mechanical alloying process would not change the predicted concentration range of amorphous phase formation in the Zr-Ni system.

In the Nb-Ni system two metastable phases, the so-called M-phase with a composition of approximately 50 at.% Ni and the η-phase with about 17 at.% Ni has to be considered besides the stable $NbNi_3$ compound as primary crystallization products of amorphous NbNi alloys. As in the Ni-Zr system a possible formation and subsequent amorphization of the $NbNi_3$ and the M-phase would not alter the concentration range predicted by the metastable equilibrium between the amorphous phase and the solid solutions. However, a precipitation of the η-phase followed by a (polymorphous) transition into the amorphous phase by further milling would result in a pronounced extension of the formation range of the amorphous phase down to 17 at.% Ni. (This assumption will be checked by corresponding experiments.)

In addition, it should be considered that mechanical energy of the ball milling can be stored by the creation of dislocations and grain boundaries (or by topological disorder, in general) in the phases. This would effect the positions of the free energy curves. However, an (upper) estimate of the relative increase in the free energy of the solid solutions with respect to the amorphous phase amounts typically a few kJ/mol which results in an extension in the homogeniety range of amorphous phase formation by only 1-2 at.%.

From the calculated free energy curves, Fig. 2, it can therefore be concluded that amorphous NbNi alloys with Nb concentrations higher than 60 at.% Nb can only be achieved if a Nb-rich intermetallic compound, such as the η-phase for example, is formed and subsequently amorphized during the mechanical alloying procedure. The proposed process is further supported by the observation that the η phase is the primary crystallization product for Nb-rich amorphous alloys prepared by mecha-nical alloying (18).

Conclusions

The amorphization process during the mechanical alloying differs from the thermal annealing of composite materials in the respect that a formed intermetallic compound can be subsequently amorphized by further milling. This can result in an enhanced range of amorphous phase formation by mechanical alloying with respect to the thermal treatment.

If the formation of intermetallic compounds can be entirely avoided in both methods, the concentration range of amorphous phase formation can be estimated by the assumption of a metastable equilibrium between the amorphous phase and the terminal solid solutions. The required free energy curves can favorably be calculated by the CALPHAD approach, if thermodynamic data of the system are available (8).

In the case where the compositions of the intermetallic compounds are **within** the metastable phase region of the amorphous alloy a formation and a subsequent amorphization does not change the predicted concentration range of amorphous phase formation. Examples of this are the Zr-Ni and the Ti-Ni system (2) (12) (14).

In the Nb-Al system the NbAl$_3$ compound cannot be transformed into the amorphous structure under usual milling conditions (24). Therefore, the amorphous phase region is limited by the formation of the NbAl$_3$ compound for Al-rich concentrations.

If the composition of the amorphized intermetallic compound is **outside** the metastable amorphous phase region the concentration range of amorphous phase formation may be extended up to the composition of the intermetallic compound. In the Nb-Ni system the formation of Nb-rich amorphous alloys up to 80 at.% Nb (5) is obviously facilitated by the precipitation and the subsequent amorphization of the metastable η-phase with a Nb concentration of approximately 83 at.% Nb (17). A similar behavior is expected for other Nb-alloys, such as Nb-Co and Nb-Fe.

Acknowledgements

We are grateful to F. Gärtner, P. Haasen, F. Petzoldt and K. Zöltzer for stimulating discussions.

References

(1) C.C. Koch, O.B. Cavin, C.G. McKamey and J.O. Scarbrough, Appl. Phys. Lett. **43** (1983), 1017
(2) R.B. Schwarz, R.R. Petrich and C.W. Shaw, J. Non-Cryst. Sol. **76** (1985), 281
(3) C. Politis and W.L. Johnson, J. Appl. Phys. **60** (1986), 1147
(4) E. Hellstern and L. Schultz, Appl. Phys. Lett. **60** (1986), 124
(5) F. Petzoldt, B. Scholz and H.-D. Kunze, Mater. Lett. **5** (1987), 280
(6) R.B. Schwarz and C.C. Koch, Appl. Phys. Lett. **49** (1986), 146
(7) E. Hellstern and L. Schultz, J. Appl. Phys. **63** (1988), 1408
(8) R. Bormann, F. Gärtner and K. Zöltzer, Proc. E-MRS Strasbourg 1988, to be publ. J. Less-Comm. Met. (special issue)
(9) L. Kaufman and H. Bernstein, "Computer Calculations of Phase Diagrams", Academic Press, New York, 1970
(10) R. Bormann, F. Gärtner and P. Haasen, Proc. LAM 6, Zt. Phys. Chem.**157** (1988),29
(11) R. Bormann, F. Gärtner and F. Haider, Proc. RQ 6, Mat. Science and Eng. **97** (1988), 29
(12) K. Zöltzer and R. Bormann, J. Less-Comm. Met. **140** (1988), 335
(13) R. Bormann, Habilitationsschrift, Göttingen 1988
(14) F. Gärtner and R. Bormann, DPG Frühjahrstagung, Verhandlungen **4** (1988), M-3.4
(15) J. Eckert, L. Schultz and K. Urban, DPG Frühjahrstagung, Verhandlungen **4** (1988), M-20.5
(16) A.W. Weeber and H. Bakker, J. Phys. F.: Met. Phys. **18** (1988), 1359
(17) R. Busch and R. Bormann, to be publ.
(18) F. Petzoldt, private communication
(19) W.J. Meng, C.W. Nieh, E. Ma, B. Fultz and W.L. Johnson, Proc. RQ 6, Mat. Science and Eng. **97** (1988), 87
(20) U. Gösele and K.N. Tu, J. Appl. Phys. **53** (1982), 325
(21) F. Petzoldt, B. Scholz and H.-D. Kunze, these proceedings
(22) P.Y. Lee, J. Jang and C.C. Koch, J. Less-Comm. Met. **140** (1988), 73
(23) W.L. Johnson, Prog. Mater. Science **30** (1986), 81
(24) E. Hellstern, L. Schultz, R. Bormann and D. Lee, Appl. Phys. Lett. **53** (1988) 1399

Amorphization of a Binary Metal-Metalloid System by Mechanical Alloying

G.Veltl, B.Scholz, H.-D.Kunze, Fraunhofer-Institut für angewandte Materialforschung (IFAM), Bremen, West-Germany

Abstract

Elemental crystalline Ti- and Si- powders were mechanically alloyed in the concentration range Ti_xSi_{100-x} (20<x<80). It is shown that it is possible to produce amorphous powders in a metal-metalloid system. The degree of amorphization was investigated by X-ray measurements and calorimetry. By rapid quenching from the melt only the alloy $Ti_{80}Si_{20}$ can be produced amorphous. In contrast mechanical alloying offers the possibility to prepare fine amorphous powders over a wide concentration range.

Introduction

During the last five years several alloy systems of the type early transition metal - late transition metal were prepared amorphous by mechanical alloying, e.g.(1,2,3). Only few investigations are known about the amorphization of metal-metalloid systems by mechanical alloying, especially of the type early transition metal - metalloid. One essential step in the amorphization process is the refining of microstructure by repeated welding and fracturing of single powder particles resulting in a fine lamellar structure. Because of their inherent brittleness the metalloids seem to be unsuitable components for the amorphization by mechanical alloying. The brittleness will shift the balance of fracturing and rewelding towards a grinding of the material without sufficient welding and alloying.

The system TiSi has a high negative enthalpy of mixing, which is a prerequisite for amorphization in the solid state. Alternately sputtered multilayers of titanium and silicon form an amorphous phase with a composition of about $Ti_{45}Si_{55}$, (6), during annealing below the crystallization temperature. Rapid quenching of TiSi melts results in the formation of an amorphous phase only near the eutectic concentration, e.g. $Ti_{80}Si_{20}$, (4,5). Aim of this investigation is to study the behaviour of TiSi powder blends during mechanical alloying.

Experimentals

Elemental titanium and silicon powder of 10μm particle size were used for these investigations. Mechanical alloying was carried out in the concentration range $Ti_{20}Si_{80}$ to $Ti_{80}Si_{20}$ in steps of 10 at%. Additionally, mixtures of the compositions of the intermetallic phases TiSi, Ti_5Si_4, Ti_5Si_3 and Ti_3Si were examined. The powder blends were canned under argon into steel vessels together with steel milling balls in a weight ratio of 1:15. The milling was performed in a planetary ball mill for 5h to 45h under argon atmosphere.

To check the degree of amorphization, the processed powders were
analysed by X-ray diffraction with CuK_α-radiation in a diffractometer in reflection geometry. To determine size, microstructure
and morphology of the produced powders SEM, Cam Scan 4, with EDX
was used. Thermal analysis was carried out with a differential
scanning calorimeter, Perkin-Elmer DSC7, with heating rates from
2.5 K/min to 40 K/min to determine the activation energy of crystallization by Kissinger-peak-shift method.

Results and discussion

One alloy from the middle of the concentration range, $Ti_{60}Si_{40}$,
was milled for 5h to 45h to determine the time necessary for
amorphization. The X-ray patterns after different milling times
are shown in <u>Fig.1</u>. With increasing milling time the intensity of

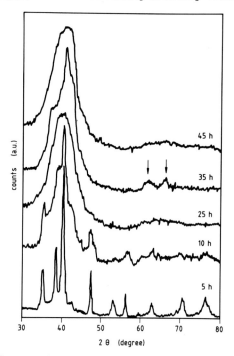

<u>Fig.1:</u> X-ray diffraction patterns of $Ti_{60}Si_{40}$ after different
milling times. The peaks of the metastable phase after
35h milling is marked by arrows

the peaks of crystalline elemental phases decreases and a typical
broad amorphous maximum develops. After 25h milling the powder is
completely amorphous. Further milling up to 35h leads to a partially crystallized powder. The peaks visible in the X-ray pattern
do not correspond to the crystalline elemental phases or to the
stable intermetallic phases of the binary phase diagram. After
additional 10h milling time this phase is no more detectable. The
powder is again completely amorphous. This effect is investigated

and explained in detail for the system NiZr by F.Petzoldt et al. (7) and will not be dicussed further in this paper.

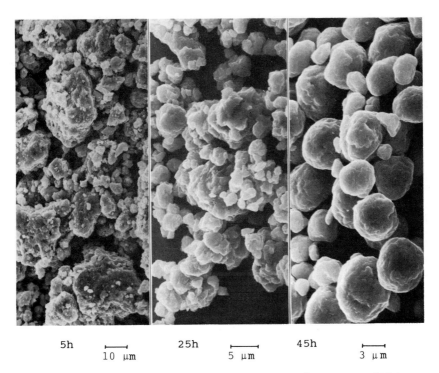

5h   ⊢———⊣   25h   ⊢———⊣   45h   ⊢———⊣
10 μm   5 μm   3 μm

Fig.2: Powder morphology of $Ti_{60}Si_{40}$ after different milling times

Fig.2 shows the development of the powder morphology with increasing milling time. After 5h milling large particles of about 20μm predominate. 25h milling leads to a refinement of the powder. The typical spheroidization of the powder particles occurs mainly during the following milling up to 45h. The mean particle size after 45h milling amounts from 3μm to 5μm. This value is only a tenth of the typical particle size of amorphous powders of a binary transition metal alloy processed with the same milling parameters,(8).

The structural refinement in the same milling intervals can be seen in   **Fig.3**.   5h milling leads to a uniform distribution of titanium and silicon. The typical lamellar structure found in the systems of two transition metals does not form. Reason for that is the brittleness of the silicon leading to a refinement of the structure mainly by fracture processes. No compositional inhomogeneities are visible in the amorphous powders after 25h and 45h milling.

The investigation of the alloy $Ti_{60}Si_{40}$ has shown, that 25h milling time is sufficient to produce completely amorphous powders.

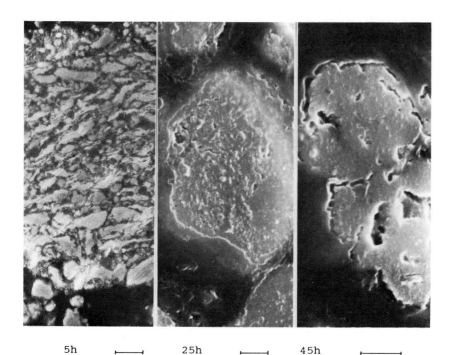

5h   ├────┤   25h   ├──┤   45h   ├────┤
10 μm   1 μm   1 μm

<u>Fig.3:</u> Microstructure of $Ti_{60}Si_{40}$ powder particles after different milling times

Therefore, the other compositions were processed for the same time with the same milling parameters. The X-ray patterns prove an amorphization range from $Ti_{75}Si_{25}$ to $Ti_{40}Si_{60}$. This composition range does not include the eutectic alloy $Ti_{80}Si_{20}$ which was produced amorphous by melt quenching, (4,5). Milling for 50h of this alloy results in complete amorphization of the powder. The alloys of the other boundary compositions investigated, $Ti_{20}Si_{80}$ and $Ti_{30}Si_{70}$, do not become amorphous during 50h milling. The X-ray patterns still reveal the peaks of the elements.

The thermal stability of the alloys milled for 25h was investigated by DSC-measurements. Typically the DSC-curves show two crystallization peaks. The intensity of the two peaks changes with composition of the alloys and with heating rate applied. Some peaks even vanished at certain heating rates which renders the evaluation of the activation energy more difficult. The highest crystallization temperatures, <u>Fig.4</u>, were reached in the middle of the concentration range. The values of the boundary concentrations represent the crystallization temperatures of the amorphous portion present after 25h milling.

The amorphous powder $Ti_{50}Si_{50}$ has two crystallization peaks in the DSC-curve, one at 650 °C, the other at 675 °C. After heating to 650 °C and rapid cooling the specimen was analysed by X-ray. The X-ray pattern shows besides the amorphous maximum some weak and broadened crystalline peaks, which do not fit to the patterns of the elements or the stable intermetallic phases.

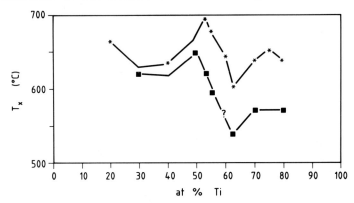

Fig.4: Crystallization temperatures of TiSi-powders
determined with a heating rate of 40 K/min

In annealing experiments of alternately sputtered layers, (6), a metastable cubic silicide was detected at a similar concentration and heating temperature, but the presented data and the few peaks measured in this investigation do not allow to affirm the correspondence of the two results. But it is worth noting that the X-ray peaks of the phase found in $Ti_{50}Si_{50}$ after heating to 650°C

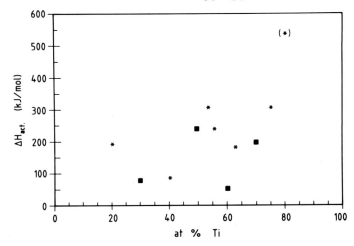

Fig.5: Activation energies of the crystallization of
TiSi-samples (■ first peak, * second peak)

correspond to the peaks visible at the alloy $Ti_{60}Si_{40}$ after 35h milling time, <u>Fig.1</u>. Samples heated above 675 °C exhibit the X-ray pattern of the stable phase TiSi. These results suggest that the crystallization proceeds according to a reaction path of the type:

$$Amorphous_1 \xrightarrow{T1} Metastable + Amorphous_2 \xrightarrow{T2} Stable$$

From the shape of the DSC-curves of the other compositions may be concluded that the crystallization follows the same reaction scheme. In <u>Fig.5</u> the activation energy of the crystallization of the amorphous phase is depicted. For the evaluation of the activation energy only the stronger DSC-peak has to be used. As there are at least two different processes involved in the crystallization further investigations have to be carried out to obtain more information for an interpretation of the data.

So far it can be concluded that the system TiSi is a candidate for amorphization by solid state reaction during mechanical alloying. It has a maximum enthalpy of mixing of about -40 kJ/mole. The structural refinement at the early stages of mechanical alloying does not lead to a lamellar structure, but to a fine distribution of the two elements. The results of Holloway,(6), - fast diffusion of silicon in titanium and amorphization of thin TiSi layers by annealing - confirm this conclusion.

Summary

Amorphous TiSi-powders were produced by mechanical alloying for 25h in the concentration range Ti_xSi_{100-x} (40<x<75). The highest thermal stability was found in the middle of the concentration range. The crystallization temperatures with values up to 650 °C are relatively high for a metalloid containing system. The crystallization follows a two stage reaction path from the amorphous state via a metastable phase to the stable intermetallic phases.

Acknowledgement

The authors would like to acknowledge the technical assistance of E.Woelki and H.Großheim.

References:

(1) C.C. Koch, O.B. Cavin, C.G. McKamey, J.O. Scarbrough
    Appl. Phys. Lett. 43 (1983), 1017.
(2) E. Hellstern, L. Schultz,
    Appl. Phys. Lett. 49 (1986), 1163.
(3) F. Petzoldt, B. Scholz, H.-D. Kunze
    Materials Letters 5 (1987), 280.
(4) D.E. Polk, A. Calka, B.C. Giessen, Acta Met., 26 (1978),1097.
(5) W.K. Wang, H. Iwasaki, C. Suryanarayana, T. Masumoto,
    J.Mat.Sci.Eng. 18 (1983), 3765.
(6) K. Holloway, R. Sinclair, J. Less Common Met. 140 (1988),139.
(7) F. Petzoldt, B. Scholz, H.-D. Kunze, this volume
(8) G. Veltl, B. Scholz, H.-D. Kunze, Proc. of the Int. Powder
    Met. Conf., Orlando 1988, in print

Compositional Dependence of Glass Formation in Mechanically Alloyed Transition Metal-Transition Metal Alloys

J. Eckert, L. Schultz, Siemens AG, Research Laboratories, Erlangen, FRG
and K. Urban, KFA Jülich, FRG

Amorphous TM-TM (TM = Ni, Mn, Cr, or V) powders have been prepared by mechanical alloying of elemental powders in a planetary ball mill. The glass-forming ranges have been determined by X-ray diffraction and differential scanning calorimetry. For Ni-V an amorphous phase appears in the middle concentration range, whereas for Ni-Cr and Ni-Mn an extended solid solution can be obtained during mechanical alloying. The extent of the glass-forming range is discussed in terms of a thermodynamic approach. The larger the driving force for the interdiffusion reaction the wider the glass-forming range.

Introduction

Since it has been shown that amorphization can be achieved by mechanical alloying [1], a lot of work has been focussed on high-energy ball milling of elemental powders or intermetallic compounds [2-6]. (For a recent review, see Ref. [7]). A large negative heat of mixing provides the driving force for the amorphization via a solid-state reaction of pure elemental powders [8-10] leading to extended glass-forming ranges in many alloy systems [3, 8, 9, 11-14]. In this paper we report on the glass-forming range in the Ni-V alloy system. Furthermore we show that extended solid solutions can be achieved for mechanically alloyed Ni-Cr and Ni-Mn.

Results and Discussion

The mechanical alloying is performed in a planetary ball mill under an argon atmosphere (for details see Ref. [12]). After milling, the powders are investigated by X-ray analysis using a Siemens D 500 diffractometer with CuKα radiation (λ = 0.154 nm). Fig. 1 shows X-ray diffractograms of several Ni_xV_{100-x} samples after 60 h of milling. For the samples in the middle concentration range the sharp crystalline lines of the elemental powders have almost completely disappeared and the typical broad diffuse maximum of the amorphous state appears. This maximum shifts slightly to higher diffraction angles with increasing Ni content as shown in Fig. 2 where the peak position of the amorphous maximum represented by the wavenumber $Q_R = 4 \cdot \pi \cdot \sin \Theta / \lambda$ (related to the mean nearest neighbour distance in these alloys) is plotted versus Ni content. Whereas the powders in the middle concentration range are perfectly amorphous, the samples with less than 40 at.% Ni and

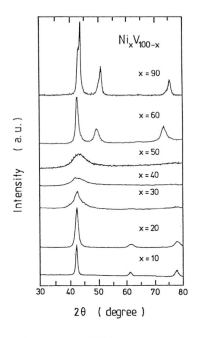
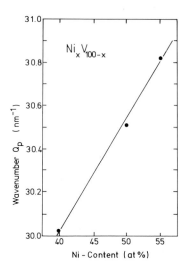

Fig. 1: X-ray diffraction patterns of several Ni_xV_{100-x} samples after 60h of mechanical alloying.

Fig. 2: Wavenumber Q_p of mechanically alloyed Ni-V versus Ni-content.

more than 55 at.% Ni show some irregularities at the amorphous peak or even crystalline lines due to the primary V and/or V_3Ni phases for V-rich samples and to crystalline Ni and/or Ni_2V for Ni-rich compositions. It is not yet clear whether the intermetallic equilibrium structures really appear during milling or whether the observed X-ray peaks are related to the strongly distorted lattices of the pure elements with extensive terminal solid solubility or perhaps to metastable intermetallic phases. For this purpose further work is still going on. A completely different behaviour is shown in Fig. 3 for the Ni-Cr system. The X-ray diffraction patterns show that 60 h of milling leads to a true alloying. But instead of an amorphous phase, as in the case of Ni-V, the primary α-Cr phase (bcc) is formed up to about 35 at.% Ni and the primary fcc Ni phase shows up for more than 45 at.% Ni. From 35 to 45 at.% Ni, a two-phase region of both phases exists which is characterized by a simultaneous appearance or by an overlapping of both sets of diffraction peaks in the X-ray patterns. This situation is also illustrated in Fig. 4, where the interlattice spacing d of the strongest X-ray peak is plotted versus Ni content. A similar result has been obtained for Ni-Mn, where the interlattice spacing also shows a strong dependence on the Ni content (see Fig. 5). Up to 70 at.% Mn the d value strongly increases from the value of pure fcc Ni to the interlattice spacing of the tetragonal

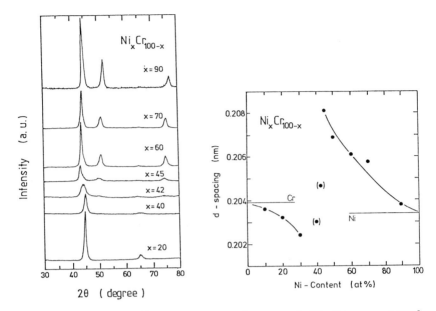

Fig. 3: X-ray diffraction patterns of several Ni_xCr_{100-x} samples after 60h of mechanical alloying.

Fig. 4: Lattice spacing d of the strongest observed peak in the X-ray diffractograms of mechanically alloyed Ni-Cr versus Ni-content.

Mn(2U) phase. For less than 30 at.% Ni the d values show a drastic decrease to a smaller interlattice spacing than observed in pure α-Mn (cubic). It can not clearly be decided from these preliminary experiments whether there exists a two-phase region of γ-Ni and Mn for less than 30 at.% Ni or whether a mixture of tetragonal and cubic Mn with dissolved Ni atoms appears in this composition range.

In order to estimate the glass-forming range, X-ray diffraction patterns as shown in Fig. 1 can give a first qualitative hint, although it is scarcely possible to determine an accurate value for the terminal composition of the amorphous phase. For this purpose it is much more reliable to measure the compositional dependence of various physical properties within the glass-forming range and within the two-phase regions between the amorphous and the crystalline phase. Accordingly, the crystallization temperatures and enthalpies were measured in a differential scanning calorimeter (Perkin-Elmer DSC 7) at a heating rate of 40 °C/min. The crystallization temperatures T_x derived from the first exothermic peak of the DSC scans are plotted versus Ni content for the Ni-V alloy system in Fig. 6. Whereas T_x shows a linear compositional dependence for less than about 35 and more than 55 at.% Ni, the values in the middle concentration range show a completely

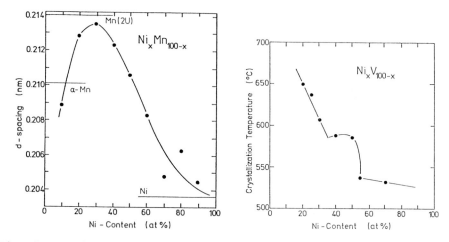

<u>Fig. 5:</u> Lattice spacing d of the strongest observed peak in the X-ray diffractograms of mechanically alloyed Ni-Mn versus Ni-content.

<u>Fig. 6:</u> Crystallization temperature T_x of mechanically alloyed Ni-V versus Ni-content.

different behaviour. From an extrapolation of the values below 35 and above 55 at.% Ni the terminal compositions of the amorphous phase can be determined to be $Ni_{35}V_{65}$ and $Ni_{55}V_{45}$. On both sides of the amorphous phase there exist two-phase regions, as mentioned before.

To consider these results from a thermodynamic point of view we calculated the enthalpy of mixing by using the Miedema model [15,16]. It is expected that the glass-forming ranges are wider in

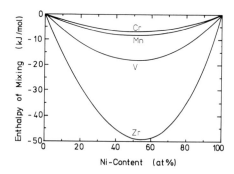

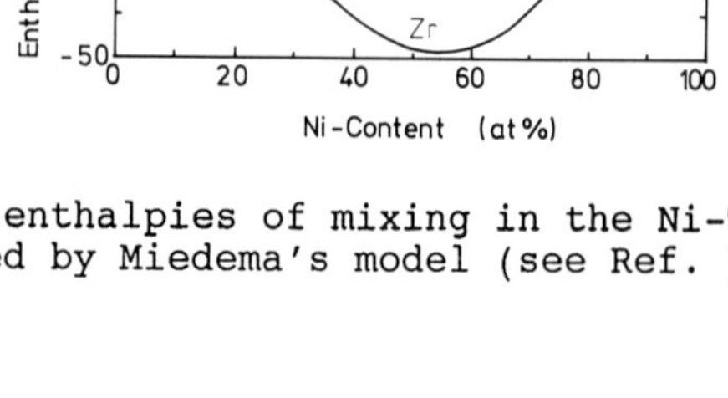

<u>Fig. 7:</u> Negative enthalpies of mixing in the Ni-TM alloys calculated by Miedema's model (see Ref. [15,16]).

alloy systems with a large negative enthalpy of mixing than in those with relatively small ones. In Fig. 7 the enthalpies of mixing calculated by the Miedema model are shown for the Ni-TM alloy systems (where TM = Mn, Cr, V and Zr). Obviously the enthalpy of mixing for Ni-V is large enough for the formation of an amorphous phase from 40 - 55 at.% Ni. For Ni-Cr and Ni-Mn the negative enthalpy of mixing as thermodynamic driving force for a solid-state reaction is not sufficient for the formation of an amorphous phase via interdiffusion, but solid solutions and intermetallic phases are produced by mechanical alloying. This behaviour may be due to a small atomic size mismatch between Ni and Cr or Mn. Therefore it is quite easy for these alloys to form solid solutions or intermetallic phases at relatively low temperatures during milling.

Conclusions

We have shown that glass formation by mechanical alloying is in general possible in the central composition range for Ni-V. On both sides the amorphous phase is enclosed by two-phase regions, consisting of elemental crystalline material with extended solid solubility and intermetallic phases. Mechanically alloyed Ni-Cr and Ni-Mn show extended solid solutions. This can be explained in terms of a thermodynamic approach, comparing the enthalpy of mixing as driving force for a solid-state reaction for the different transition metals. The formation of the solid solutions is also due to a relatively small atomic size mismatch of Ni and Cr or Mn allowing an easy formation of these phases during milling at relatively low temperatures.

Acknowledgement

The authors are grateful to M.I. Cooper and F. Gaube for tech-nical assistance, and K. Wohlleben for stimulating discussions.

References

[1] C.C. Koch, O.B. Cavin, C.G. McKamey and J.O. Scarbrough, Appl. Phys. Lett. 43 (1983) 1017.
[2] E. Hellstern and L. Schultz, Appl. Phys. Lett. 48 (1986) 124.
[3] C. Politis and W.L. Johnson, J. Appl. Phys. 60 (1987) 1147.
[4] E. Hellstern and L. Schultz, Phil. Mag. B. 56 (1987) 443.
[5] E. Hellstern and L. Schultz, Mat. Sci. Eng. 93 (1987) 213.
[6] A. W. Weeber, H. Bakker and F.R. de Boer, Europhys. Lett. 2 (1986) 445.
[7] L. Schultz, Mat. Sci. Eng. 97 (1988) 15.
[8] R.B. Schwarz, R.R. Petrich and C.K. Saw, J. Non-Cryst. Solids 76 (1985) 281.
[9] E. Hellstern and L. Schultz, Appl. Phys. Lett. 49 (1986) 1163.
[10] E. Hellstern and L. Schultz, J. Appl. Phys. 63 (1988) 1408.

[11] J. Eckert, L. Schultz and K. Urban, Proc. E-MRS Spring Meeting, Strasbourg 1988, in: J. Less-Common Met. (in the press).

[12] J. Eckert, L. Schultz, E. Hellstern and K. Urban, J. Appl. Phys. $\underline{64}$ (1988) 3224.

[13] F. Petzold, B. Scholz and H.D. Kunze, Mat. Sci. Eng. $\underline{97}$ (1988) 25.

[14] A. W. Weeber and H. Bakker, Physica B (in the press).

[15] A.K. Niessen, F.R. de Boer, R. Boom, P.F. de Chatel, C.W.M. Mattens and A.R. Miedema, CALPHAD $\underline{7}$ (1983) 51.

[16] A.R. Miedema, P.F. de Chatel and R.R. de Boer, Physica $\underline{100\ B}$ (1980) 1.

Amorphization of Gd-Co Powders by Mechanical Alloying

D. Girardin[*] and M. Maurer, Laboratoire Mixte CNRS SAINT-GOBAIN, Pont-à-Mousson (France)

Summary

We have investigated the solid state reaction of Gd-Co powder mixtures by high-speed ball milling. The Gd-Co system, which offers a large variety of magnetic properties has already been extensively studied in conventional amorphous forms like liquid or vapor quenched alloys. On the other hand, few is known today about the possibility of solid state amorphization in this system. In this preliminary work, we have successfully prepared a nearly complete amorphous phase by mechanical alloying in argon flow. A metastable fcc phase forms in an intermediate stage, which could not be identified by reference to the equilibrium phase diagram.

Introduction

The solid state reactions are now one of the most studied routes to prepare amorphous alloys (1). The mechanical alloying by high-speed ball milling (2-6) is a very attractive method because it is versatile and it is also able to produce significant amounts of amorphous powders, thus opening possibilities for new applications.

Two leading criteria are now recognized in order to identify couples of metals which can be obtained as metastable amorphous phase by solid state reaction (1): the first one is the thermodynamic instability of an inhomogeneous crystalline mixture as compared to a homogeneous amorphous alloy. The second condition is that one of the components is an anomalously fast diffuser into the other metal. This feature is usually observed when one of the two elements is much smaller than the other one. In this context one of the most studied families of alloys is that involving an early transition metal (Ti, Zr or Hf) and a late transition metal (Fe, Co, Ni, Cu, Pd). Some investigations have also been carried out with rare earth based systems (La-Au, Er-Ni, Y-Co) where amorphous alloys were obtained by interdiffusion of multilayer (1,7), cold rolling (8) or milling of an intermetallic compound (2,3).

The system Gd-Co which is selected here, is believed to fulfill both the thermodynamic (9) (Fig. 1) and the kinetic criteria in spite of the fact that only few is known on the diffusion of Co in Gd. Nevertheless, Co or Ni are known to be anomalous fast diffusers in other rare earth metals (10) and then the same situation is expected in the peculiar case of Co in Gd.

Experimental procedures

Commercial powders of Co (99.9 %) and Gd (99.8 %) were used as base material. The initial particle sizes were respectively 5-10 μm and 200-400 μm for the two metals. Ball milling was achieved in tungsten carbide mortars with 1 ball in a volume of

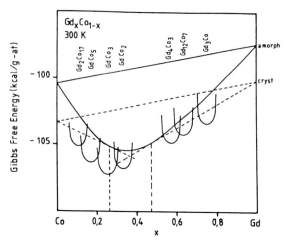

Fig. 1: Diagram of Gibbs free energy of formation of Gd-Co alloys
(9). For each compound, the stoichiometry dependence is
hypothetical.

about 1.5 cm^3. The typical speed of the ball was evaluated to
about 1 m/s. In order to avoid compaction, several droplets of cy-
clohexane were added. The volume occupied by the powders was typically 10 % of the volume of the mortar. All the processing was
carried out without exposure to air. This is an important point
due to the high reactivity of rare earth to oxygen. X-ray diffraction experiments were performed with an automated Philips X-ray
diffractometer using Co-Kα radiation. Powders were embedded in
collodion and then deposited onto a tilted sapphire single crystal
in order to avoid any spurious contribution to the diffracted
intensity.

Results

We focus here on the structural investigations of mechanically
alloyed powders (Fig. 2). All the results concern equiatomic mixtures of Gd and Co. A first interesting observation is that in
spite of a rather large initial size of Gd particles (200-400 µm)
the X-ray patterns of the two elements disappear after about 2-3
hours of milling at room temperature (Fig. 3).

This is a proof that chemical mixing occurs rather easily, as a
consequence of the fast diffusion of Co into Gd. The amorphous
phase does not form directly. Indeed, we argue that a metastable
crystalline phase forms in the second step. It could not be identified as one or several of the numerous crystal phases of the
equilibrium phase diagram. A comparison with oxide and tungsten
carbide patterns also failed.

This intermediate phase could not be obtained with equiatomic alloys obtained by arc melting or by sintering at 630 °C. In both cases a mixture of $GdCo_2$ and Gd_4Co_3 (11) is obtained. In the mechanically alloyed materials the intermediate crystalline phase, which was systematically and reproducibly observed, could be identified as a fcc structure with a lattice parameter of a ≈ 5.3 Å. Longer milling times result in fully amorphous phase with a major diffraction halo centered at d = 2.90 Å (2Θ = 36°) and two secondary bumps at 2Θ ≈ 65° and 2Θ = 95°.

Discussion

The formation of an amorphous alloy from multilayer interdiffusion or from mechanically alloyed powders is believed to be promoted by the fast interdiffusion of the small metal (here Co) at clean interfaces. Our results must be compared with the widely studied Zr based materials. Here an intermetallic compound forms in an intermediate stage. However metastable Ni-Zr crystalline phases have been observed in some instances during mechanical alloying (12). It is now established that milling conditions can dramatically influence this formation of an intermediate crystalline compound during amorphization by mechanical alloying. Our findings agree quite nicely with the

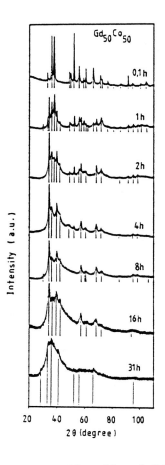

Fig. 2: X-ray diffraction patterns of mechanically alloyed Gd-Co mixtures as a function of the milling time (Co-Kα radiation).

works of Yermakov et al. (2,3). The reason why rare earth-cobalt intermetallic phases can be so easily amorphized remains still unclear. Extensive comparison of the crystallization process in mechanically alloyed Gd-Co amorphous alloys as compared to rapidly quenched glasses (13) are now needed to understand the back-transformation into a crystalline compound.

Conclusion

We have prepared amorphous Gd-Co alloys be mechanical alloying of an equiatomic mixture. Amorphization occurs via the formation of a metastable intermetallic compound, which is contrasting with most of Zr based materials. Studies of crystallization of our Gd-Co powders are now in progress and we have undertaken the investigation of magnetic properties of mechanically alloyed Gd-Co amorphous alloys with the aim to compare them with their rapidly quenched counterparts.

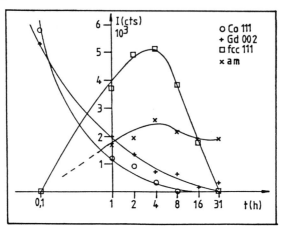

Fig. 3: Variation of the diffracted intensity by the various phases versus milling time.

Acknowledgement

*) This work is supported by a "Convention CIFRE" with the "Compagnie de Saint-Gobain".

References:

(1)   W.L. Johnson: Mat. Sci. and Eng. 97 (1988) 1.
(2)   A.YE. Yermakov, YE.YE. Yurchikov and V.A. Barinov: Fiz. metal. metalloved. 52 (1981) 1184.
(3)   A.YE. Yermakov, V.A. Barinov and YE. YE. Yurchikov: Fiz. metal. metalloved. 54 (1982) 935.
(4)   C.C. Koch, O.B. Cavin, C.G. McKamey and J.O. Scarbrough: Appl. Phys. Lett. 43 (1983) 1017.
(5)   E. Hellstern and L. Schultz: Appl. Phys. Lett. 48 (1986) 124.
(6)   L. Schultz: Mat. Sci. and Eng. 97 (1988) 15.
(7)   R.B. Schwarz and W.L. Johnson: Phys. Rev. Lett. 51 (1983) 415.
(8)   M. Atzmon, K. Unruh and W.L. Johnson: J. Appl. Phys. 58 (1985) 3865.
(9)   C. Colinet, M. Pasturel and K.H.J. Buschow: Met. Trans. A 18 (1987) 903.
(10) M.P. Dariel: Handbook on the Physics and Chemistry of Rare Earth. Chap. 12, 847, eds K.A. Gschneider and L. Eyring (North Holland).
(11) W.G. Moffatt: Handbook of Binary Phase Diagrams (G.E.).
(12) A.W. Weeber, A.J.N. Wester, W.J. Haag and H. Bakker: Physica 145B (1987) 349.
(13) K.H.J. Buschow, H.A. Algra and R.A. Henskens: J. Appl. Phys. 51 (1980) 561.

An Electron Microscopy Study of the Formation of Amorphous Alloys by Mechanical Alloying in the Ni-Zr System

E. Gaffet, N. Merk, G. Martin*), J. Bigot
Centre d'Etudes de Chimie Métallurgique/C.N.R.S.,
Vitry/Seine, France

Summary

The morphology, microstructure and microchemistry of mixtures of Ni and Zr powders after ball-milling are studied by SEM, TEM and EDX. After a transient agglomeration stage, the particles are split down to a steady state size of about 20 µm. Under steady-state conditions, amorphous phases coexist with intermetallic compounds on a submicron scale. Amorphous phases, homogeneous on the micron scale, occur with a composition not far from the nominal one. Two of the observed intermetallic compounds have a composition in the range of two of the latter amorphous phases.

I. Introduction

In 1970, J.S. Benjamin [1] used a high energy driven ball mill in order to improve superalloys by dispersion strengthening. A new aspect of mechanical alloying was revealed when the first work on amorphization by ball-milling has been reported by C.C. Koch et al. [2]. Since then, many other amorphous binary alloys have been prepared by such a technique, starting from crystalline powders or from a mixture of intermetallic compounds (for a review, see [3]). In the present work, we report on the evolution of mixtures of elemental Ni and Zr powders during the early stages of ball milling. SEM and TEM investigations show the morphological and structural evolutions which are discussed in terms of the different parameters involved during the ball-milling process.

II. Experimental procedure

Two series of experiments (labeled I and II) were performed on elemental Ni and Zr powder mixtures containing respectively 61 and 63.7 at% Ni. The milling proceeded in dry air atmosphere, in sealed steel containers (oxygen to Zr ratio is less than 1 %). Series I was processed sequentially: 15 min milling time followed by 10-15 min rest periods, series II was processed continuously with milling of one to two hours. Every one or second hour of milling, 0.3 to 0.6 g of powder are taken away for further electron microscopy investigations. More experimental details are given in [4]. The particle morphology and size distribution after various milling times have been characterized using a digital scanning electron microscope (Zeiss DSM 950) in the secondary electron mode. TEM observations (BF images, HREM and SAD pattern) and EDX analyses with an electron beam probe diameter of 0.5 to 2 µm have been performed with a Jeol 2000 FX TEM equipped with a Si-Li detector and a Tracor microanalyzer system (SMTF program). TEM samples have been prepared by two different techniques: either by sectioning, with an ultramicrotome, powder particles previously embedded in epoxy resin, or by simply spreading some particles on the TEM grid and then observing the areas which are sufficiently thin.

III. Results

III.1. SEM investigations

Fig. 1 shows typical SEM micrographs of Ni powders and Zr sheets prior to any milling. The Ni particles (Fig. 1a) with an average size of abour 5 µm exhibit a typical dendritic solidification morphology. The Zr sheets are thin saw-cut ribbons, 20 µm thick and

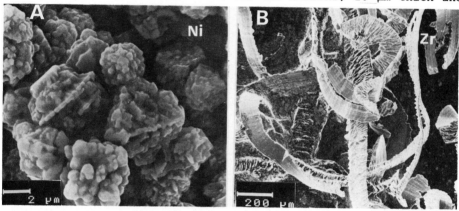

Fig. 1: Scanning electron micrographs of a) initial elemental Ni powder, b) initial elemental Zr sheets.

about 100 µm wide. During ball-milling, particle morphology and size distribution evolved as shown in Fig. 2. An aggregation has

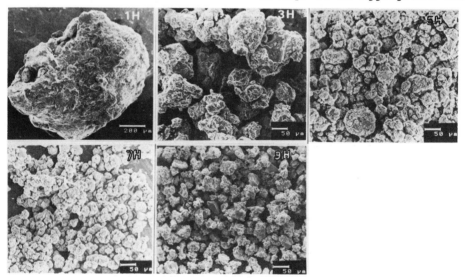

Fig. 2: Particle morphology of the $Ni_{63.7}Zr_{36.3}$ alloy after 1, 3, 5, 7 and 9 hours of ball-milling.

obviously occurred during the first hour of processing. The particles, subjected to larger milling times experience a rapid decrease in average size followed by a stabilization of the particle size at about 20 μm for milling time larger than 5 hours.

III.2 TEM Investigations

III.2.1. TEM Observations

The structural state of the particles after 12 hours of milling is revealed by the SAD patterns (Fig. 3). In some areas, only the amorphous structure is found (Fig. 3a); in other areas, crystalline and amorphous structures coexist (Fig. 3b). Furthermore the

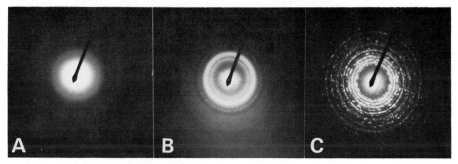

Fig. 3: Typical SAD patterns obtained after 12 hours of ball-milling for $Ni_{61}Zr_{39}$.

HREM observations reveal the presence of crystalline phases (Fig. 4). After calibration on graphite, the interplanar distances have been evaluated. They do not correspond to those of the Ni and Zr elements, but rather to some intermetallic compounds like Ni_7Zr_2 and $Ni_{10}Zr_7$, (see Table 1). Notice that the latter are stable for a range of composition which allows for a fluctuation of the interplanar distances.

<u>Table 1</u> : Measured interplanar distances, calibrated with respect to graphite interplanar distance and phases identified (series I)

Measured distance (nm)	Correction	Identified phase	Theoretical distance (nm)
0.2899	-1.40% → 0.286	$Ni_{10}Zr_7$	[222] 0.288
0,3086	-2,44% → 0,301	Ni_7Zr_2	[311] 0.299
0.3106	-1.40% → 0.306	Ni_7Zr_2	[400] 0.303
0.4065	No - 0.407	$Ni_{10}Zr_7$	[021] 0.411
0.4120	-2.44% → 0.402	$Ni_{10}Zr_7$	[021] 0.411
0.6100	No - 0.610	Ni_7Zr_2	[200] 0.611
0.6170	-2.44% → 0.602	Ni_7Zr_2	[200] 0.611

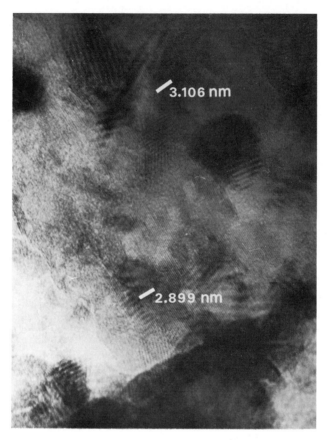

<u>Fig. 4:</u> HREM micrograph showing the presence of some intermetallic precipitates.

III.2.2. Chemical microanalysis

Fe-Cr contamination

Based on 60 EDX microanalyses, an average value of 2 at% Fe and 1 at% Cr may be given for the contamination of the powder after 12 hours of ball-milling. A definite correlation appears between the Zr and Fe-Cr concentrations: at a given Ni fraction, the lower the Zr, the higher the Fe-Cr concentrations. This may be related to the preferential sticking of Zr on the container walls.

Composition and structural heterogeneousness

A large number of EDX microanalyses has been performed on the micron scale after 12 hours of ball-milling [5]. In series I, the amorphous phase observed at this latter scale has a Ni content in the range of 57.4 to 59.7 at%. In series II, three distinct amorphous phases are observed on the micron scale with Ni contents of 56.6±1.2, 61.5±1.5, 72.2±0.7 at%. One amorphous area was found with 67.2 at% Ni. Moreover, in series I, EDX measurements at particles chosen at random, showed that the local chemical compositions clustered in well defined fields: 56.5±0.6, 59.4±0.7, 63.3±0.4 at% Ni.

It is worth noting that the amorphous phases observed do not necessarily correspond to eutectic compositions, and do sometimes have the composition of intermetallic compounds ($Ni_{11}Zr_9$, $Ni_{10}Zr_7$, Ni_5Zr_2), some of which are actually observed by HREM. All the above intermetallic compounds have been previously observed by X-ray diffraction as described in [4].

IV. Discussion and conclusions

The morphological and compositional evolutions of Ni and Zr mixtures during ball-milling have been studied by SEM, TEM and EDX analyses. The conclusions are as follows:

- <u>Particle size distribution:</u> the evolution we observe (agglomeration followed by subdivision before reaching a steady state where interparticle welding compensates the splitting of agglomerates) is typical of milling in dry atmosphere [6,7]. The occurrence of a dynamical steady-state is in agreement with [7]. The atomic structures and compositions observed thus correspond to that dynamical state.

- <u>Microstructure and microchemistry:</u> complementary to our previous work, where the same samples were studied by X-ray diffraction [4], we have here a microstructural information which confirms and complements our previous observations. We confirm the simultaneous occurrence of several amorphous phases and of intermetallic compounds. The amorphous phase (in series I) or phases (in series II) observed here are homogeneous on the micron scale. As shown in Fig. 3b, amorphous phases also exist, mixed with crystalline phases on a finer scale. In ref [4], amorphous phases with compositions far away from the nominal composition of the initial load (10 and 80 Ni at%) were identified by X-ray diffraction, together with those observed here. We must conclude that the former are distributed on a submicron scale, while the latter with a composition close to the nominal composition have grown above the micron scale.

Finally, we confirm that some of the amorphous phases we observed have a composition typical of intermetallic compounds which are also observed. The later stages of ball-milling (72 hours) are now under study.

It is then quite clear that the type of compounds to form under ball milling is very sensitive to the milling conditions (compare series I and II in the present work and [4], see also [8] for a comparison between vibrating frame and planetary milling). Moreover, since ball milling induced demixing has been reported recently [9], it is clear that the formation of compounds under ball milling is not the mear result of enhanced diffusion, and is not driven by the usual thermodynamical potentials. A similar situation is well known for alloys under irradiation where, in some cases, the appropriate driving force for phase changes has been identified and implies the alloy composition and temperature plus a dynamical variable which scales with the power injected in the alloy [10,11]. We propose by analogy that such should be the case under ball milling.

References

*): Now at: CENS-DTech-SRMP, F-91190 Gif/Yvette Cedex, France.

[1] J.S. Benjamin: Met. Trans. $\underline{1}$ (1970) 2943.
[2] C.C. Koch, O.B. Cavin, C.G. McKamey, J.O. Scarbrough: Appl. Phys. Lett. $\underline{43}$ (1983) 1017.
[3] L. Schultz: E-MRS Strasbourg 1988: J. Less Comm. Met., to be published.
[4] E. Gaffet, N. Merk, G. Martin. J. Bigot: ibid [3].
[5] The EDX compositions reported here must be understood as relative to a large set of measurements done during one working session at the microscope.
[6] P.S. Gilman, J.S. Benjamin: Ann. Rev. Mat. Sci. $\underline{13}$ (1983) 279.
[7] J.S. Benjamin, T.E. Volin: Met. Trans. $\underline{5}$ (1974) 1929.
[8] A.W. Weeber, W.J. Haag, A.J. Wester, H. Bakker: J. Less Comm. Met. $\underline{140}$ (1988) 119.
[9] H. Bakker, P.I. Loeff, A.W. Weeber: DIMETA 1988, in the press.
[10] G. Martin, P. Bellon: J. Less Comm. Met. $\underline{140}$ (1988) 211.
[11] P. Bellon, G. Martin: Phys. Rev. B, in the press; Phys. Rev. $\underline{B38}$ (1988) 2570.

Amorphization of Intermediate Phases by Mechanical Alloying/Milling

C. C. Koch, J. S. C. Jang, and P. Y. Lee, Materials Science and Engineering Department, North Carolina State University, Raleigh, NC, U.S.A.

Introduction

Recent interest in synthesis of amorphous alloys from the solid crystalline state resulted in the first International Conference on Solid State Amorphizing Transformations which was held in Los Alamos, NM, August 10-13, 1987 (1). Of the several methods used for solid state amorphization, which include thin film diffusion couples, hydrogenation of intermetallic compounds, and irradiation of intermetallic compounds, amorphization by high energy ball milling of powder - mechanical alloying (MA) or mechanical milling (MM) - is least well understood. Weeber and Bakker (2) have recently reviewed the experimental evidence for amorphization by high energy ball milling. The process may be classified into two major categories: 1) mechanical alloying (MA) and, 2) mechanical milling (MM). Mechanical alloying involves material transfer-alloying-between dissimilar powders which may be pure elements or alloys. MA has resulted in the amorphization of many binary systems (2) since first observed in Nb_3Sn by White (3) and studied in $Ni_{60}Nb_{40}$ by Koch et al (4). The mechanical alloying process consists of a competition between cold-welding and fracturing of the component powders which first results in a lamellar structure (5) and, depending on the alloy system, ultimately alloying on the atomic level to produce a solid solution, intermediate phase, or in certain systems an amorphous alloy. The kinetic energy of the milling media is partly converted to heat on impact with the powders, and typically temperature increases of 100-200K are possible during milling (6). Thus some thermal activation is often present to aid the alloying process.

The process of amorphization of elemental powders by MA has been explained by Schwarz et al (7) and Hellstern et al (8) as the synthesis of an ultrafine composite in which a solid-state amorphizing reaction takes place, i.e. a method for producing fine-scale diffusion couples. The driving force for this reaction is the composition-induced destabilization of the crystalline phases in systems with large negative heats of mixing (9). A kinetic constraint on formation of the equilibrium phase(s) is aided by asymmetry of the component diffusivities. However, in MA the severe plastic deformation creates defects - dislocations, planar defects, point defects - which can alter the reaction both thermodynamically and kinetically. The stored energy of cold work introduced by plastic deformation is usually a small fraction (<10%) of the difference in free energy between the amorphous and crystalline states, but might influence the composition range for amorphization in a minor way. The effect of defects introduced by the deformation might, however, have a significant effect on the kinetics of amorphization by providing diffusion short circuits and/or nucleation sites. The latter effects might favor or oppose amorphization. At least one alloy system, Nb-Ge, has been reported to undergo amorphization by MA (10) but not by solid state diffusion couples (11). While the defects introduced by MA may modify the amorphization reaction, most of the experimental evidence, including transmission electron microscopy by Petzoldt (12), points to solid state diffusion between the crystalline components as the major factor.

The case of amorphization of an intermediate phase by milling (MM), as first observed by Ermakov et al (13), is even less well understood. Here, the starting material is the equilibrium phase. As in the case of irradiation-induced amorphization of intermetallics (14), the free energy of the equilibrium crystalline intermediate phase must be raised to that of the amorphous phase. This must come about by the increase of free energy due to defects introduced by plastic deformation (or irradiation). The kinds of defects responsible and the nature of the crystalline-to-amorphous phase transition are not yet well defined.

An additional class of components which have been found to undergo amorphization on milling are mixtures of intermetallic compounds (15,16). In this case two possible mechanisms for amorphization are: 1] material transfer raises the free energy of each compound as they are moved off the stoichiometric composition such that the mixture of non-stoichiometric compounds has a free energy greater than the amorphous alloy, or 2] each compound is amorphized separately by accumulation of defects, as above, and the amorphous alloys' composition becomes homogeneous on subsequent material transfer during milling. Experiments have not yet clearly differentiated the two possible mechanisms.

In this paper we will describe some recent experimental results on the amorphization of intermediate phases by MM and discuss their implications regarding possible mechanisms for the crystalline to amorphous phase transition.

Experimental

The intermediate phases (Nb_3Sn, Nb_3Ge, Ni_3Al) and "line" intermetallic compounds ($NiZr_2$, $Ni_{11}Zr_9$) were prepared by arc-melting the pure components in a partial pressure of Zr gettered argon on a water-cooled copper hearth. The buttons were remelted at least six times to ensure homogeneity. Weight losses were negligible and nominal compositions are assumed. The buttons were crushed to powder in a mortar and pestle and the powder was screened to a size of less than 45 μm.

Milling was carried out in either a Spex Mixer/Mill, model 8000 or in an Invicta vibratory mill, model BX920/2. In each case a hardened tool-steel vial was used with 440C martensitic stainless-steel balls. The vial was sealed with an elastimer O-ring in an argon-filled glove box. The ball-to-powder weight ratio was 10:1. The vibratory mill required longer milling times for a given reaction but also minimized contamination of iron from the milling media.

The milled powders were characterized by x-ray diffraction measurements in either a Debye-Scherrer powder camera (CuKα radiation and nickel filter) or a GE XRD5 diffractometer with a singly bent graphite diffracted-beam monochromator. The stability of the amorphous phases formed was tested by studying their crystallization behavior in a computerized differential scanning calorimeter (DSC) DuPont model 9900.

The microstructural evolution during milling was followed with both SEM and TEM. The polished surface of powder compacts, metallographically prepared and etched, was examined in a Hitachi-530 SEM. Specimens for TEM were prepared by making powder pellet compacts by compressing the powder under a pressure of 2 GPa. The compacted powder disks were thinned electrochemically in a solution of 35 ml perchloric acid, 165 ml butanol, and 200 ml methanol at -55°C. TEM was performed in a Hitachi-800 microscope operating at 200 kV and with an energy dispersive x-ray analysis (EDS) capability. A probe size of about 10 nm was used for EDS analysis.

Experimental Results

Line Compounds: $NiZr_2$, $Ni_{11}Zr_9$

The compounds $NiZr_2$ and $Ni_{11}Zr_9$ are "line" compounds which exhibit a negligible range of composition away from stoichiometry. These compounds have been amorphized by milling them separately or by milling mixtures of them (15). The TEM microstructures after relatively short milling time indicate crystalline intermetallic particles embedded in an amorphous matrix. Figure 1 illustrates the TEM micrographs for $NiZr_2$ milled 75 minutes. A characteristic feature of these micrographs is an amorphous "halo" around the crystals. This effect has been observed previously in milled $NiZr_2$ + $Ni_{11}Zr_9$ mixtures (16) and by Kamenetzky et al

(17) for amorphization by isothermal cold rolling of $CuTi_2$ and Cu_3Ti_2 particles embedded in a glassy $Cu_{35}Ti_{65}$ matrix. The growth of the amorphous phase in these cases is observed to proceed from the crystal/amorphous interface.

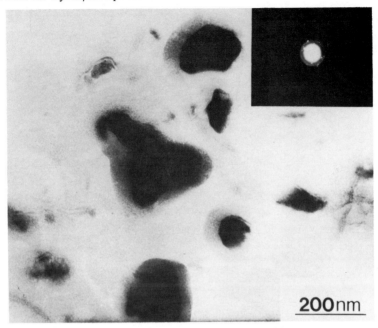

Fig. 1:   TEM micrograph, $NiZr_2$ particles in an amorphous matrix. $NiZr_2$ milled 75 minutes. Bright field.

The early stages of amorphization of mixtures of $NiZr_2$ + $Ni_{11}Zr_9$ give a unique TEM micro-structure. This is illustrated in Figure 2 for 15 minutes of milling. The dark field micrograph using a $NiZr_2$ reflection, reveals an irregular crystalline shell - resembling the surface of a cauliflower - with an amorphous interior. This structure is only observed in the early stages of milling; at longer times the same morphology as seen in Figure 1 is observed with amorphi-zation proceeding from the crystalline/amorphous interfaces.

Intermediate Phases: Nb_3Ge, Nb_3Sn, Ni_3Al

Intermediate phases are distinguished from line compounds by the significant composition range over which they may exist. The A-15 structure intermediate phases Nb_3Ge and Nb_3Sn have been amorphized by either MA of the pure component powder mixtures, or by MM of the compound powders. In the former case, the crystalline A-15 phase forms initially and then transforms to the amorphous structure on continued milling (18,19). Thus MA first re-sults in the formation of the equilibrium, or highly defected metastable, crystalline compound which is then destabilized by subsequent milling. The nature of this destabilization in Nb_3Sn and Nb_3Ge is the subject of ongoing research.

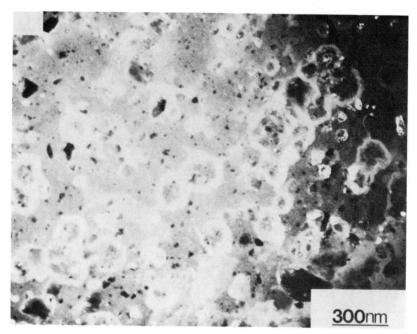

<u>Fig. 2:</u>   TEM micrograph of $NiZr_2$ + $Ni_{11}Zr_9$ powders milled 15 minutes. Dark field
using $NiZr_2$ reflection.

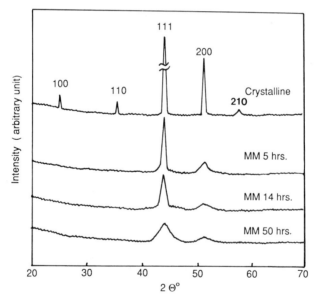

<u>Fig. 3:</u>   X-ray diffraction patterns for Ni_3Al for selected milling times.

Intermediate phases in the Ni–Al alloy system provide a range of behavior with respect to solid state amorphization. The line compound, Al_3Ni, with a complex orthorhombic structure, has been amorphized by ion bombardment (20,21). The very stable NiAl compound with the CsCl (ordered bcc) structure has not been amorphized by ion bombardment (20,21), high energy electron irradiation (14) nor even disordered at 10K by high voltage electron irradiation (22). The Ll_2 structure (ordered fcc) Ni_3Al exhibits intermediate behavior with regard to solid state amorphization. High voltage electron irradiation does not produce amorphization (14), but can lead to complete disorder at irradiation temperatures $\leq$ 190K (22). Partial amorphization was claimed for the composition of Ni_3Al by ion beam irradiation of Al/Ni multilayers (21). Recently, Ivanov et al (23) have mechanically alloyed mixtures of pure Ni and Al powders and obtained an extended fcc solid solution at compositions where the equilibrium Ni_3Al phase occurs and an amorphous structure at compositions where the Ni_3Al + NiAl two-phase field exists in the equilibrium diagram. We have been studying the structures of the Al_3Ni, NiAl, and Ni_3Al intermediate phases as a function of milling time. Preliminary results of milling on the structure of the Ni_3Al compound are presented below.

The x-ray diffraction patterns of Ni_3Al for selected milling times are presented in Figure 3. The unmilled powder exhibits the superlattice lines of the Ll_2 structure. After milling for 5 hours the superlattice lines disappear leaving only the fundamental fcc lines of a Ni_3Al solid solution. Continued milling serves to further broaden the lines but even after 50 hours at least some crystalline material remains. It is not possible to determine from x-ray diffraction whether some fraction of amorphous phase forms at the longer milling times. The relative long-range-order parameter (LRO), S, was calculated from the relative integrated intensities of the superlattice/fundamental intensity ratio of I_{100}/I_{111}. The relative LRO parameter was defined, after Carpenter and Schulson (24) as $S = [(I_s/I_f)\Delta t/(I_s/I_f)\Delta t=0]^{0.5}$. Here $(I_s/I_f)\Delta t$ is the intensity of the superlattice to the fundamental line at a given milling time, Δt. This relative LRO parameter, S, is plotted against the logarithm of the milling time in Figure 4. S exhibits a monotonic decrease with milling time, reaching S = 0 after 5 hours. Microhardness was measured on polished surfaces of mounted powder compacts and is also plotted in Figure 4. The microhardness increases rapidly to a maximum at the milling time when $S \approx 0.5$, and then decreases and becomes essentially constant after S = 0. A similar dependence of flow

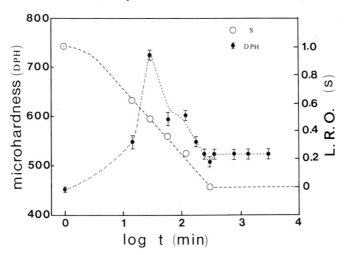

Fig. 4:   Relative long-range-order parameter, S, and microhardness as a function of milling time.

stress on S has been observed in FeCo-V and Fe$_3$Al ordered intermetallics where a peak in flow stress occurs at intermediate (S = 0.4 - 0.6) values of the LRO parameter (25). The small dip in hardness at the time where the material disorders is reproducible.

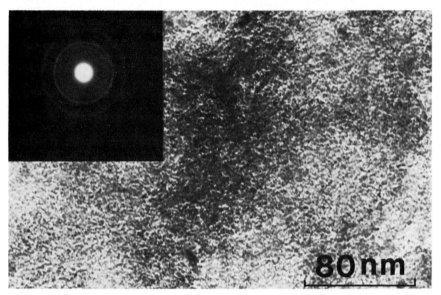

<u>Fig. 5:</u>   TEM micrograph, Ni$_3$Al powder milled for 28 hours. Bright field.

Transmission electron microscopy was used to study the microstructure at the longer milling times where x-ray diffraction indicates broad lines but still crystalline patterns. If the Scherrer formula is applied to the breadth of the major x-ray diffraction peak, the effective scattering length decreases with milling time and saturates at a value of about 4.2 nm. At milling times > 5 hours a fine microcrystalline microstructure is observed as illustrated in Figure 5 for a sample milled 28 hours. The selected area electron diffraction pattern is clearly crystalline, and the microcrystalline grain size appears to be about 2 to 4 nm in diameter. This microcrystalline structure is also observed for longer milling times (i.e. 50 h) over parts of the sample, but in some areas an image and a selected area electron diffraction pattern consistent with the amorphous structure is observed, as illustrated in Figure 6.

Differential scanning calorimetry revealed two exothermic peaks for the milled powders, a small peak at about 140°C and a much larger peak at about 320°C. Annealing at temperatures just above the first peak did not result in any apparent change in structure. The fundamental x-ray diffraction peaks remained broad, and no superlattice lines were observed. Annealing above the higher peak, however, served to reorder the structure such that the fcc → Ll$_2$ transition was completed and the grain size also increased, although only to about 5 nm after annealing 0.5 h at 450°C.

Enthalpy values obtained for the integrated area of the peak are illustrated in Figure 7 for the large exothermic peak as a function of milling time. The values for ΔH increase to a maximum at about 15 hours milling time and remain constant for longer times. This is the same milling time where the effective scattering length from the Scherrer formula also saturates. A change in the slope of the monotonically increasing ΔH vs t curve is evident at about 5 hours,

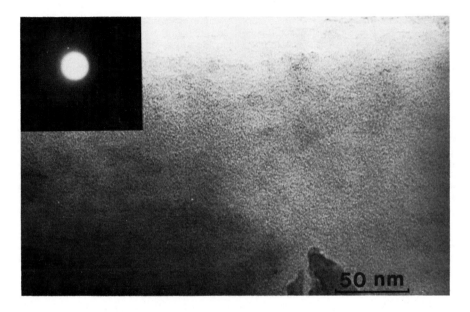

Fig. 6:   TEM micrograph, Ni$_3$Al powder milled for 50 hours; area exhibiting amorphous structure. Bright field.

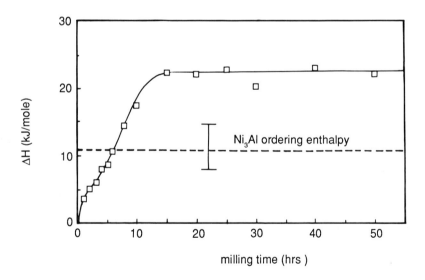

Fig. 7:   Enthalpy of large exothermic peak in Ni$_3$Al as a function of milling time. Estimated ordering enthalpy shown by the horizontal line.

the milling time where the LRO disappears. A value estimated for the ordering enthalpy, shown on the Figure, is consistent for the ΔH value and time where the disordered fcc phase begins. The ordering enthalpy was estimated from literature values for the heat of formation of Ni_3Al (20,26) and the observation of Luzzi and Meshii (14) that the ordering energy for several ordered compounds is 21-37% of the enthalpy of formation. The continued rise in ΔH for milling times after disorder is complete may represent the release of stored energy of cold work in the form of reduced defect densities, and "grain" growth. It may also include crystallization of amorphous regions.

Discussion

The thermodynamic requirement for amorphization by mechanical milling of intermediate phases is the increase in free energy of the equilibrium intermediate phase by the milling reaction to that of the corresponding amorphous alloy. As first suggested by Swanson et al (27) for amorphization of Si and Ge by irradiation, this is likely to occur when the free energy of the crystalline phase (G_C) plus the free energy increase due to defects produced by milling (G_D) is greater than the free energy of the amorphous phase (G_A): that is $G_C + G_D > G_A$. The central question in this regard is what defects can be responsible for the needed energy changes in a given system. The stored energy of cold-work as conventionally obtained appears to be too small for the above criterion. However, under the severe deformation conditions which may exist during the milling of powder this is still an open question.

Luzzi and Meshii (14) have presented evidence from amorphization by electron irradiation that chemical disordering is responsible for amorphization in several intermetallic compounds and that the amorphous phase nucleates heterogeneously at antiphase boundaries. It is possible that disordering is the source of amorphization in the line compounds $NiZr_2$ and $Ni_{11}Zr_9$. The crystalline to amorphous transition apparently proceeds from the crystalline/amorphous interfaces. TEM at earlier stages of the transformation is required to identify the sites in the crystalline powders where the amorphous phase nucleates.

The partial amorphization of Ni_3Al presents a different observation regarding the defects responsible for the crystal/amorphous transition. Disorder is complete in Ni_3Al at milling times far short of those where an amorphous phase is observed in TEM. Apparently the disordered fcc solid solution transforms in certain regions by the accumulation of defects.

Johnson (28) has proposed an intriguing theory for destabilization of crystalline solids which, analogous to melting, may in certain cases occur by a homogeneous transformation that he has termed a "shear strain spinodal". The shear modulus for pure metals can be extrapolated to zero at temperatures above T_M at a temperature Johnson defines as T . This shear strain spinodal temperature may be a function of composition in binary alloys and even fall below T_g, the glass transition temperature. It has been shown that the introduction of interfaces (surfaces, or phase boundaries) on a fine scale can markedly depress T_M of pure metals such as gold (29) and tin (30). In the case of these pure metals the ΔT_M still lies well above T_g, which is near zero K for pure metals. However, it is possible that in an alloy such as Ni_3Al the large interfacial area due to the fine microcrystalline microstructure introduced by milling destabilizes the crystalline structure, but since the milling temperature (nominally ambient) is less than T_g, (which is presumably near the crystallization temperature of ~ 320°C) instead of "melting", an amorphous solid is formed. More experimental work is needed to determine whether the crystalline to amorphous transition in Ni_3Al is a nucleation and growth reaction or a homogeneous transformation.

Summary

Preliminary results of TEM studies on the evolution of the crystalline to amorphous transition in certain mechanically milled intermediate phases were presented. The line compounds $NiZr_2$ and $Ni_{11}Zr_9$ exhibit a microstructure that suggests growth of the amorphous phase at the crystalline/amorphous interface. Whether disordering is responsible for nucleation of the amorphous structure has yet to be determined.

Partial amorphization of Ni_3Al is observed at milling times well after the $Ll_2 \to$ fcc order/disorder transformation is complete. A fine fcc microcrystalline structure with "grain" diameters of 2-4 nm apparently transforms to the amorphous structure in certain areas, perhaps where the density of interfaces ("grain" boundaries) reaches a critical value for destabilization of the crystalline solid solution.

Acknowledgements

The authors' research reported in this paper was supported by the U. S. National Science Foundation under contract DMR-8620394.

References

(1) "Solid State Amorphizing Transformations", Proc. of Conf. on Solid State Amorphizing Transformations, Los Alamos, NM, August 10-13, 1987, edited by R. B. Schwarz and W. L. Johnson, also in J. Less Common Metals 140, Elsevier Sequoia S. A., Lausanne 1988.
(2) Weeber and Bakker: to be published in Physica B.
(3) R. L. White: Ph.D. Dissertation, Stanford University 1979.
(4) C. C. Koch, O. B. Cavin, C. G. McKamey, and J. D. Scarbrough: Appl. Phys. Lett., 43 (1983) 1017.
(5) J. S. Benjamin and T. E. Volin: Metall. Trans. 5 (1974) 1929.
(6) R. M. Davis, B. T. McDermott, and C. C. Koch: to be published in Metall. Trans.
(7) R. B. Schwarz, R. R. Petrich, and C. K. Saw: J. Non Cryst. Solids 76 (1985) 281.
(8) E. Hellstern and L. Schultz: Appl. Phys. Lett. 48 (1986) 124.
(9) R. B. Schwarz and W. L. Johnson: Phys. Rev. Lett. 51 (1983) 415.
(10) C. C. Koch and M. S. Kim: J. Phys. (Paris), Colloq., 46 (1985) C8-573.
(11) R. Bormann: Private communication, 1987.
(12) F. Petzoldt: J. Less-Common Metals, 140 (1988) 85.
(13) A. E. Ermakov, E. E. Yurchikov, and V. A. Barinov: Phys. Met. Metall. 52 (1981) 50.
(14) D. E. Luzzi and M. Meshii: Res. Mechanica, 21 (1987) 207.
(15) P. Y. Lee and C. C. Koch: Appl. Phys. Lett., 50 (1987) 1578.
(16) P. Y. Lee, J. C. S. Jang, and C. C. Koch: J. Less-Common Metals, 140 (1988) 73.
(17) E. A. Kamenetzky, P. D. Askenazy, L. E. Tanner, and W. L. Johnson: in M. Tenhover, L. E. Tanner and W. L. Johnson (eds.), Proc. Symp. on Science and Technology of Rapidly Quenched Alloys, Boston, MA, December 1986, Vol. 80, MRS, Pittsburgh, PA p. 83.
(18) M. S. Kim and C. C. Koch: J. Appl. Phys., 62 (1987) 3450.
(19) E. A. Kenik, R. J. Bayuzick, M. S. Kim, and C. C. Koch: Scr. Metall., 21 (1987) 1137.
(20) J. L. Brimhall, H. E. Kissinger, and L. A. Charlot: Radiation Effects, 77 (1983) 273.
(21) L. S. Hung, M. Nastasi, J. Gyulai, and J. W. Mayer: Appl. Phys. Lett. 42 (1983) 672.
(22) H. C. Liu and T. E. Mitchell: Acta Metall. 31 (1983) 863.
(23) E. Ivanov, T. Grigorieva, G. Golubkova, V. Boldyrev, A. B. Fasman, S. D. Mikhailenko, and O. T. Kalinina: to be published in Mater. Lett.
(24) G. J. C. Carpenter and E. M. Schulson: J. Nuclear Mater. 23 (1978) 180.

(25) N. S. Stoloff and R. G. Davies: Progress in Materials Science, 13 (1966) 3.
(26) R. P. Santandrea, R. G. Behrens, and M. A. King: in High Temperature Ordered Intermetallic Alloys II, ed. N. S. Stoloff, C. C. Koch, C. T. Liu, and O. Izumi, MRS Conf. Proc. Vol. 81 MRS, Pittsburgh, PA (1987) p. 467.
(27) M. L. Swanson, J. R. Parsons, and C. W. Hoelke: Radiation Effects 9 (1971) 249.
(28) W. L. Johnson: Progress in Mater. Science 30 (1986) 81.
(29) Ph. Buffat and J-P. Borel: Phys. Rev. A 13 (1976) 2287.
(30) C. C. Koch, S. S. Gross, and J. S. C. Jang: submitted for publication, 1988.

Amorphization and Phase Transformations during Mechanical Alloying of Nickel-Zirconium

F. Petzoldt, B. Scholz, H.-D. Kunze
Fraunhofer Institut für angewandte Materialforschung (IFAM)
Bremen, West-Germany

Abstract

Two commercially available powders with nominal compositions $Ni_{61}Zr_{39}$ and $Ni_{78}Zr_{22}$ were amorphized by mechanical alloying. The starting powders are mixtures of intermetallic compounds. The time dependence of the amorphization process and of subsequent phase transformations was monitored by x-ray analysis and calorimetry. During further milling of the completely amorphized powders reversible phase transformations occured. These effects are discussed in relation to milling parameters and thermodynamic considerations.

Introduction

In 1983 Koch et.al.(1) were the first who published a new method to synthesize amorphous powders. They were successful in amorphizing $Ni_{60}Nb_{40}$ by mechanical alloying (MA) a mixture of the pure crystalline elements. In the last few years some groups of scientists started working in that field. A lot of late transition metal / early transition metal alloy systems were synthesized amorphous by MA (2-4).

One of the interesting questions in all these examinations is to understand the mechanism of amorphization during MA. At present researchers believe that the amorphization process during MA is similar to a solid state interdiffusion reaction in alternating thin films (5). There is direct evidence for such a process found in transmission electron microscopic examinations of not completely amorphized NiNb powders showing a sequence of layers: crystalline Ni - amorphous NiNb - crystalline Nb (6). This observation suggests that the formation of an amorphous phase starts at the boundary between the two elements. This way of looking at the amorphization process during MA is compatible with the negative heat of mixing of an amorphous phase in all considered systems.

Recently Schwarz and Koch (7) showed the possibility to amorphize NiTi and NiNb intermetallic compounds by milling the powders. They suggest a defect mechanism to raise the free energy of the intermetallic compound above that of the amorphous phase.
In this paper we present results on the proceeding amorphization process during MA a mixture of intermetallic NiZr-phases and on subsequent phase transformations occuring during further milling the amorphous powder.

Experimental

The starting powders were commercial powders with nominal compositions $Ni_{61}Zr_{39}$ and $Ni_{78}Zr_{22}$ supplied by Ventron. These powders are mixtures of different intermetallic phases. The MA process was carried out in a planetary ball mill. The material of the milling vessels and milling balls is the steel 100 Cr 6. The vessels were filled and sealed under argon atmosphere with a powder to milling ball weight ratio of 1 : 15. The powders were processed between 1 and 60 hours. The progress in MA was determined by x-ray analysis using $Cu-K_\alpha$-radiation. Crystallization temperatures and the enthalpy of crystallization were measured by differential scanning calorimetry (DSC 7, Perkin Elmer) at a heating rate of 40 K/min.

Results

The starting powder $Ni_{78}Zr_{22}$ consists of at least the two intermetallic compounds Ni_5Zr and Ni_7Zr_2 as detected by x-ray analysis. Additionally there are some unidentified peaks in the x-ray diffraction patterns. The progress of amorphization during MA of this powder is shown in Fig.1. The early stage of MA (5 hours)

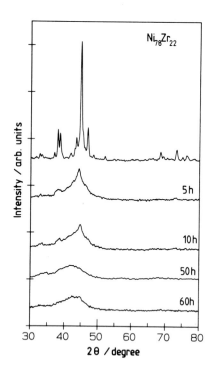

Fig. 1
X-ray diffraction patterns of $Ni_{78}Zr_{22}$ after different milling times

effects a decreasing intensity of the crystalline peaks combined with line broadening. With proceeding milling time a broad maximum typical for an amorphous phase becomes clearly visible in the diffraction patterns. At longer milling times the maximum is shifted to smaller 2-Theta-values. This is a hint for a concentration variation of the amorphous phase caused by the formation of an unknown crystalline phase. This new crystalline phase seems to consist of very small crystallites finely dispersed in the amorphous matrix. This effect and the probably small amount of crystalline phase results in a very weak intensity in its diffraction patterns. This supposition is directly confirmed by calorimetric measurements. Fig. 2 depicts the crystallization temperature T_x and the crystallization enthalpy ΔH for different milling times. T_x is an intrinsic property being expected to depend on composition variations only. The measured T_x increases

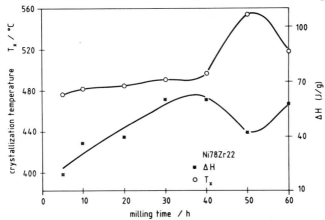

Fig.2 Crystallization temperatures and crystallization enthalpies of $Ni_{78}Zr_{22}$ versus milling time

only slightly between milling times of 5 and 40 hours. That means the concentration of the amorphous phase is nearly constant. The clearly higher T_x after 50 hours milling implies a concentration shift of the amorphous phase. In contrast the extrinsic property ΔH varies with the amount of amorphous phase. During MA the crystallization enthalpy increases until the powder becomes completely amorphous. The decrease in ΔH between 40 and 50 hours milling can be explained by a smaller portion of amorphous phase. These results also suggest that a new crystalline phase is formed during further milling of the amorphous powder. The data points at 60 hours milling (Fig. 2) indicate that the formation of the crystalline phase is a reversible process.

To prove these results the same experiments were carried out with another mixture of intermetallics (nominal compostion $Ni_{61}Zr_{39}$). After MA 10 to 15 hours the powder was completely amorphous as detected by x-ray. In the diffraction patterns of a sample milled 30 hours peaks of a crystalline phase are clearly visible (Fig.3). This phase could not be identified as one of the stable intermetallic phases. After 40 hours milling this crystalline phase could not be detected any more by x-ray. The reversibility of this process is confirmed by x-ray analysis (Fig.3, 50 hours)

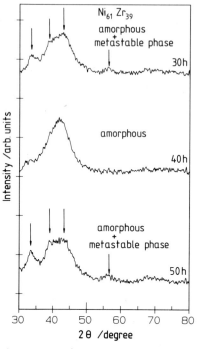

Fig.3
$\overline{\text{X-ray}}$ diffraction patterns of reversible phase transformations in $Ni_{61}Zr_{39}$

and by measuring the crystallization temperature T_x. T_x is nearly constant until the MA process is completed (Fig.4). The variation in T_x during further milling reflects changes in composition of the amorphous phase similar to $Ni_{78}Zr_{22}$.

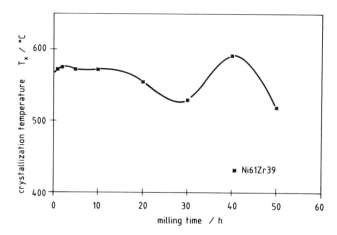

Fig.4 Crystallization temperature of $Ni_{61}Zr_{39}$ vs. milling time

The crystallization enthalpy could not be determined exactly for

$Ni_{61}Zr_{39}$ because of a superimposed endothermal effect.

Discussion

To understand these results it is useful to consider a principal
free energy diagram. The phases involved in this process are at
least two intermetallic phases of the starting powder, the amorphous phase and a metastable crystalline phase. The mechanism of
amorphization of an intermetallic compound differs from that of
MA of elemental powders. The amorphization by MA of elemental
powders is explained with a solid state reaction driven by the
large negative heat of mixing (5). In contrast the amorphization
by milling of an intermetallic phase cannot be explained in this
way because its free energy is lower than that of the amorphous
phase. To raise the free energy of the compound a defect mechanism is suggested (7,8). In case of the amorphization of $Ni_{61}Zr_{39}$
and $Ni_{78}Zr_{22}$ the reaction path seems to be more complicated. In
Fig.5 the free energy of the starting powder is defined as G_{12}.

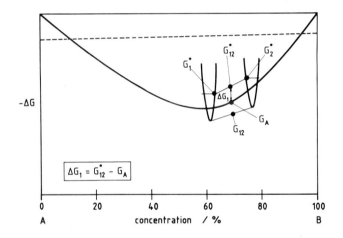

Fig. 5 Schematic free energy diagram for the amorphization process of two intermetallic compounds. The amorphous phase
and the two intermetallic phases are represented by solid
lines. The dashed line represents the mixture of crystalline elements.

This value has to be increased above that of the amorphous phase
to get a driving force for the amorphization. In this case two
intermetallic compounds are involved in the MA process. So the
free energy can be raised by a combination of a defect mechanism
and by composition changes to off-stoichiometry of the intermetallic compounds. The points G_1^* and G_2^* represent the free energy of these non-equilibrium intermetallic phases. G_{12}^* corresponds to the free energy of the nominal composition. The driving
force for the amorphization reaction ΔG_1 is the difference between G_{12}^* and the free energy of the amorphous phase, G_A. One
can imagine such a reaction path for the amorphization of both
considered NiZr-compositions. The results presented in the previ-

116

ous chapter of this paper concerning with MA of intermetallic
phases (the expression "MA" is used only for the process until
the powder is completely amorphous) fit to such thermodynamic
considerations. In the following the reversible phase transformation amorphous - crystalline - amorphous, occuring during continuous milling of the amorphous NiZr-powders, is discussed.

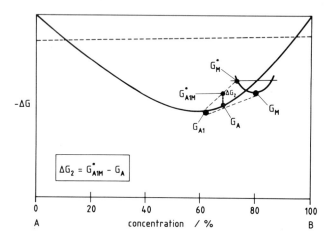

Fig. 6 Schematic free energy diagram of reversible phase transformations. The free energy of the amorphous and the metastable phase are represented by solid lines.

In a principal free energy diagram (Fig.6) the amorphous phase is
defined as G_A. During MA and the milling process the average
temperature of the milling vessel rises to about 100 °C. That
means the amorphous powder is heat treated for a very long time
and simultaneously milled. For NiZr this results under our special milling conditions in the crystallization of a metastable
phase, G_M, and the amorphous phase shifts in composition to G_{A1}.
This behaviour was determined by the x-ray measurements as well
as by calorimetric measurements. It is necessary to review the
crystallization temperatures of the amorphous phase in the NiZr
system to understand the changing crystallization temperatures
presented in Fig.2 and Fig.4. In Fig.7 the T_x of amorphous NiZr
prepared by melt spinning (9), mechanical alloying of elements
(10) and by MA of mixtures of intermetallics are plotted. There
is a maximum in T_x near 65 at.% Ni. The two considered concentrations of this work are on both sides of this maximum. The amorphous $Ni_{78}Zr_{22}$ powder has a T_x of 480 °C. As measured by x-ray
the concentration of the amorphous phase changes during further
milling to about 70 at.% Ni. This concentration shift is accompanied by an increasing T_x up to 555 °C. In the amorphous $Ni_{61}Zr_{39}$
sample the T_x is 595 °C being shifted to 518 °C if there are crystalline peaks detected by x-ray. The crystalline metastable phase seems to be the same for both concentrations but could not be
identified. The measured enrichment in Zr of the amorphous phase
indicates that the metastable phase has to be rich in Ni.

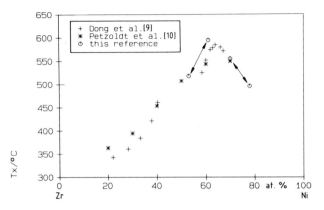

Fig. 7 Crystallization temperatures of amorphous NiZr samples

So far the process of amorphization and the formation of a crystalline phase can be understood and fit well to the crystallization data plotted in Fig.7. To understand the observed reversibility of this process the schematic free energy diagram, Fig.6, is considered again. The mechanism leading to a rise of the free energy of the metastable phase G_M above that of the amorphous phase can be explained similar to the amorphization process of an intermetallic compound (7). By that process the metastable crystalline phase reaches a free energy G_M^*. The value $G_{A_1M}^*$ represents a mixture of off-stoichiometric metastable crystalline phase and the amorphous phase. For a renewed amorphization ΔG_2 can act as a driving force. If this occurs the crystallization temperature and crystallization enthalpy must change again. Indeed all experiments presented in this paper confirm such a thermodynamic conception. It is indicated by the arrows in Fig. 7 that the T_x varies corresponding to the portion of metastable crystalline phase.

It may be expected that the formation of stable intermetallic compounds during MA is favoured if the milling process is carried out in an equipment with too high energy input (11,12). In our experiments the energy input is high enough to amorphize the mixture of intermetallic compounds and too long milling only results in the described reversible phase transformations but not in complete crystallization of the amorphous phase.

Conclusions

Two mixtures of intermetallic compounds $Ni_{61}Zr_{39}$ and $Ni_{78}Zr_{22}$ were amorphized by MA. Further milling of the amorphous phases caused the partial formation of a metastable crystalline phase. This transformation was detected to be reversible with proceeding milling time. The reversibility can be explained related to existing thermodynamic considerations describing possible amorphization processes.

Acknowledgements

The authors wish to thank G. Veltl for helpful discussions and
critical reading of the manuscript. Technical assistance of
E. Woelki and H. Großheim is gratefully acknowledged. Financial
support is given by the Deutsche Forschungsgemeinschaft.

References

(1) C.C. Koch, O.B. Cavin, C.G. McKamey, J.O. Scarbrough
     Appl. Phys. Lett., 43 (1983) 1017
(2) R.B. Schwarz, R.R. Petrich, C.K. Saw
     J. Non-Cryst. Sol. 76 (1985) 281
(3) L. Schultz, E. Hellstern
     in: Tenhover, Tanner, Johnson (eds.)
     Science and Technology of Rapidly Quenched Alloys
     MRS Symp. Proc. Pittsburgh, PA, 1987. p.3
(4) C. Politis, W.L. Johnson
     J. Appl. Phys., 60 (1986) 1147
(5) R.B. Schwarz, W.L. Johnson
     Phys. Rev. Lett., 51 (1983) 415
(6) F. Petzoldt, B. Scholz, H.-D. Kunze
     Mater. Lett., 5 (1987) 280
(7) R.B. Schwarz, C.C. Koch
     Appl. Phys. Lett., 49 (1986) 146
(8) P.Y. Lee, C.C. Koch
     J. Mater. Sci., 23 (1988) 2837
(9) Y.D. Dong, G. Gregan, M.G. Scott
     J. Non-Cryst. Sol., 43 (1981) 403
(10) F. Petzoldt, B. Scholz, H.-D. Kunze
     Mat. Sci. Eng., 97 (1988) 25
(11) L. Schultz
     Mat. Sci. Eng., 97 (1988) 15
(12) A. Weeber, W.J. Haag, A.J.H. Wester, H. Bakker
     J. Less Common Metals, 140 (1988) 119

Structural Transformation and Decomposition of Binary Alloys by Grinding

P.I. Loeff, H. Bakker and F.R. de Boer
Natuurkundig Laboratorium der Universiteit van Amsterdam
Amsterdam, the Netherlands

Introduction

Since its discovery in 1983, the subject of amorphization by ball milling has been intensively studied. A large number of binary alloy systems that transform to the amorphous state by mechanical alloying or by grinding the crystalline intermetallic phases has been discovered. Especially the Zr-TM and Ti-TM systems have been examined (1). However, the criteria that govern the glass forming ability (GFA) for this technique are not fully understood. A tentative criterion for the GFA of Zr-TM systems has been presented by Weeber and Bakker (2). In their model a Zr-TM alloy is expected to exhibit amorphization upon high energy ball milling when a) the heat of mixing of the system is negative, and b) the volume ratio v^{TM}/v^{Zr} of the atomic volumes is smaller than about 0.6. (The atomic volumes v^{TM} are corrected for the volume change by alloying, following the Miedema model (3).) Nearly all zirconium-TM alloys with a negative heat of mixing, and a volume ratio of less than 0.6 appeared to exhibit amorphization by ball milling. So far, the only exception seemed to be the Cr-Zr system, which did not amorphize in spite of its volume ratio of 0.45 and mixing enthalpy of -12 kJ/mol.

To test the applicability of these criteria to binary systems that contain other elements than zirconium, we examined the GFA of several La-TM alloys. Lanthanum seemed to be a good candidate to replace Zr, since the difference in atomic volumes between La and the late transition metals is large. In a previous paper (4), we reported the amorphization of $La_{45}Au_{55}$ by milling a mixture of crystalline intermetallic compounds. Since V_{Co}/V_{La} and V_{Ni}/V_{La} are smaller than V_{Au}/V_{La}, and both mixing enthalpies ΔH^{mix} are negative, La-Co and La-Ni alloys were expected to exhibit amorphization by ball milling. This paper reports the results of the ball milling of these alloys and also of La-Ag. A tentative explanation of the results will be given in terms of energy absorption and the thermodynamics of the various systems.

Experimental procedures

La-TM alloys were prepared by arc melting the pure elements together in an argon atmosphere. The melting was repeated several times to ensure the homogeneity of the alloys. The arc-melted buttons were brittle enough to be pulverised in the first stage of the ball milling procedure, so that no preceding crushing in a press was necessary. After the arc melting the alloys were transported directly to a glove box filled with purified argon, in which the milling took place.

The milling apparatus consisted of a hardened steel vial with a tungsten carbide bottom and an inner diameter of 6.5 cm, in which the material was milled by a hardened steel ball (diameter 6 cm). The ball was kept in motion by a vibrating frame (Fritsch: Pulverisette 0) upon which the vial was mounted. The hopping amplitude of the ball was 1-2 mm.

After the milling, the resulting material was examined by means of X-ray diffraction in a Philips vertical diffractometer, employing Cu Kα radiation. To prevent oxidation, the material was mixed with high vacuum grease inside the glove box, prior to the diffraction measurements.

Magnetic measurements were carried out at 4.2 K in fields up to 14 T in the High Magnetic Field Installation at the University of Amsterdam (5).

Results and discussion

An arc melted button of $La_{43}Ni_{57}$ was pulverized in the ball milling apparatus, and examined by means of X-ray diffraction. The X-ray diffractogram is displayed in fig. 1a. The Bragg peaks can be identified as being from the equilibrium compounds LaNi and La_2Ni_3. Figure 1b shows the diffractogram of a similar La-Ni alloy of the same composition, ball milled for 136 hours.

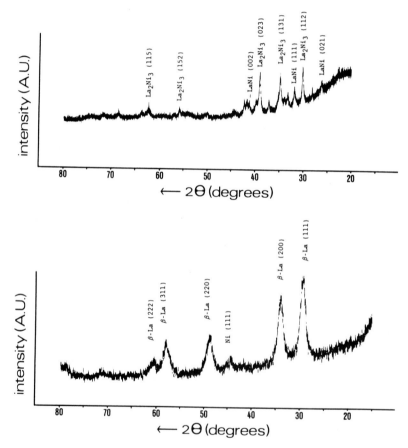

Fig.1: (a) X-ray diffractogram (Cu Kα radiation) of $\overline{La_{43}Ni_{57}}$ before the milling. The observed Bragg peaks are from NiLa and from La_2Ni_3. (b): X-ray diffractogram of the material, ball milled for 136 hours. The peaks are identified as being β-La and Ni peaks. (For clarity, not every peak has been marked.)

Clearly, the equilibrium compounds have transformed into another phase, which we identify to be a mixture of elemental Ni and β-La (a high temperature f.c.c. phase of La, in the equilibrium phase diagram present above about 330 °C). The identified peaks have been marked in the diffractogram. The lattice parameter of the β-La, derived from the position of the peaks, is 5.29 ($\pm$ 0.01) Å, which is in very good agreement with the literature value of 5.291 Å (6). Although some mutual solution of the elements into each other (especially of Ni into La) can not be fully excluded, we attribute the Bragg peaks to the pure elements because the lattice parameter, derived from the reflections, is in such good accordance with the literature value. So, the mixture of equilibrium compounds has decomposed into the elements by the milling process. By using the Scherrer formula and the width of the peaks in the diffractogram, we estimate the particle size of the mixture of the elements to be 100 - 200 Å.

The same results as for the $La_{43}Ni_{57}$ alloy were found in the case of $La_{34}Ni_{66}$, $La_{38}Co_{62}$, $La_{40}Co_{60}$, $La_{50}Ag_{50}$ and $La_{45}Ag_{55}$. The milling resulted in elemental β-La and Ni, Co or Ag, respectively.

To investigate the decomposition process further, we studied the magnetic behaviour of $La_{40}Co_{60}$ as a function of the milling time. As an example, in figure 2 the X-ray diffractogram of $La_{40}Co_{60}$, ball milled for 216 hours is displayed. Present are the reflections from β-La, as well as a slight rise in the background where the strongest Co line is to be expected. Cobalt is hard to see with the used Cu $K\alpha$ radiation, because Co has a large absorption coefficient for this wavelength. Figure 3 shows a plot of the magnetization of the powder at 4.2 K with an applied field of 14 tesla, versus the milling time. It is clear that in the early stages of the milling the magnetization increases rapidly, which is attributed to the formation of regions of pure cobalt. The maximum observed magnetization is 1.23 bohr magneton per Co atom. This is lower than the bulk magnetic moment of Co, being 1.7 μ_B/atom. It is, however, not reasonable to expect to find the bulk value of the Co magnetic moment, even if the La-Co alloy has decomposed completely into pure Co and La. Because of the very small particle size of 100-200 Å the magnetization of the Co is expected to be significantly reduced with regard to the bulk value (7). In the later stages of the grinding there appears to be some decrease in the magnetization.

The sample with maximum magnetization is the sample with a milling time of 72 hours. This suggests that by that time the decomposition reaction has completed. This is supported by the results of the X-ray diffraction measurements, the diffractograms do not change from 72 to 216 hours of milling time, whereas in the earlier stages we observe reflections from La_2Co_3.

In contrast to the above results, we mention the results of the grinding of intermetallic La-Au alloys, reported previously (4). Figure 4 is an X-ray diffractogram of $La_{45}Au_{55}$, ball milled for 130 hours. The material clearly is not crystalline but amorphous, which was also confirmed by means of Differential Scanning Calorimetry. The difference with the above results is striking, ball milling of crystalline La-Au alloys produces an amorphous alloy, whereas the same technique, applied to a comparable system like La-Ag, causes the alloy to decompose into the elements.

On the basis of the experience in Zr-TM systems, we had expected La-Ni and La-Co to undergo an amorphization reaction, since the criteria of Weeber and Bakker for Zr-TM (2) are, as we already mentioned in the introduction, that the system should exhibit a negative heat of mixing, and that the ratio of the atomic volumes should be smaller than 0.6. There was no indication that the magnitude of the heat of mixing was of any importance, since both Ni-Zr (ΔH^{mix} = -49 kJ/mol) and V-Zr (ΔH^{mix} = -4 kJ/mol) could be made amorphous by the milling process. Consequently, the main criterion for the glass forming ability seemed to be the volume criterion.

Applying the same criteria to La-TM systems, one would expect the formation of an amorphous phase in the La-Ni case and in the La-Co case, because of the

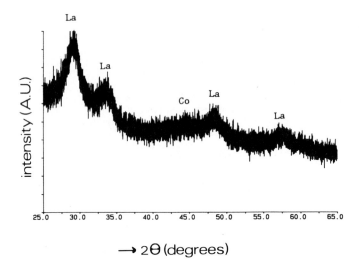

Fig.2: *X-ray diffractogram (Cu Kα radiation) of* $La_{40}Co_{60}$, *ball milled for 216 hours.*

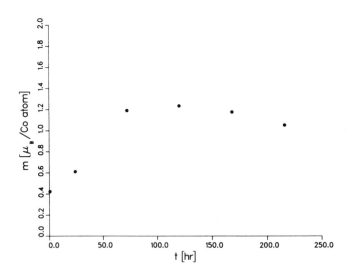

Fig.3: *Magnetization of* $La_{40}Co_{60}$ *versus the milling time, with an applied field of 14 T.*

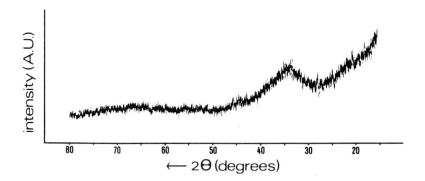

Fig.4: X-ray diffractogram (Cu Kα radiation) of La$_{45}$Au$_{55}$, ball milled for 130 hours.

small volume ratio and the negative heat of mixing. The volume ratio in both cases is smaller than in La-Au, which did amorphize by ball milling. Based on the same criterion, it was not obvious what the result of the milling of La-Ag would be.

The demixing of La-Ag, La-Ni and La-Co lead us to the conclusion that the magnitude of the heat of mixing must be an important factor in glass formation, since ΔH^{mix} of La-Au is about twice as large as ΔH^{mix} of La-Ag, La-Ni and La-Co. Apparently for La-TM systems a minimum $|\Delta H^{mix}|$ is required for glass formation by ball milling.

To give a tentative explanation of the observed phenomena, we consider the enthalpy of formation of the crystalline and the amorphous alloys, in connection with the energy that is absorbed during the milling process. The enthalpies are calculated by means of the Miedema model (8,9). Although the enthalpy values, given by the (semi-empirical) Miedema model are not claimed to be exact values, they have proved to be very reasonable estimates, when compared with experimental evidence. Therefore, we will use these values in the following enthalpy considerations.

From the fact that in the cases of decomposition the equilibrium compounds are transformed into a mixture of the elements, we must conclude that the energy that is absorbed by the material, must be in the order of the formation enthalpy of the stable compounds. For instance, the formation enthalpy ΔH^{form} of the compound AgLa is -43 kJ/mol. This compound shows demixing upon grinding, which means that at least this amount of energy can be stored in the material by milling in our equipment. This is of course a surprisingly high amount, it is for example about four times as large as the heat of fusion of most metals. However, the idea that the energy absorption can be this high is supported by our experiments on Zr-Pd alloys, in which we succeeded to amorphize the intermetallic compounds (for instance the composition Pd$_{49}$Zr$_{51}$) (10). The enthalpy difference between the crystalline phase and the amorphous phase for Pd$_{49}$Zr$_{51}$ is 53 kJ/mol, so the conclusion has to be drawn that we can store about 50 kJ/mol in these materials.

With this estimate of the absorbed energy we give a tentative explanation of the difference in the end products of the milling of La-Au and for example La-Ni alloys. The formation enthalpy ΔH_c^{form} of the crystalline La$_{43}$Ni$_{57}$ compound is -38 kJ/mol, ΔH_a^{form} of the amorphous La$_{43}$Ni$_{57}$ alloy is -23 kJ/mol. Assuming that an amount of energy of 50 kJ/mol is stored in the material, we suppose that the milling process thus brings the La-Ni into an unstable state with an energy content somewhat higher than that of a mechanical mixture of the elements. This unstable state of course can not be maintained, so the end product of the ball milling apparently becomes the state with an energy level which lies directly under the energy of the unstable state. In the La-Ni case the end product would therefore be a mixture of the pure components. On the other hand, when 50 kJ/mol is stored into crystalline $La_{45}Au_{55}$ (ΔH_c^{form} = -105 kJ/mol) the unstable state in which the material is brought lies directly above the enthalpy level of the amorphous $La_{45}Au_{55}$ alloy, for which ΔH_a^{form} = -68 kJ/mol (-105 + 50 = -55 kJ/mol). So, the end product of the ball milling of $La_{45}Au_{55}$ is an amorphous alloy. The results on the La-Co and the La-Ag alloys can be explained by similar enthalpy considerations.

Thus, in general, we speculate that when we ball mill crystalline compounds in our milling apparatus, we bring the material in an unstable state that lies a certain amount of energy above the stable equilibrium state. The exact amount of energy by which the material is raised above the state of stable equilibrium, will depend on the milling apparatus and on the ability of the material to absorb the energy. In the present cases, this amount would be about 50 kJ/mol. The result of the milling is speculated to be the (metastable) state with an energy level which lies directly under the energy of the unstable state. In this way, the result of the milling is determined by the (non-equilibrium) thermodynamics of the system, or in other words by the location of the various energy levels belonging to a certain configuration of the alloy, e.g. amorphous, demixed etc.

In figure 5 we show an energy diagram for all binary systems which we ball milled starting from (mixtures of) crystalline intermetallic compounds. In this diagram the enthalpies (50-50 at.%) for the crystalline state and the amorphous state are presented, calculated following the Miedema model. Obviously the demixed state has zero enthalpy. We add to the enthalpy level of the crystalline state the amount of 50 kJ/mol that we store in the material by ball milling. If we then take as the end product of the milling the (metastable) state that lies directly below the unstable state, we find that in this way we amorphize Ni-Zr, V-Zr, Pd-Zr, Co-Zr and Au-La, which is exactly in conformity with our experiments. The systems Co-La, Ag-La and Ni-La would decompose according to fig. 5, which also confirms our experiments. But also the Cr-Zr case appears to be clarified. We already mentioned before that we could not prepare amorphous Cr-Zr in our ball milling equipment. Instead, according to the X-ray diffraction, the result of the milling was a two phase mixture of elemental Cr and an f.c.c. phase which we suppose to be an f.c.c. Zr phase. This supposition is supported by the fact that regardless of the fraction of Cr in the starting alloy, the lattice parameter of the f.c.c. phase is always the same, namely 4.53 Å. Also, with this lattice parameter, the molar volume of this phase is the same as the molar volume of pure Zr. So, we believe that in the case of Cr-Zr demixing occurs as a consequence of the ball milling, in agreement with what we expected above on the basis of thermodynamic considerations.

Thus, all our experimental results agree with our speculative model. However, much more experiments (e.g. TEM investigations) have to be carried out to confirm the above explanation. Moreover we have to discuss the 'old' criterion for the glass forming ability based on the volume sizes of the components. It would be quite imaginable that such a difference in size is a condition for the ability of the system to store the energy. We also emphasize the role of the milling apparatus that is used. It was reported earlier, that different milling equipments can give very different results (11), so it is reasonable to assume that the amount of energy that is stored in a material is dependent on the particular milling device.

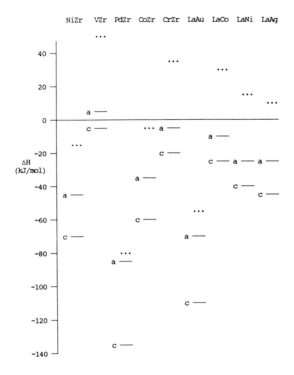

NiZr VZr PdZr CoZr CrZr LaAu LaCo LaNi LaAg

Fig.5: *The formation enthalpies of crystalline (c),* *amorphous (a), demixed (zero enthalpy) and unstable* (•••) *state of various alloys.*

Conclusion

We report the decomposition of La-Ni, La-Co and La-Ag alloys into the elements by ball milling. The decomposition is observed by means of X-ray diffraction and magnetization measurements. From our ball milling experiments we conclude that we can store about 50 kJ/mol into these alloys by grinding them in our ball mill. La-Au alloys do not decompose by the same treatment, since the enthalpy of formation of La-Au compounds is much larger in absolute value than the amount of energy that is absorbed. La-Au alloys transform to an amorphous phase instead. The observations are explained by assuming a model in which the ball milled material is brought to a configuration with an energy content a certain amount of energy above the equilibrium state. The model also explains results on Zr-TM systems, obtained previously.

Acknowledgements

The authors wish to thank Mr A.J. Riemersma for technical assistance, and Mr A.C. Moleman and Mr W.F. Moolhuyzen for their help with the X-ray analysis. Dr Ir F.J.A. den Broeder helped us with some useful suggestions about the results of the magnetic measurements. This work was financially supported by the Stichting voor Fundamenteel Onderzoek der Materie (Dutch Foundation for Fundamental Research on Matter).

References

(1) A.W. Weeber and H. Bakker, accepted for publication in Physica B.
(2) A.W. Weeber and H. Bakker, Proc. LAM6 (august 1986, Garmisch
    Partenkirchen)
    Z.Phys.Chemie 157,   221 (1988).
(3) A.R. Miedema and A.K. Niessen, Physica 114B,   367 (1982).
(4) P.I. Loeff and H. Bakker, Scr. Metall. 22,   401 (1988).
(5) R. Gersdorf, F.R. de Boer, J.C. Wolfrat, F.A. Muller and L.W. Roeland, in
    High Field Magnetism, ed. M. Date, p. 277 (Amsterdam: North-Holland 1983).
(6) P. Villars and L.D. Calvert, Pearson's Handbook of Crystallographic Data
    for Intermetallic Phases, vol. 3, p.2624 (American Society for Metals,
    1985).
(7) F.J.A. den Broeder, private communication.
(8) A.R. Miedema, F.R. de Boer and R. Boom, CALPHAD 1(4),   341 (1977).
(9) P.I. Loeff, A.W. Weeber and A.R. Miedema, J.L.C.M. 140,   299 (1988).
    (Proc. Int. Conf. on SSAT, Los Alamos, august 1987.)
(10) P.I. Loeff and H. Bakker, acc. for publ. in Proc. E-MRS Spring Conf.,
    Strasbourg, june 1988.
(11) A.W. Weeber, W.J. Haag, A.J.H. Wester and H. Bakker, J.L.C.M. 140 (1988),
    119-127 (Proc. Conf. on SSAT, Los Alamos, NM, August 10-13, 1987).

B:
PROCESSING OF MECHANICALLY ALLOYED MATERIALS

Consolidation of Powders under High Pressures - Cold Sintering

E.Y. Gutmanas, Department of Materials Engineering, Technion, Haifa, Israel.

Introduction

Advanced powder metallurgy methods offer means to produce high performance materials with improved physical and mechanical properties (1). The mechanical behaviour of materials with complex microstructures is controlled by superposition of various strengthening mechanisms, namely grain and subgrain boundary hardening, solid solution hardening and precipitation/dispersion hardening. Refinement of grain size and of second phase precipitates or dispersoids, homogenization of microstructure and bonding integrity between second phase particles and the matrix provide a combination of high strength and ductility. Fine scale microstructures may be achieved using advanced powder metallurgy processing technologies. Rapid solidification processing of powders or ribbons results in a high degree of supersaturation of solid solutions, fine scale precipitates, a fine grain size, and a high degree of homogeneity of composition and microstructure, a retention of metastable phases including amorphous metallic glasses (1-3). Alternatively, mechanical alloying or high energy attrition milling of metal powders and ceramic particles gives rise to a uniform distribution of fine scale dispersoids in the metal matrix (4-6). In some systems attrition milling results in formation of amorphous phases (7,8).

To produce high performance materials from rapidly solidified powders/ribbons or from attrition milled powders, the latter should be consolidated to full density. Hot processing, hot extrusion, hot forging and hot isostatic pressing (HIP) are usually used for consolidation of rapidly solidified or mechanically alloyed powders. The refined microstructures obtained by rapid solidification or by mechanical alloying may coarsen and metastable constituents may dissolve in the matrix during hot consolidation of the powders (9). Actual consolidation temperatures for hot extrusion, forging or HIP are frequently higher than the service temperatures. Thus, in order to retain fine scale microstructures and metastable constituents, it is necessary to shorten the time at temperature during hot consolidation or to develop consolidation methods which are capable of achieving full density at temperatures which are lower than the service temperatures. Dynamic compaction (10,11) or explosive compaction (13,14) have certain shape limitations. Consolidation using these methods can result in surface melting and rapid solidification, providing interparticle welding and retention of fine scale microstructures. High amplitude rarefaction waves may break the bonds between powder particles obtained during consolidation and cause cracking. Spray deposition of metal alloys offers the possibility to combine rapid solidification with consolidation and is successfully used to produce various high performance materials (14-16).

An alternative ambient temperature consolidation method for rapidly solidified powders and for metal-ceramic composites is cold sintering or high pressure consolidation (17-19). It was shown that when the flow stress of cold worked powder particles is lower than pressure applied for compaction, plastic flow of powder (see Fig.1) at ambient temperature leads to high density, bonding between freshly formed oxide free surfaces and good mechanical properties, or to cold sintering. High strength levels were obtained for refractory metal powders (17,20), materials with ionic and ionic-covalent bonds and various metal alloys, including high speed steels, superalloys and aluminium alloys for elevated temperatures (19,21,22) as well as for metal-ceramic composites (23-25). High pressure consolidation was successfully used for joining of high speed steel to high strength steel (26).

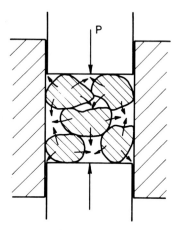

Fig.1: A schematic drawing of plastic flow of powder during high pressure consolidation.

Subsequent heat treatment of cold sintered material at temperatures significantly lower than those used in conventional sintering and often at temperatures lower than those used in hot processing of powders, results in excellent combination of strength and ductility.

During the last few years the main efforts were made in the field of design of high performance materials by combining high pressure consolidation of rapidly solidified powders or of attrition milled fine elemental powders with surface treatment of powders prior to consolidation to full density. Such kind of treatment may improve the bonding integrity between powder particles and reduce or even eliminate dissolution of second phase particles in the matrix.

This paper describes experimental techniques used in design of high performance materials by cold sintering of powders, discusses the experimental results and considers possible applications of this method.

Experimental Techniques

High Pressure Consolidation. The results obtained on high pressure consolidation of Fe, Cu, Al, Ni and Co powders and some of their alloys show that density close to theoretical (>99%) may be achieved at ambient temperatures at compacting pressure $P \approx 3GPa$. With careful design, tool steel dies and punches can be used at pressures up to 3GPa for hundreds of cycles. In the present work, dies and punches were made from T15 P/M high speed steel (CPM, Crucible). Provided alignment of the tooling is maintained, it is possible to compact powders at pressures up to about 3.5 GPa. To withstand these pressures, the outer diameter of the die is approximately three times larger than the part diameter. Typically the compact is loaded to $P \approx 2GPa$ in several seconds, loading is then increased more slowly over 10–15 second intervals to achieve a compaction pressure of 3GPa. Pressures as high as 4.5 GPa are possible utilizing cemented carbide dies. These are more brittle and high pressure consolidation of composite materials containing hard or superhard particles results in failure of cemented dies after several cycles. Cylindrical dies with diameters from 3mm to 50mm and rectangular dies with cross sections from 2x18 mm^2 to 25x70 mm^2 and to 40x42 mm^2 were used. A 5MN Toni-

Technic provides P=2.5-3GPa for the largest dies. A large number of smaller size specimens may be produced in a die with a large opening or in parallel each in a separate rigid cell (see Fig.2). Recently an upscaling to a larger size pressure cell was made, employing a press with more than 20 MN capacity. Densities close to theoretical were obtained for Al, Fe and T15 high speed steel powders (27) in a pressure cell with 88x88 mm² opening. The large cell with two anvils, punches and with some of the specimens produced is shown in Fig.3. The density of compacted specimens is measured by the Archimedes method. Samples for transmission electron microscopy (TEM) were prepared from cold sintered discs 0.1-0.15 mm thickness. Densities less than 99% of theoretical are achieved after high pressure consolidation at P=3GPa for some of the highly alloyed rapidly solidified powders, including amorphous and microcrystalline powders, for some of the commercially available elemental powders (W,Mo,Zr), and for metal-ceramic composites with a large volume concentration of hard particles. Annealing of consolidated specimens, followed by repressing at the same pressure P=3GPa at ambient temperature or warm sintering at 200-400°C at pressures up to 3GPa results in most cases in full densification. Dies and punches made from P/M high speed steel may work at P≈3GPa at temperatures up to 500°C.

Fig.2: High pressure die and punches for preparation of ten specimens.

Fig.3: Large pressure cell for high pressure consolidation with punches and examples of prepared flat specimens, opening 88x88mm².

Materials. Water atomized (WA) iron and 4600 steel powders from Hoeganaes Corporation were used, stainless steel and high speed steel WA powder were supplied by SCM Metal Products, WA nickel, Ni-Cr and Nimonic 80A powders were supplied by BSA Metal Products and WA Co and Co alloys by Pfizer, Inc. Gas (air) atomized aluminium alloys were supplied by Alcoa, inert gas atomized (IGA) aluminium alloy was from Pratt and Whitney, IGA cobalt alloys (Stellites) were from Cabot Corporation and IGA high speed steel and superalloys were from Crucible, Colt Industries. Amorphous and microcrystalline powders were from Allied Corporation and from Marco Materials and were made by crushing of splat cooled ribbons. Water and air atomized powders were normally screened to give particle sizes less than 44μm (-325 mesh). IGA powders were screened to give a particle size less than 100μm and less than 44μm. Water atomized powders and air atomized aluminium alloy powder were irregular in shape. Such shape provides better conditions for plastic deformation of powder particles during consolidation, resulting in physical contact and chemical bonding of freshly formed oxide free surfaces. IGA powders are spherical in shape. Consolidation to full density of spherical shape powder results in a relatively low amount of lateral flow. Elemental fine (1-2μm) carbonyl iron powder (Poudmet) and fine cobalt powder (1.5-3μm) prepared by a combination of pyrolysis and reduction of cobalt oxalate (Hoboken) and fine (0.5μm) oxidereduced tungsten powder (H.C. Starck) were used for preparation of alloys using ball and attrition milling. Carbide, nitride and oxide powders were purchased from H.C. Starck and from Cerac Co. Coarser (~ 20μm) W powders were from BDH and were used for preparation of tungsten based alloys. Some of the alloyed powders were purchased from Alfa Ventron.

Milling. A number of iron and cobalt based alloys was prepared by milling of elemental metal powders with additions of hard particles (carbides, nitrides, oxides). To obtain an homogeneous distribution of fine particles some of the hard particles were dry milled and, only after such treatment, added to metal powders. High energy attrition milling was much more effective compared to ball milling. A laboratory attritor (Model 1-S, Union Process) was used in the experiments. Examples of VC powder ball milled (20h) and high energy attrition milled (1hr) are shown in Fig.4.

Fig.4: SEM micrographs of VC powder: a) ball milled for 20hrs and b) high energy attrition milled for 1hr.

Reduction Studies. Treatment of metal powders before and simultaneously to sintering in a reducing atmosphere leads to an improvement in the properties of the sintered product. In (18) it was shown that reduction of the surface oxide layer of pure iron powder particles provides very high "green" strength of cold sintered material. Transverse rupture strength, $\sigma_{TRS} \approx 600$MPa, was obtained for samples cold sintered (P=3GPa) from compacts with interconnected systems of pores which were treated in hydrogen at T=600°C. Similar results were obtained for Cu, Ni and Co, having oxides which can be removed in hydrogen flow at relatively low temperatures. Oxides of other metals require higher reducing temperatures or are not reducible at all. Since removal of surface oxide layers can considerably improve the bonding integrity of consolidated powder particles reduction treatment prior to consolidation to full density may be beneficial in achieving high mechanical strength. On the other hand heat treatment at high temperatures may result in coarsening of the microstructure: grain growth, coarsening of fine precipitates etc. Thus it is important to know what temperatures and exposures are necessary for removal of oxide layers and what will be the effect of such treatment on microstructure and mechanical behaviour of consolidated material.

Reduction diagrams were used to establish the temperature limits within which oxides present on the surface of metal powders or oxides, carbides, nitrides, etc. added as hard particles to the metal matrix are reducible. The apparatus used in the experiments is shown schematically in Fig.5. A specimen in a quartz or alumina boat is placed inside a quartz reactor in the stream of hydrogen, supplied at a flow rate of 2-5 dm^3/h. The reactor is placed in a tube furnace, the temperature of which can be controlled automatically within $\pm 5°C$. The temperature of the furnace can be increased at a constant rate, so that the reduction can be studied with a steady increase in temperature. The temperature of a sample is measured by means of a thermocouple placed near the sample and covered in a quartz tube. During treatment of metal powders or metal-ceramic composites in hydrogen the reduction of oxides, sulfides, carbides, etc. may take place as well as the reaction of hydrogen with dissolved carbon (or other elements).

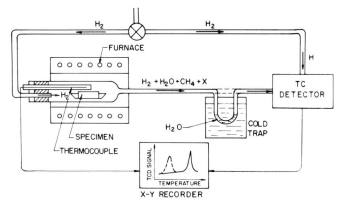

Fig.5: A schematic diagram of the apparatus used for the study of reactions of hydrogen with powders and compacts.

As a result of these reactions water vapour, hydrogen sulfide or methane are formed. The amount of the gaseous product of the reaction is in direct proportion to the amount of the compound reacted with hydrogen and this means, for example, that the amount of water is proportional to the amount of oxides reduced. The kinetics of a reaction can be studied by measuring the amount of gas produced in a reaction as a function of time and temperature. To determine the amount of a gas produced in reaction, the reactor is connected to a thermal conductivity (TC) detector. The signal from the detector is proportional to the concentration of gases evolved as a result of chemical reactions of hydrogen with oxides, carbon, nitrogen, sulphur etc. The signal is recorded versus temperature (or time) on an X-Y recorder. The amount of gases evolved is proportional to the amount of the compounds (oxides, carbides, sulphides, etc.) or dissolved elements (carbon, nitrogen, sulphur, etc.) which reacted with hydrogen. The starting temperature for the reaction depends on the strength of the bond between the element removed and the compound. Thus for each reaction a peak of the TC signal versus temperature may be found within a specific range (28). The product of the reaction of oxides with hydrogen is water and can be removed by using a cold trap placed between the reactor and TC detector. A dry ice cold trap was used in the present work. Peaks which belong to the reactions of hydrogen with carbon (CH_4), nitrogen (NH_3), etc. remain. Using a number of cold traps at different temperatures, peaks corresponding to different reactions can be determined. The apparatus shown in Fig.5 was also used to study the decomposition of hydrides in Al alloy powders (29,30).

A new approach to the P/M processing of fully dense metal-ceramic composites has been developed (23,25). A metal-ceramic compact containing interconnected pores is treated in hydrogen (Fig.6a). The open interconnected pore structure provides access for hydrogen in order to remove oxide layers from the surface of the metal particles and for partial reduction or metallization of the surface layers of the ceramic particles, at elevated temperature (Fig.6b). After treatment in hydrogen the reduced compacts were cold sintered to full density (Fig.6c).

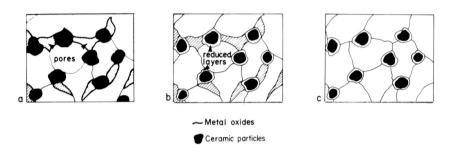

— Metal oxides

⬤ Ceramic particles

Fig.6: A schematic of the processing sequence for metal-ceramic composites (a) compact to ≈80% of theoretical density; (b) treatment in hydrogen, flow-removal of oxide layers from metal particles and metallization of surface layers of ceramic particles, (c) cold sintering to full density.

Mechanical Testing. The transverse rupture strength, σ_{TRS}, was measured in three point bending using usually a sample size 44x4.5x18mm³ and a distance between the outer points of 16mm. Yield stress was determined at different temperatures on compression specimens 4x4·5x8mm³ at a cross-head speed of 0.1mm·s⁻¹ on a standard Instron machine. Tensile specimens were prepared by machining of round or rectangular shape specimens.

Microstructure Characterization. Microstructure of cold sintered specimens, as sintered and after different thermal treatments was studied employing scanning electron microscopy (SEM) with energy dispersive analysis (EDS), X-ray diffraction analysis, transmission electron microscopy, Auger spectroscopy and optical microscopy.

Results and Discussion

Effect of shape, size and composition of powders on densification and properties. As mentioned earlier, irregular shape of powder particles provides better conditions for plastic flow during consolidation, resulting in physical contact of oxide free surfaces and in chemical bonding. This was proved directly for powders of pure Al (31). The broken oxide layers exist on the interface of powder particles as islands (Fig.7a). Similar results were obtained later for cold sintered Cu (see Fig.7b), Fe and stainless steel powders.

For spherical shape powders consolidated to full density a relatively small amount of plastic flow results in contact of oxide layers along interfaces of powder particles and in relatively poor bonding integrity. The green strength of spherical powders consolidated to full density is typically 5-10 times lower than that of irregular shape powders of the same or similar composition (18,32). The mode of fracture of consolidated spherical powders is transgranular. For irregular shape powders the mode of fracture is partially intergranular. Good bonding integrity may be achieved for cold sintered spherical shape powders which were treated in a reducing atmosphere. This relates to Fe, Cu, Ni and Co powders; oxides of all these metals can be easily removed. For spherical shape aluminium alloy, superalloy and stainless steel powders, containing oxide layers, which are difficult or impossible to remove, a substantial increase in mechanical properties can be obtained after heat treatments of cold sintered specimens. Such treatments lead to spheroidization of surface oxide layers (32).

Coarser powders can be consolidated to full density at lower pressures than fine powders of the same composition (17,18,29). This is specially applied to rapidly solidified powders, in which fine size is related to a finer scale microstructure. Grain size can be related to the particle size and in this case the Petch-Hall relation should determine the flow stress of powder under consolidation. Fine (micron size) pure powders of spherical carbonyl iron, of irregular shape cobalt and nickel can be cold sintered to full density at P=3GPa.

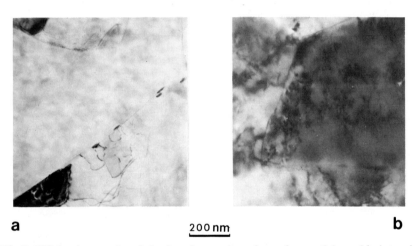

a   200 nm   b

Fig.7: TEM micrographs of the interface region of powder particles cold sintered at P=3GPa: a) pure Al; b) pure Cu.

The effect of hardening by solute atoms and by fine precipitates/dispersoids sees to be much more pronounced than that of grain size. This is the reason that rapidly solidified highly alloyed powders in most cases are hot consolidated at relatively high temperatures. Rapid solidification results in residual thermal stresses and in martensitic transformation in steels. To achieve higher compressibility the powders are usually annealed prior to consolidation. This applies also to high pressure consolidation. As already mentioned, to achieve full density for highly alloyed powders (high speed steels, stainless steels, superalloys) repressing of cold sintered and annealed specimens is employed (18,26,29,34). Some of the alloyed powders, such as for example austenitic stainless steel, may undergo martensitic phase transformations during high pressure consolidiation resulting in hardening (35). Annealing, followed by repressing or warm sintering (200-400°C), provides consolidation to full density.

Fine structures may be obtained by attrition milling or even mixing of fine elemental powders, followed by high pressure consolidation and heat treatment. In this case compressibility
is determined mainly by the matrix or main component powder. If pure ductile powder is the
main component, full densities may be achieved for high pressure consolidated material containing relatively large amounts of homogeneously distributed non-compressible hard particles. Full density was achieved for carbonyl iron-carbides and for Co-WC composites, containing upto 30 volume percent of carbide (36,37).

Examples of microstructures obtained for attrition milled (1 hour) and cold sintered steel
composition are shown in Fig.8 and compared with the microstructure of cold sintered rapidly
solidified powders. The structure of attrition milled and cold sintered specimens is as fine as
that of rapidly solidified powders. Further refinement may be achieved by using fine scale
starting powders or by prolonged milling. Crushing of hard particles and use of submicron
size metal powders should provide finer microstructures. Steel compositions prepared by attrition milling and cold sintering exhibit high mechanical properties, exceeding those of cold
sintered rapidly solidified powders of the same composition after the same thermal treatments.
For example, hardness Rc=66-67 and σ_{TRS}=2600 MPa was achieved for T15 composition
processed via attrition milling from elemental powders after standard heat treatments.
Crushing of carbides and high energy attrition milling results in much finer microstructures as
compared with ball milling. An example is shown in Fig.9 for Co-20w/o WC. Refinement of
microstructure results in short diffusion distances. Thus consolidation of the pure ductile
matrix material to full density may be followed by a relatively short heat treatment, providing
homogeneous distribution of solute atoms, prior to quenching, aging etc. If several components are involved, the diffusion of one of them is not affected by others at least in the initial
period. This is shown schematically in Fig.10 for a specimen cold sintered to full density from
elemental powders.

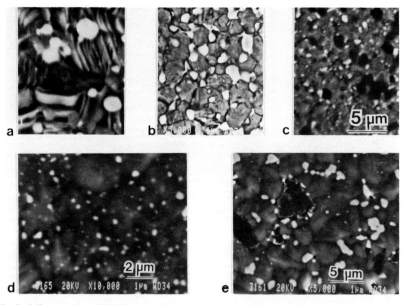

Fig.8: Micrographs (SEM) of polished and thermally etched cold sintered Fe-base
materials. (a) T15, IGA powder; (b) T15, WA powder; (c) T15, attrition milled (AM);
(d) Fe-12w/o W-1w/o C (AM); (e) Fe-10w/o WC-5w/o VC; (c-e) carbonyl iron
matrix material.

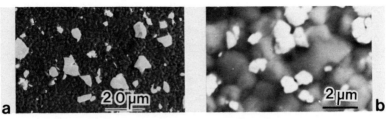

Fig.9: Micrographs (SEM) of cold sintered and thermally etched Co-20w/o W-20w/o WC (a) ball milled for 20hrs, (b) high energy attrition milled for 1hr.

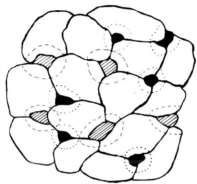

Fig.10: A schematic of a specimen prepared by cold sintering to full density of attrition milled elemental powders.

A number of iron and cobalt based composites containing a large volume fraction of hard particles was prepared using attrition milling, followed by cold sintering. Examples of the microstructure for Fe-VC composites are shown in Fig.11. Hardness of Rc=60 and σ_{TRS}=1700MPa was achieved for Fe-50w/o VC after heat treatments. In (38) attrition milling was used to obtain a high speed steel with a high concentration of NbC by liquid phase sintering. Liquid phase sintering results in partial coarsening of the microstructure. Compositions containing up to 20w/o NbC were successfully cold sintered from elemental attrition milled powders and exhibit fine microstructures and high mechanical properties.

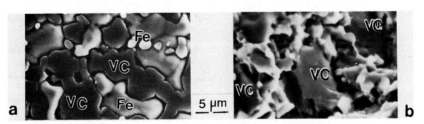

Fig. 11: Micrographs (SEM) of attrition milled, cold sintered and heat treated Fe-50w/o VC: (a) polished and thermally etched; (b) fracture surface.

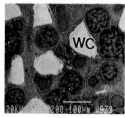

<u>Fig. 12:</u> Micrograph (SEM) of attrition milled, cold sintered and thermally etched Fe-40w/o WC-30w/o Stellite (Co alloy).

Attrition milling may be beneficial in homogenization of materials containing coarse hard particles. An example is shown in Fig.12 for an iron based alloy containing coarse WC and hard cobalt alloy (spherical) particles.

<u>Effect of Reduction Treatment</u>. It was shown that removal of surface oxides from powder particles considerably improves the bonding integrity of cold sintered elemental and alloy powders, as well as metal-ceramic composites, when the ceramic particles can be partially reduced/metallized (18,21,23-26, 33,34,36). The effect of the reduction temperature on subsequent room temperature transverse rupture strength of Fe and Fe-30w/o VC is shown in Fig.13a. Treatment in hydrogen was for 1hr at each temperature. In the cold sintered Fe, the σ_{TRS} begins to increase at about 400°C; this correlates with the peak corresponding to the removal of oxygen, Fig.13b. There is a two-step increase in σ_{TRS} for the Fe-30w/o VC composite. The first step at about 400°C is associated with the reduction of the oxide from the iron powder particles. The second-step increase above about 900°C correlates with the reduction peak observed for VC, Fig.13b. After treatment in hydrogen at the highest temperature examined (1100°C) a decrease in the intensity of the main X-ray diffraction peak of VC is observed. This is believed to reflect the partial reduction (surface metallization) of the VC particles, with attendant improvement in bonding at carbide particle-metal particle interfaces. For the Fe and Fe-VC composites prepared from fine powders and reduced in hydrogen at temperatures above 900°C prior to cold sintering, crack propagation occurs through the matrix. Fig.13c shows the metallized (decarbonized) surface of a cleaved coarse VC particle in Fe-VC composite that was treated at 950°C (1hr) prior to cold sintering. The response of composites prepared from Co-WC composites is similar to that of Fe-VC but the reduction temperature is higher (36).

Removal of oxides from elemental powders enhances the diffusion process that may be beneficial when homogenization is considered but may also lead to undesirable grain growth (33). Reduction and dissolution of carbides in high speed steels results in grain coarsening (26,34,39).

Treatment of a compact in a hydrogen atmosphere may prevent dissolution of carbon in iron or cobalt based matrices (34,40) and prevent undesirable reactions when hydrogen forms stable bonds with free carbon (graphite diamond).

Homogeneous distribution of fine particles results in some cases in interesting diffusion effects of alloying elements. In Fig.14 X-ray diffraction data are presented for Co-20w/o W and Co-20w/o W-20w/o WC composition prepared by cold sintering from attrition milled (1 hr) powders. It can be seen that for Co-W (Fig.14a) W is dissolved in Co already during the treatment of the compact in H₂ at 600°C (surface diffusion followed by bulk diffusion). For the Co-20W-20WC composite, if the compact is treated in hydrogen at high temperature

(1000°C), WC is dissolved in the matrix preventing dissolution of W particles. If the compact is treated at 900°C, W is dissolved and in specimens cold sintered from such compacts, WC remains stable at higher heat treatment temperatures. The microstructure of Co-20W treated in H_2 at 600°C and cold sintered is shown in Fig.14c. The effect of reducing treatment on dissolution, diffusion of various components should be taken into account in design of materials with complex microstructures.

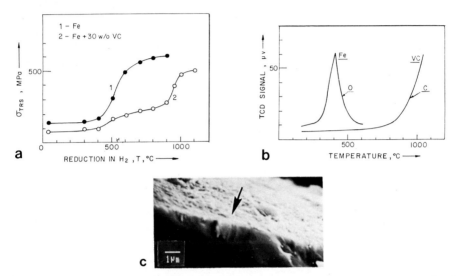

Fig.13: Reduction treatment of Fe-VC composites: (a) room temperature σ_{TRS} of cold sintered materials as a function of green compact reduction temperature, (b) reduction diagrams for Fe and VC powders; (c) SEM fractographs of reduced at 950°C (1 hr) and cold sintered composite (coarse cleaved VC particles with metallized surface).

On structure stability during thermal treatments. For rapidly solidified aluminium alloys for elevated temperatures it was shown that processing via cold or warm (300°C) sintering provides the means of retaining fine scale microstructures of as-atomized powder, resulting in a substantial increase (~ 30%) of the yield stress for consolidated material as compared with hot processed material of the same composition (29). Fine scale structures (grain size and precipitates) can be retained also in high speed steels and stainless steels (25,26,33,34). Dispersions of fine precipitates prevent grain coarsening in RSP hot processed 303 stainless steel (41). Cold sintered WA powder exhibits even finer grain size (see Fig.15a) where the data for cold sintered material are compared with hot processed RSP powder and with conventionally produced 303 steel. In Fig.15b, a TEM micrograph of cold sintered material annealed for 1 hr at 1200°C is presented.

Grain growth may be prevented and full density achieved by a combination of cold sintering with activated sintering. Thin layers of Ni (or Co) produced on the surface of tungsten powder by chemical deposition of salts and reduction procedure enhance the densification of W powder by accelerating the diffusion rate of W (42,43). Such activated sintering does not provide full density and material produced in this way is brittle. Combination of cold sintering to full density of W powder particles covered by a thin Ni layer provides reasonable strength after thermal treatments. In Figs.16a,b fracture surfaces of such specimens are shown. Some of the coarser W grains are cleaved, the fracture mode being mainly intergranular. In Fig.16c data for σ_{TRS} of such specimens after different heat treatments are presented.

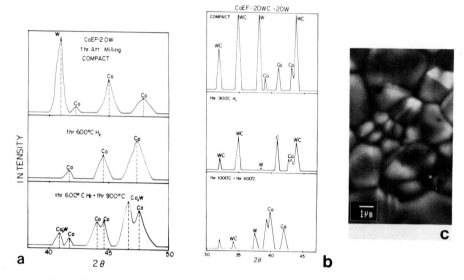

Fig.14: Reduction treatment of attrition milled Co-20w/o W and Co-20w/o W-20w/o WC composites (a) X-ray diffraction spectra after various treatments of (a) Co-20 W; (b) Co-20W-20WC; (c)SEM micrograph of Co-20W reduced at 600° C (1 hr) and cold sintered (thermal etching).

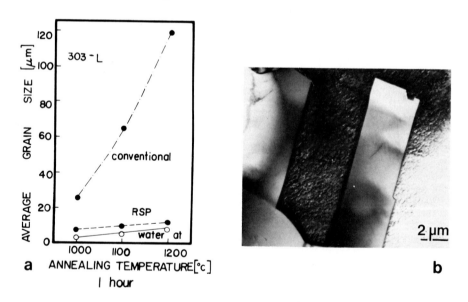

Fig.15: (a) Grain growth characteristics of 303 stainless steel after conventional processing, rapid solidification processing (RSP) and cold sintering of WA powder, (b) TEM micrograph of cold sintered specimen annealed for 1 hr at 1200° C.

<u>Production of final shape and size parts</u>. Since in high pressure consolidation rigid dies are used and full densities are achieved, very close tolerances may be obtained (<0.01mm for 10mm opening) (18). Thus the method can be used for production of final shape and size parts with high accuracy. Wear resistant parts (metal-bonded diamond tools, high speed steel tools etc.) and parts for elevated temperature service are of first consideration. Research in these directions is on the way.

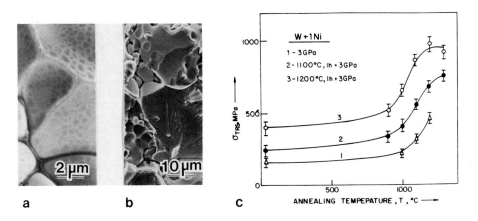

<u>Fig.16:</u> Cold sintered W-1% Ni (chemically deposited layer): (a)(b) fractographs (SEM); (a) for compact treated at 1100°C 1 hr in H_2, P=3GPa and annealed 1 hr at 1300°C, (b) at 1200°C 1 hr in H_2, P=3GPa, annealed 1 hr at 1300°C, (c) σ_{TRS} as function of annealing temperature.

References

(1) M. Cohen, "Advancing Materials Research", National Academy Press, Washington, D.C. (1986) 52.
(2) N.J. Grant, "Advances in Powder Metallurgy", ASM, Metals Park, OH (1982) 1.
(3) W.J. Boettinger, Mater. Sci. Eng. 98 (1988) 123.
(4) J.S. Benjamin, Met. Trans., 1A (1970) 2913.
(5) J.S. Benjamin and M.J. Bomfort, Met. Trans., 8A (1977) 813.
(6) P.S. Gilman and W.D. Nix, Met. Trans., 12A (1981) 813.
(7) L. Schultz, Mater. Sci. Eng. 97 (1988) 15.
(8) R. Sundarsen, A.G. Jackson, S. Krishnamurthy and F.H. Froes, Mat. Sci. Eng. 27 (1988) 115.
(9) E.Y. Gutmanas, M. PremKumar and A. Lawley, "Progress in Powder Metallurgy", MPIF, Princeton, N.J. 39 (1984) 669.
(10) D. Raybold, D.G. Morris and G.A. Cooper, J. Mater. Sci. 19 (1979) 2523.
(11) D. Raybold, "Progress in Powder Metallurgy", MPIF, Princeton, N.J. 41 (1986) 95.
(12) A.K. Bhalla and J.D. Williams, Powder Met. 19 (1976) 31.
(13) R. Prummer, Mater. Sci. 98 (1988) 461.
(14) A.R.E. Singer, J. Inst. Met. 100 (1972) 195.
(15) D. Apelian, "Progress in Powder Metallurgy, MPIF, Princeton, N.J. 42 (1986) 17.
(16) E.J. Lavernia and N.J. Grant, Mater.Sci.Eng. 98 (1988) 461.
(17) E.Y. Gutmanas, A. Rabinkin and M. Roitberg, Scripta Met. 13 (1979) 11.
(18) E.Y. Gutmanas, Powder Met. Int. 15 (1983) 129.

(19)   E.Y. Gutmanas and A. Lawley, "Progress in Powder Metallurgy" MPIF, Princeton, N.J. 39 (1984) 653.
(20)   E.Y. Gutmanas, "Strength of Metals and Alloys", Aachen-5, Pergamon Press, Vol.1 (1978) 577.
(21)   D.B. Goldman, J.B. Clark, S. Hart and E.Y. Gutmanas, "Modern Developments in Powder Metallurgy", MPIF, Princeton, N.J., 17 (1985) 427.
(22)   O. Botstein, E.Y. Gutmanas and A. Lawley, "Modern Developments in Powder Metallurgy", MPIF, Princeton, N.J., 15 (1985) 761.
(23)   E.Y. Gutmanas, D.B. Goldman, J.B. Clark and S. Hart, "Progress in Powder Metallurgy, MPIF, Princeton, N.J., 41 (1986) 631.
(24)   E.Y. Gutmanas, "Modern Developments in Powder Metallurgy", MPIF, Princeton, N.Y. 15 (1985) 175.
(25)   E.Y. Gutmanas, D.B. Goldman, S. Hart and D. Zak, Powder Met. Int. 18 (1986) 401.
(26)   I. Gotman and E.Y. Gutmanas, Powder Met.Int. 19,4 (1987) 11.
(27)   E.Y. Gutmanas and G. Vekinis, to be published.
(28)   D.B. Goldman, J. Amer. Cer. Soc. 66 (1983) 147.
(29)   O. Botstein, E.Y. Gutmanas and A. Lawley, "Progress in Powder Metallurgy", MPIF, Princeton, N.J., 41 (1986) 123.
(30)   O. Botstein, E.Y. Gutmanas and D. Zak, "Horizons of Powder Powder Metallurgy", Verlag Schmid, Freiburg, V.2 (1986) 961.
(31)   D. Shechtman and E.. Gutmanas, Praktische Metallographie 18 (1981) 587.
(32)   E.Y. Gutmanas and D. Shechtman, "Materials Engineering Conference", Freund Publ., Tel-Aviv (1981) 76.
(33)   D.B. Goldman and E.Y. Gutmanas, Powder Met. Int. 17 (1985) 269.
(34)   E.Y. Gutmanas and D. Zak, "Modern Developments in Powder Metallurgy", MPIF, Princeton, N.J., in press (1988).
(35)   E.Y. Gutmanas, Powder Met. Int. 12 (1980) 178.
(36)   E.Y. Gutmanas and A. Lawley, "Modern Developments in Powder Metallurgy", MPIF, Princeton, N.J., in press (1988).
(37)   E.Y. Gutmanas and I. Mitrani, to be published.
(38)   F. Thummler, Powder Met. Int. 20,2 (1988) 39.
(39)   I. Gotman and E.Y. Gutmanas, J. Mater. Sci. Lett. 6 (1987) 1303.
(40)   E.Y. Gutmanas and E. Levin, to be published.
(41)   T.F. Kelly and J.B. Vandersande, "Rapid Solidification Processing", Claitor's Publ. Baton Rouge, Vol.II (1980) 100.
(42)   G.H. Gessinger and H.F. Fischmeister, J. Less Common Metals, 27 (1972) 129.
(43)   R.M. German and Z.A. Munir, High Temp. Sci. 8(1976) 267.

Microstructural changes during consolidation - a comparison of rapidly solidified and mechanically alloyed materials.

D.G.Morris and M.A.Morris, Institute of Structural Metallurgy, University of Neuchatel, Switzerland.

Abstract

Both rapid solidification (RS) and mechanical alloying (MA) are able to produce powders having refined microstructures and metastable phases. Apart from retaining an extremely intense deformation microstructure, the two techniques create materials having similar microstructures, but differing sometimes in subtle ways. The techniques differ also in creating powders having different overall and surface morphologies, and this may affect the compaction behaviour of the materials. The compaction technique chosen for a particular material will depend, amongst other factors, on the thermal stability of the alloy microstructure. For the highly stable materials, for example the coarsening-resistant dispersoid-containing materials, an easy, high temperature consolidation technique can be chosen: extrusion is frequently used. For materials in highly metastable states, for example the glassy materials, a more difficult, lower temperature technique must be used. In this presentation some of the techniques which may be used to consolidate such materials are discussed.

Introduction-Material classes and characteristics

Mechanical alloying has been used mostly to create alloys having highly stable dispersed particles, for example oxides or carbides in nickel-base or aluminium-base matrices (1,2,3,4). Such materials are conceived to operate at high temperatures, near about half of the melting temperature, and can withstand high temperatures during consolidation without excessive loss of properties. Alloys resistant to high temperatures are also produced by rapid solidification, but here the selection of dispersoids of high stability must be compromised by the need to achieve a sufficiently high solubility in the liquid state before quenching. The high-temperature aluminium alloys are typically those containing transition metals (5), such as iron or chromium, and in these cases alloy stability is determined by the diffusion of these metals through the aluminium matrix. Such alloys are clearly not as stable as the mechanically-alloyed materials and property degradation during consolidation is a serious concern. From this point of view mechanical alloying is capable of producing more stable alloys simply because of the nature of the dispersoid particle.

Amorphisation during or subsequent to mechanical alloying is a recent discovery (6-12). The creation of an amorphous state can be regarded as an extreme case of metastable phase or structure creation, since the finely distributed or lamellar morphologies of the dispersion-strengthened alloys can also be regarded as metastable. Comparisons of the detailed structures of amorphous materials prepared by the two techniques (6,7,9) confirm that there are no important differences between materials prepared either way. Since this is the case, it is likely that the conditions for consolidating the mechanically amorphised powders will be similar to those used for melt-quenched glasses. These conditions must necessarily involve short times at as low temperatures as possible in order to maintain significant metastability.

In comparing the consolidation conditions for MA and RS powders it is important to note the different powder morphologies obtained by the two techniques. These are illustrated in Fig. 1. Powders prepared by the gas atomisation techniques are generally nearly spherical with smooth surfaces. Such powders flow freely to high loose-packing densities and are somewhat difficult to consolidate because of the limited mechanical interlocking and shear-induced bonding. Powders prepared by comminuting foils or ribbons obtained from quenching onto a substrate are generally blocky, the aspect ratio depending on the degree of comminution, and tend to have angular surface facets of high smoothness. These powders are consolidated similarly to the atomised powders. Powders prepared by MA are globally of equiaxed morphology and can flow to relatively high loose-packing densities. The density achieved is lower than for the gas atomised powders, partly because the milling process may leave imperfectly packed layers within the powders themselves. In addition, the surfaces of the powders are rough, such that during compaction there will be extensive mechanical interlocking and shear-induced bonding.

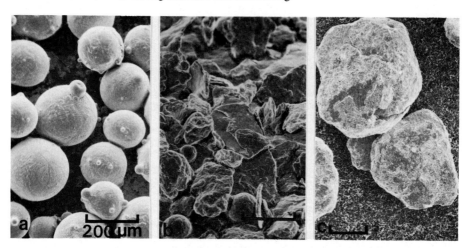

Fig. 1: Morphologies of typical RS and MA powders. The powders shown are (a) spherical gas atomised nickel alloy, (b) blocky, comminuted splat aluminium alloy, and (c) mechanically alloyed copper alloy. Global morphology is similar for each, but the surface roughness varies according to the fabrication method.

Compaction techniques

It is interesting to compare the various techniques which are used to consolidate RS and MA powders on the basis of the Stress or Pressure used to achieve compaction and the Time-Temperature excursion associated with the requirement of complete consolidation. Such a comparison for the more commonly used consolidation techniques is shown in Fig. 2. Long hold times at high temperatures are useful for obtaining full consolidation by diffusion bonding and by voidage removal, but will lead to significant loss of any metastability.

A second comparison of the consolidation techniques can be made on the basis of the types of forces imposed, specifically on the pressure and shear components of these forces. This is important because the shear component will tend to bring about intense cleaning and then bonding of the powder particle surfaces, whilst the pressure component will press the powder surfaces into contact without itself bringing about bonding. The ideal consolidation technique will use a sufficient hydrostatic pressure to achieve densification, for the temperature-time condition, whilst a large shear component will perform an important cleaning operation on the powder particle surfaces leading to effective bonding. Such a comparison of consolidation techniques is made in Fig. 3. It is clear that techniques such as Hot Isostatic Pressing and Dynamic Compaction, which use hydrostatic pressing conditions with limited shear, are not ideal from this point of view. The extrusion and forging processes make use of extensive shear, with the material under certain pressure, and appear to be among the best processes. The precise technique chosen for a given application will clearly depend on many factors: cost of material;final form required; the cleanliness of the powders and the extent of shear cleaning needed; acceptable force and temperature values to achieve the required degree of plasticity. The more interesting techniques would appear to be those of extrusion, when such shear extension is desirable or possible, forging and ROC (Rapid Omnidirectional Compaction).

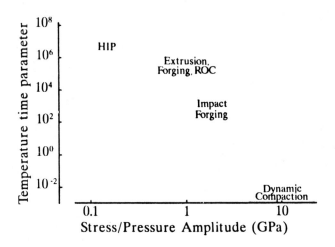

Fig. 2: Schematic comparison of techniques for consolidation of RS and MA powders on the basis of the Time-Temperature excursion and the Stress or Pressure used. The Time-Temperature parameter is simply the product of the time (s) and the temperature (K) used.

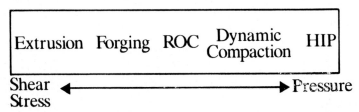

Fig. 3: Schematic comparison of techniques for consolidation of RS and MA powders on the basis of whether shear or pressure forces are applied.

The ROC process is described briefly here, on the grounds that the technique is poorly known and may be of significant interest for the consolidation of the MA powders. This technique essentially uses a forging press to create high pressures within a thick-walled, hot, solid die (13,14). The temperatures and forces are sufficient for the die to flow as a fluid, thus creating a nearly hydrostatic pressure. Several modifications of the basic process exist (15,16,17), to produce shaped parts using simple and cheap fluid dies. The process is being evaluated for several RS materials, for example for titanium alloys and superalloys (18,19).

Consolidation of a thermally-stable,dispersion-strengthened alloy system: comparison of RS and MA materials

In the following section, reference will be made to the copper-chromium system, specifically to the Cu-5at%Cr alloy. This alloy has been extensively studied both in the form of RS ribbons (20) as well as MA powders after extrusion (21) and after various heat treatments. More recent work on other copper-dispersoid systems has confirmed the essential results (22). The experimental conditions are described only briefly here, and further information may be obtained in the previously cited references.

The melt spun ribbon was cast under helium on a copper wheel rotating at a speed of 36m/s giving a ribbon 30 microns thick and 3mm wide. Mechanical alloying was performed by ball milling the elemental components in a Fritsch Mini-Planetary mill (Pulverisette 7) for 20 hours, when it is judged by metallographic examination that the chromium is sufficiently finely dispersed. Powders, both by ribbon comminution and by mechanical alloying, were consolidated by hot extrusion. The ribbon-powder was pressed to about 60% density in mild steel capsules which were evacuated for 15mins at the extrusion temperature before sealing, and further preheated for 15mins before extrusion. MA powders were cold pressed at 200MPa, giving about 75% density, and heated directly to the extrusion temperature for 25mins under an argon atmosphere. Extrusion was performed on a direct-acting press with the dies and chamber heated to 500 C using a semi-conical die with a reduction ratio of 10/1. Immediately after extrusion the product was water quenched. Mechanical properties were evaluated on bar and ribbon samples using standard techniques, and microstructures evaluated by optical and electron microscopy.

Fig. 4 shows the microstructures of Cu-5Cr ribbons and extruded MA material. The alloy is of hyper-eutectic composition, and it has not been possible to maintain all the chromium in solution by rapid quenching. The as-cast ribbon shows many fine chromium particles within a columnar-

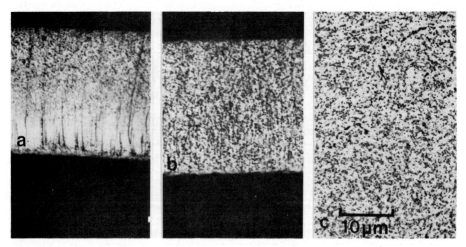

Fig. 4: Optical micrographs showing the microstructures of Cu-5Cr materials:(a) as-cast melt spun ribbon; (b) ribbon after heating for 1h at 800°C;(c) MA alloy extruded at 700°C.

grained microstructure. After ageing at high temperature, precipitation and coarsening of the chromium occurs. The microstructure of the extruded MA material is similar to that of ribbon heat treated for the same time and temperature, namely showing a similar distribution of particles.

It is interesting to compare the microstructures of the materials at a finer level, as shown in Fig. 5, which shows the structures of the Cu-5Cr alloy in ribbon form after heat treatment, in extruded ribbon form, and in the form of extruded MA material. In the extruded ribbon samples, there are some large particles but most of the particles remain much finer than in the ribbon heated for similar times and temperatures. In the MA materials there are no (or very few) large particles and the small particles have about the same size as in the extruded ribbons. It is thus seen that the fine-scale microstructure of each material differs slightly from the others, both in terms of grain size and retained dislocation structure, but also in terms of the extent of particle coarsening which has taken place.

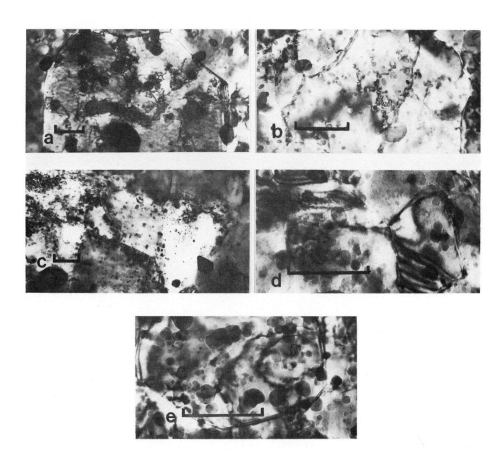

Fig. 5: Transmission electron micrographs;Cu-5Cr (a) ribbon after 1h 800°C; (b) ribbon-powders extruded 700°C; (c) ribbon-powders extruded 800°C; (d) MA powder extruded 700°C; (e) MA powder extruded 700°C, heated 1h at 900°C.

This difference in microstructure is also reflected in the mechanical properties of the different materials, as seen in Fig. 6. In each case the mechanical strength falls at the high temperatures because of coarsening of the microstructure, controlled by coarsening of the chromium particles. It is likely that the MA material as well as the extruded ribbon-powder is also somewhat strengthened by retained dislocation structures or by fine grain size, but as seen in Fig. 7 the essential cause of the additional strength of these materials over the heated ribbon is the finer particle size distribution maintained in this material. These factors are examined in more detail elsewhere (21,22).

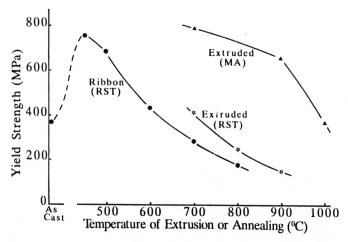

Fig. 6: Yield strength of Cu-5Cr in the form of heat treated ribbon, extruded ribbon-powder, and extruded MA powder.

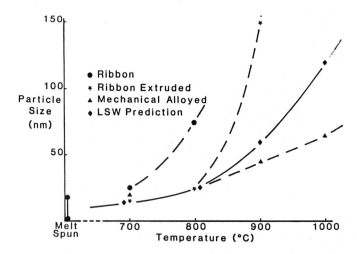

Fig. 7: Particle size in Cu-5Cr, heat treated ribbon, extruded ribbon-powder, and extruded and heat treated MA materials. Predictions (21) of particle size based on coarsening are shown for comparison.

It is clearly seen in Fig. 7 that the predicted and observed particle sizes are very similar for low annealing temperatures, but differ at the high temperatures. The point where the predicted and observed particle sizes begin to differ depends on the particular material examined, and notably is at a higher temperature and the difference is least marked for the MA materials. The reason for this difference between extruded ribbon-powders and MA powders is to be found in the size and distribution of the chromium particles present in the powders before compaction, and in the important role played by the few, large particles present. The RS process here has been unable to keep more than 2% chromium in supersaturated solid solution, and the remaining chromium is present in the form of particles about 50nm in size (20). The MA process here has managed to disperse almost 4% chromium in the form of very small particles, such that only about 1% is left as large particles, about 250nm in size. These large particles play a very important role at high temperature as sinks for solute from the nearby strengthening particles. This principle is illustrated in Fig. 8. The presence of many large particles (as is found in the RS material) means that there are many sinks available, close to each fine,strengthening particle, such that dissolution will readily occur, as soon as diffusion is able to transport the solute over these distances. As such, this RS material is less stable than the MA material at the high temperatures, over the time and temperature ranges where such diffusion occurs.

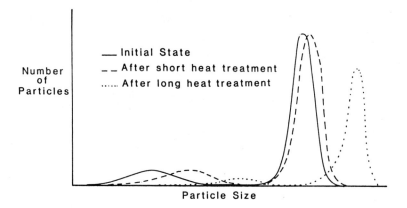

Fig. 8: Schematic particle size distributions during heat treatment of a material having an initial bimodal size distribution. The large particles act as sinks for the fine ones which gradually disappear.

Finally, a few comments should be made about the effectiveness of the consolidation achieved here by extrusion. The best judgement of this is made by examination of the fracture surfaces of broken test samples, see Fig. 9. It has been found that extrusion of the dispersion-strengthened copper alloys at temperatures of 700-900°C provides sufficient shear to achieve complete bonding, provided that the strength of the material does not exceed about 700-800MPa (22). When the material is stronger than this, the fracture surfaces show many cracks parallel to the extrusion direction where the individual powder particles have torn apart. Even under these conditions the material shows a global ductility (a few % before necking). In order to achieve full interparticle bonding it has been found advisable to heat treat, for example for 1h at 800-900°C. For those materials where the as-extruded strength does not exceed 700-800MPa, the fracture surfaces are always typical of ductile materials, with no sign of interparticle tearing.

150

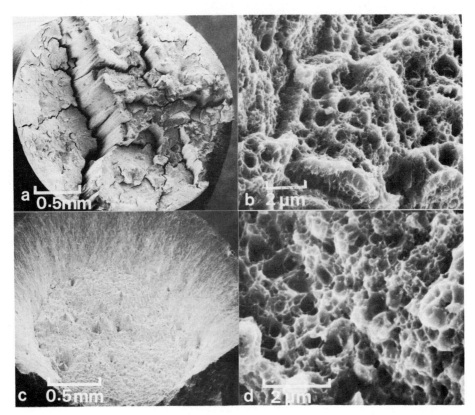

Fig. 9: Fracture surfaces of tensile samples of Cu-2Nb prepared from MA powder: (a) and (b) as-extruded at 700°C; (c) and (d) extruded at 700°C and annealed 1h at 800°C.

Consolidation of metastable, layered or amorphous materials prepared by MA

The experiments described here refer mostly to fine, layered powders of nickel and titanium prepared by ball milling the elemental powders. The objective has been to achieve consolidation of the intermetallic NiTi maintaining an extremely fine microstructure (23). Some additional experiments on amorphous powders relate to RS ribbons and flakes (24,25).

The powders were prepared by ball milling to different degrees of mixing and refinement, see Fig. 10. Initially the powders consist of coarse layers of the elements, and gradually the layer thickness is reduced, eventually down towards the atomic level of mixing. Milling must be continued at least to the level of mixing where diffusion during consolidation will be able to complete the homogenization process, but it may not be necessary to achieve atomic-level mixing by milling. Attempts were made to consolidate these different powders here, essentially using a hot uniaxial pressing or hot extrusion technique. Billets of milled powders were first cold pressed at 500MPa, preheated for 20mins to the desired temperature, and then consolidated. In some cases extrusion has been performed using a die reduction ratio of 10/1; in most cases the material is hot-pressed at

1.5GPa with a hold time of 1min under pressure. Some experiments at ROC or fluid-die pressing have been attempted for comparison, using a copper die. The powder is pressed at 100MPa and sealed into a copper can with wall thickness one third that of the die chamber: as such, during the subsequent pressing operation the powder is initially squeezed to full density and then held within the quasi-hydrostatic pressure created by the flow of the soft copper. A hold time of 1min is used and temperatures of 500 to 600°C.

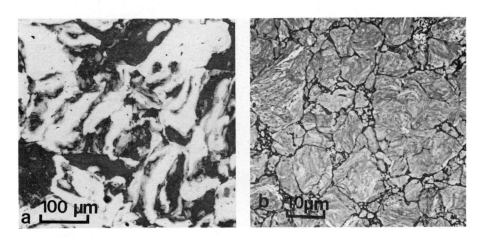

Fig. 10: Typical powders of NiTi produced by ball milling elemental powders for (a) 1h; and (b) 20h. Layered powders are created during milling and the layer thickness gradually decreases towards atomic dimensions.

One of the characteristics of the highly-metastable materials, such as those having heavily worked, fine lamellar or amorphous structures, is their high hardnesses. This makes compaction more difficult, since the pressure and temperature must be chosen such that the materials can flow readily under these conditions. Furthermore, it is only when the powder aggregate has a very high density that shear or diffusional bonding becomes effective. The difficulty associated with densification of hard materials is illustrated in Fig. 11. This figure shows the percentage density obtained when hot or cold uniaxial compacting powders of different initial hardnesses (measured cold). It is seen that to fully densify the hard microcrystalline or amorphous powders it is necessary to use high uniaxial pressures and temperatures, for example 1.5GPa and temperatures near or above 800°C: lower temperatures require significantly greater increases in pressure.

The use of temperatures as high as 800°C implies that fine-scale microstructures may be maintained in dispersion strengthened materials, but that highly metastable or amorphous structures will not be retained. The MA amorphous materials compacted by Krauss et al (26,27), for example, had crystallization temperatures typically about 400-500°C and showed signs of partial crystallization after compaction at 350°C. On the other hand, the retention of very fine grain sizes (about 10nm) in a dispersion strengthened intermetallic has been shown to be possible (23) after such high temperature consolidation.

The microstructure of a typical Fluid die compact is shown in Fig. 12. The powder used was the same NiTi powder prepared by 20h milling as that shown in Fig. 11 with a hardness of 1050Hv. The density obtained by the fluid die technique is about 92%, compared with 97% for the uniaxial compaction. Nevertheless, the fluid die sample appeared better consolidated, despite the large difference of temperature (500°C instead of 800°C), and showed no tendency for cracking across the direction of consolidation as did the uniaxially compacted samples. Even though further experiments are required to evaluate this technique, it appears to be possible to achieve fairly high densities in well-bonded powder samples while retaining highly metastable microstructures.

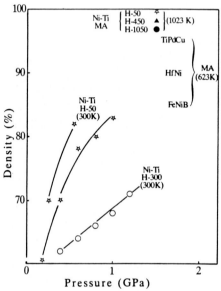

Fig. 11: Density achieved by cold or hot uniaxial compaction of powders of different hardnesses. Hardness is shown (H) as well as the temperature used for compaction (K). Some of the data on NiTi is taken from ref 28, and the data on MA amorphous materials from refs 26 and 27.

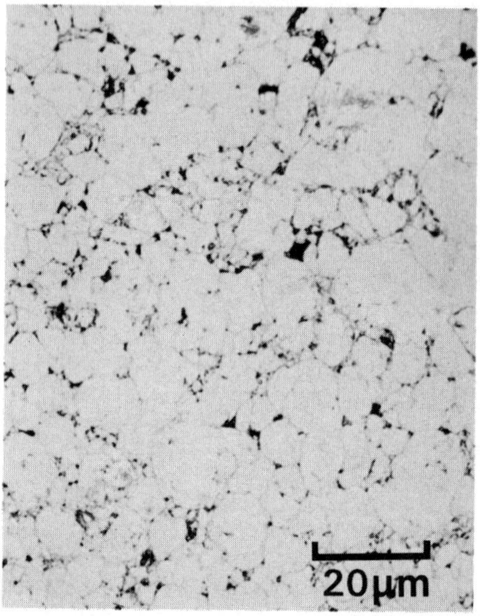

Fig. 12: Microstructure of Fluid die compaction of NiTi powders prepared by MA for 20h. Compaction took place in a copper die at 500°C with a pressure of 1.5GPa applied for 1min.

The use of shorter consolidation times, particularly with low temperatures implies that much higher pressures must be used. This was the initial reasoning behind the application of the dynamic compaction technique. This technique uses the high pressures created by a shock wave to densify the powders, while the shear occurring during compaction both cleans the powder surfaces and adiabatically heats these such that interparticle bonding is achieved. Compaction is usually carried out using impact techniques from gas gun launchers (24,25,29-31) or by explosives (32-35).

There are several difficulties associated with this technique. Extremely high shock pressures are required to consolidate the hard materials of interest here. Associated with these are very high relief waves, which can cause cracking in the solid sample unless they are controlled very carefully. Recent developments in the modelling of shock wave consolidation and new techniques for powder containment may lead to improvements here (35). A second difficulty arises from the need to utilize sufficient energy to bond powder particles but insufficient energy to cause microstructural degradation. On the one hand, high pressures are required to densify the powder aggregate, see Fig. 13. It has been shown that pressures about 2 to 3 times the material flow strength are required to achieve good densification (24). However, the total energy input may be sufficient to cause such a temperature rise that crystallization of an amorphous material can occur (29). On the other hand, it has been shown that a measurable amount of melting must occur, for typical powders and morphologies, in order for good bonding to be obtained giving a solid, tough material (24). In between these extremes of no melting, no bonding nor densification, and some melting, bonding and overheating, there are consolidation conditions where good bonding can be obtained but local regions of melting and slow cooling occur, leading to local crystalline regions, as illustrated in Fig. 13c. A slightly different example is shown in Fig.13d, where the extent of melting has been limited to an extremely low value but where crystallization has occurred along many of the interparticle surfaces as a result of solid state heating.

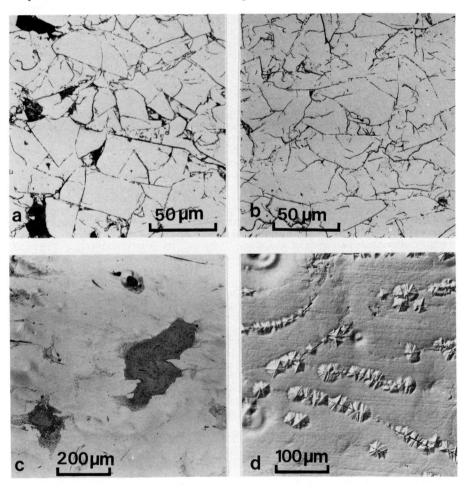

Fig. 13: Micrographs showing dynamically compacted glassy powders.(a) and (b) show Ni-7.5%Si-15%B glassy powders compacted at pressures of 2 and 4 GPa, respectively. (c) and (d) show Fe-40Ni-14P-6B glassy powders compacted at 5 and 4 GPa, respectively.

These examples illustrate the difficulties of finding the process "window" where good densification and bonding is obtained, but where the extent of global as well as localised heating and microstructural modification is limited to acceptable values. Despite the difficulties in defining this "window", it has been shown on several occasions that well-bonded, glassy compacted samples can be obtained. In view of the more attractive morphology of the MA powders, that is the overall equiaxed shape and the rough surfaces, it is likely that these may be somewhat more conducive to dynamic compaction than the typical RS spherical or plate powders.

Summary and Conclusions

All compaction techniques make use of a combination of temperature, pressure and shear. The temperature helps to diffusion bond the particles and to remove some of the porosity. The pressure provides densification, but does not itself help bonding. Shear helps to clean the powder surfaces and to bring about interparticle bonding. The choice of technique depends on the material as well as the application.

If the material is highly thermally stable, as for example the oxide dispersion strengthened alloys, high temperature techniques can readily be used. Such materials are clearly produced by the MA route since it is impossible to start with a suitable homogeneous liquid for RS.

For the medium-thermally-stable materials, for example the alloys dispersion strengthened by metallic particles (eg the Cr in Cu here), it is important to restrict the consolidation to sufficiently low temperatures to avoid excessive property degradation. Techniques of high shear, for example extrusion or forging are preferred for consolidation. In essence the MA and the RS powders are equivalent. The MA powders, however, are approximately equiaxed with rough surfaces, and both factors favour an easier consolidation. In addition, the milling operation ensures a more uniform microstructure than is often the case by RS: in the example described in this report it was impossible by RS to prevent the formation of large hyper-eutectic particles of chromium; during RS there is often a gradation of the scale of the microstructure (from one side to the other of the powder or ribbon) as the degree of undercooling changes. As shown in this report the initial uniformity of the microstructure can play an important role in determining the ultimate stability of the material.

For the highly metastable materials, such as the amorphous materials, it is the avoidance of excessive loss of metastability by high temperature excursions which is the most important criterion. Low temperature consolidation techniques such as dynamic compaction, ROC or forging seem to offer the best chances of consolidation. Based on the experiments reported here, in particular the ROC and the uniaxial pressing of the MA NiTi powders, it is clear that the high pressures and shears associated with ROC are highly beneficial for consolidation. Dynamic compaction offers the advantage of very high pressures and densification of hard materials at low temperatures, but obtaining sufficient shear and bonding without metastability loss is extremely difficult. Again, the more suitable morphology and surface roughness of the MA powders over that obtained by RS techniques may lead to easier consolidation of such highly metastable materials.

References

(1) P.S.Gilman and J.S.Benjamin, Ann. Rev. Mater. Sci. 13 (1983) 279.
(2) R.Sundaresan and F.H.Froes, J. Metals, Aug. (1987) 22.
(3) A.Layous, S. Nadiv and I.J.Lin, Powd. Met. Int. 19 (1987) 11.
(4) R.Schelleng, Metal Powd. Report, April (1988) 239.
(5) H.Jones, J. Mater. Sci. 19 (1984) 1043.
(6) C.C.Koch, O.B.Calvin, C.G.McKamey and J.O.Scarbrough, Appl. Phys. Lett. 43 (1983) 1017.

(7)   E.Hellstern and L.Schultz, Appl. Phys. Lett. 48 (1986) 124.
(8)   E.Hellstern and L.Schultz, Appl. Phys. Lett. 49 (1986) 1163.
(9)   C.Politis and W.L.Johnson, J. Appl. Phys. 60 (1986) 1147.
(10)   R.B.Schwarz and C.C.Koch, Appl. Phys. Lettl 49 (1986) 146.
(11)   E. Hellstern and L.Schultz, Phil. Mag. 56 (1987) 443.
(12)   E. Hellstern and L.Schultz, Mater. Sci. and Eng. 93 (1987) 213.
(13)   J.M.Wentzell, U.S.Patent No 3,982,934 (1976).
(14)   W.J.Rozmus, U.S.Patent No 4,142,888 (1979).
(15)   W.Smarsly and W.Bunk, Powd. Met. Int. 17 (1985) 63.
(16)   R.L.Anderson and J.Groza, Metal Powd. Report, April (1988) 272.
(17)   R.L.Anderson and J.Groza, Proc. Int. Conf. on PM Aerospace Materials,
        MPR Pub. Services, Shrewsbury, (1988) paper 37.
(18)   Y.R.Mahajan, D.Eylon, C.A.Kelto, T.Egerer and F.H.Froes, Powd. Met.
        Int. 17 (1985) 75.
(19)   R.F.Singer, Powd. Met. Int. 17 (1985) 284.
(20)   M.A.Morris and D.G.Morris, Acta Metall. 35 (1987) 2511.
(21)   D.G.Morris and M.A.Morris, Mater. Sci. and Eng. in press.
(22)   M.A.Morris and D.G.Morris, Met. Trans. in press.
(23)   D.G.Morris and M.A.Morris, Mater. Sci. and Eng. in press.
(24)   D.G.Morris, Mat. Res. Soc. Symp. Proc. 28 (1984) 145.
(25)   D.G.Morris, Rapidly Quenched Metals, Eds S.Steeb and H.Warlimont,
        North-Holland Pub., Amsterdam (1985) 1751.
(26)   W.Krauss, C.Politis and P.Weimar, Metal Powd. Report, April (1988)   231.
(27)   W.Krauss, C.Politis abd P.Weimar, Proc. Int. Conf. on PM Aerospace
        Materials, MPR Pub. Services, Shrewsbury, (1988) paper 49.
(28)   M.Igharo and J.V.Wood, Powd. Met. 29 (1986) 37.
(29)   D.G.Morris, Metal Sci. 14 (1980) 215.
(30)   D.Raybould, Proc. Conf. on High-Strain-Rate Phenomena in Metals,
        Eds M.A.Meyers and L.E.Murr, Plenum Press, NY (1981) 895.
(31)   T.Negishi, T.Ogura, T.Masumoto, T.Goto, K.Fukuoka, Y.Syono and
        H.Ishii, J. Mater. Sci. 20 (1985) 399.
(32)   C.F.Cline and R.W.Hopper, Scripta Met. 11 (1977) 1137.
(33)   M.A.Meyers, B.B.Gupta and L.E.Murr, J. Metals, 33 (1981) 21.
(34)   R.Prummer, Z. Werkstofftech. 13 (1982) 44.
(35)   R.N.Wright, G.E.Korth and J.E.Flinn, Adv. Mater. and Proc., Oct.
        (1987) 56.

Recrystallization of ODS Superalloys

C.P.Jongenburger and R.F.Singer, Asea Brown Boveri, Corporate Research, CH-5401 Baden, Switzerland.

1. Introduction

Oxide dispersion strengthened (ODS) superalloys exhibit a combination of excellent high temperature strength and good corrosion and oxidation resistance. This makes them attractive for application in environments where highly stressed components are exposed to high temperatures (>850°C) and aggressive gasses such as burning chamber components, gas turbine blades, etc. In the last 20 years a number of commercial ODS superalloys have been introduced that were made via mechanical alloying. MA 6000, MA 758, MA 754 and MA 760 (all nickel-base) and MA 956 (iron-base) are among the well known representatives of the class of mechanically alloyed ODS alloys (all MA-alloys are trademark of the INCO family of companies). After mechanical alloying the powder is consolidated by hot extrusion. In order to make the material suitable for use at high temperatures, the as-extruded, fine-grained microstructure needs to be transformed into a coarse grained microstructure. This is possible with a high temperature anneal during which secondary recrystallization occurs. For some of the alloys the recrystallization is performed under the influence of a high temperature gradient, which causes the formation of long grains with a high grain aspect ratio (GAR). This process is called directional recrystallization or zone annealing. The high grain aspect ratio imparts an additional strengthening effect upon the material.

Recrystallization in ODS superalloys has been subject to a number of recent investigations (1-8). In Section 2 some important aspects with regard to secondary recrystallization in ODS alloys are summarized. The aim of the present work was to obtain a better understanding of (i) what causes the onset of secondary recrystallization (the triggering mechanism, Section 3) and (ii) what influences the secondary grain size and morphology in isothermal and directional recrystallization, Section 4. In this work mainly the commercial alloys MA 6000 and MA 760 were studied, while some work was done on a number of experimental alloys, Table 1. All alloys belong to the class of nickel-base ODS alloys with a high γ'-content ($\sim$50%). ODS alloys belonging to other classes, (for instance, low γ' nickel-base like MA 754 and MA 758, or iron-base like MA 956) display a number of similarities in their recrystallization behaviour, but also some important differences (1). The conclusions of this study are therefore restricted to the high γ' nickel-base ODS alloys.

2. Secondary Recrystallization

2.1 Nature of Recrystallization

The MA 6000 microstructure after extrusion of mechanically alloyed powder is depicted in Fig.1. The characteristic features of the microstructure are (i) an extremely fine grain size of approx. 0.2 μm, (ii) a low average dislocation density, (iii) the absence of a texture, (iv) a fine dispersion of oxide particles and (v) γ'-precipitates ($\sim$50 vol-%) with a bimodal size distribution of which the larger precipitates have a size of about 0.2 μm. Upon heating to a temperature above approx. 1200°C, a few grains start to grow rapidly and consume the fine grains surrounding them. These coarse grains grow until they impinge upon each other and grain growth stops completely when all the fine grains have been consumed. This process is called secondary recrystallization or abnormal grain growth. Characteristic for secondary recrystallization is the fact that the starting microstructure is already recrystallized (low dislocation density), that grain boundary energy is the driving force and that the grain size distribution during secondary recrystallization is bimodal. The existence of a distinct temperature that has to be exceeded before abnormal grain growth can occur, is typical for recrystallization in high γ', ODS nickel-base alloys. This temperature is defined as the secondary recrystallization temperature, T_{sRx} (section 3.1). After T_{sRx} is exceeded, the recrystallization process is completed in a few seconds (section 3.2).

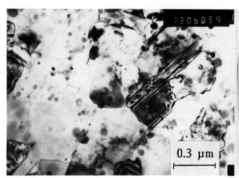

0.3 µm

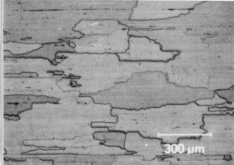

300 µm

Fig.1: TEM-Micrograph of as-extruded MA 6000

Fig.2: Grain structure of MA 6000 after isothermal recrystallization

Element [5] Alloy	Ni	Cr	Al	Ti	Ta	W	Mo	Co	B (ppm)	Zr	C	Y_2O_3	T_{sRx} (°C)	Remarks
MA 6000 [1]	bal.	15	4.5	2.5	2.0	4	2	-	100	0.15	0.05	1.1	1204	nominal composition
MA 760 [1]	bal.	20	6	-	-	3.5	2	-	100	0.15	0.05	0.95	1192	nominal composition
J2 [2]	bal.	20	6	-	-	1	2	-	100	0.15	0.05	1.1	1169	nominal composition
J3 [2]	bal.	17	6	-	2	3.5	2	-	100	0.15	0.05	1.1	1190	nominal composition
Alloy 10 [3]	bal.	19.4	5.8	-	-	3.6	2.0	-	nd	nd	0.042	0.66	1201	measured composition
TMO-2 [6]	bal.	5.9	4.5	2.5	2	12.4	2	9.7	100	0.05	0.05	1.1	1245	Ref. [7,8]
IN 738 + [6] Y_2O_3	bal.	15.8	3.4	3.4	1.75	2.5	1.75	8.4	-	0.1	0.08	1.5	1235	Nb:0.9% Ref. [6]

1) Trade Mark of the INCO family of companies
2) Experimental alloy produced by Metallgesellschaft
3) Experimental alloy produced by INCO
4) The iron content of the alloys varies generally between 1 and 2 weight-%
5) All values in weight-%
6) Not investigated in this study

Table 1: Composition of ODS Nickel-base superalloys with high volume fraction γ' (~ 50%).

2.2 The Secondary Grain Structure

The grain structure of MA 6000 after isothermal secondary recrystallization is depicted in Fig.2. The grain size is about 300 μm, which is about three orders of magnitude larger than the starting grain size. The grains show a slight elongation parallel to the extrusion direction, which is caused by stringers of inclusions that hamper transverse grain growth during secondary recrystallization (3).

2.3 Normal Grain Growth prior to Secondary Recrystallization

At temperatures below the secondary recrystallization temperature, T_{sRx}, normal grain growth takes place, driven by the grain boundary energy. During normal grain growth the grain size distribution stays unimodal. Contrary to abnormal grain growth, normal grain growth takes place in a wide range of temperatures; at low temperatures with correspondingly low growth rates. Normal grain growth is hindered by oxide particles and at temperatures below the γ'-solvus, mainly by γ'-precipitates. If too much normal grain growth takes place the primary grain size reaches a critical grain size, L_{crit}, beyond which no subsequent secondary recrystallization can take place (2,3). Apparently the driving force has become too low and the material has lost its potential for secondary recrystallization. Normal grain growth still takes place above L_{crit}, until the primary grain size has reached the Zener grain size (Section 2.4).

2.4 The Role of Dispersoids in Grain Growth

The grain boundary velocity can, in general, be described by the equation

$$G = M (P - P_P) \tag{1}$$

where G is the average grain boundary velocity, M the grain boundary mobility, P the driving pressure for grain growth and P_P the pinning pressure that the dispersoids exert upon the grain boundaries (9). Zener was first to give an expression for the maximum pinning pressure that a fine dispersion of stable particles can exert upon the grain boundaries (10):

$$P_{P,max} = 3/2 \, f\gamma/r_0 \tag{2}$$

where f is the volume fraction of dispersoids, γ the specific grain boundary energy and r_0 the radius of the dispersoids. The driving pressure is equal to the total amount of grain boundary energy:

$$P = 2 \, \gamma/L \tag{3}$$

where L is the mean grain intercept length. It may be noted that the driving pressure decreases during grain growth. Eq.(1) shows that grain growth will cease completely if $P = P_{P,max}$. The average grain size (mean intercept length) at which this happens is called the Zener grain size L_Z:

$$L_Z = 4/3 \, r_0/f \tag{4}$$

and thus,

$$P_{P,max} = 2 \, \gamma/L_Z \tag{5}$$

An important consequence of Eq.(4) is that a grain which has a mean intercept length in excess of L_Z cannot be consumed anymore by other grains.

The interaction of grain boundaries and particles has subsequently been treated more rigorously by Hillert (11), Gladman (12,13) and others (14-19). They found expressions for L_Z that differ slightly from Eq.(4). Work on MA 6000 shows that the Zener grain size is approx. 0.6 μm (3). Larger grains than 0.6 μm can only be obtained by abnormal grain growth or if the factor r_0/f changes, for instance, due to Ostwald ripening during long time exposure at high temperatures.

3. Triggering Mechanism of Secondary Recrystallization

3.1 Observation of T_{sRx}

Differential Thermal Analyis (DTA) was performed on several fine grained ODS alloys. Fig.3 shows a typical DTA curve. After an endothermal peak between about 1000°C and 1200°C, that represents γ'-dissolution, a sharp, exothermal peak can be observed at about 1230°C. The exothermal peak represents secondary recrystallization and is caused by grain boundary energy being released during secondary recrystallization. This is confirmed by the fact that a repetition of the DTA run does not show the exothermal peak again. The secondary recrystallization temperature, T_{sRx}, is the temperature at which the peak starts to rise. Table 1 gives the compositions and DTA results of a number of ODS alloys. Analysis of the results shows that the alloy composition seems to have a small influence on T_{sRx}. Although there are major differences in the composition, the value for T_{sRx} always lies close to 1200°C. Attempts to correlate T_{sRx} with the major elements all failed, which indicate that the major elements do not strongly affect T_{sRx}.

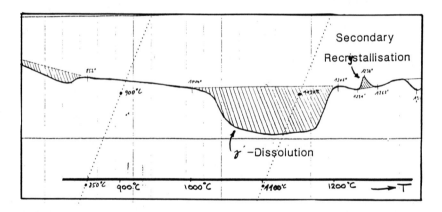

Fig.3: DTA curve of MA 6000

A comparison of the MA 760 and Alloy 10 gives valuable information about the influence of dispersoid content on T_{sRx}. Alloy 10 has the same chemical composition as MA 760, except for the dispersoid content which is nearly half of that of MA 760. The secondary recrystallization temperature, however, is virtually the same for both alloys. This would indicate that the dispersoids play little or no role in the recrystallization process. Another indication that the dispersoids have little influence on the recrystallization behaviour is found in the case of PM nickel base superalloys without dispersoids (20). Secondary recrystallization in these alloys takes place in a manner similar to the high γ' ODS alloys, although no (deliberately added) dispersoids are present.

The element boron has a strong influence on T_{sRx}. Qualitatively, this effect has been observed before (8,34) in experiments where an increased boron level was obtained via diffusional processes. In the present investigation the effect is quantified, Fig.4. It shows that T_{sRx} decreases with increasing boron content up to 200 ppm. Beyond 200 ppm boron T_{sRx} stays constant. Based on literature (21) it can be expected that at about 200 ppm the fine-grained microstructure is saturated with boron (mainly segregated at the grain boundaries) and that any extra boron will precipitate and form borides. This suggests that only boron in solid solution or segregated at the grain boundaries affect recrystallization. A similar observation has been made for secondary recrystallization in the alloy Fe-3Si (22,23).

3.2 The Grain Boundary Velocity

An experiment was carried out in which a specimen of unrecrystallized MA 6000 was rapidly heated to 1220°C using induction heating and immediately cooled down. The total time above T_{sRx} was 2.5 seconds. Metallographic investigation of the specimen revealed that secondary recrystallization had taken place and that large parts of the specimen had recrystallized completely. The largest secondary grain was 750 μm long. This implies that the grain boundary velocity, G, during abnormal grain growth was at least 150 μm/s. For comparison, the grain boundary velocity, G, during normal grain growth just before secondary recrystallization (fine grained MA 6000 just below T_{sRx}) is in the order of 0.1 nm/s (4), about six orders of magnitude lower, Fig.5. This drastic increase of the grain boundary velocity at the onset of secondary recrystallization is one of the characteristic features of recrystallization in high γ', ODS nickel-base alloys.

3.3 γ'-Dissolution

It has been suggested that for MA 6000 the dissolution of γ'-precipitates triggers secondary recrystallization (1,2). The disappearance of the grain growth impeding precipitates would suddenly allow grain boundaries to increase their velocity which, in turn, would allow the formation of coarse grains. In order to check this explanation the DTA results were used to compare the γ'-solvus and T_{sRx} for the investigated alloys, Fig.6. The results show that there is no correlation between the γ'-solvus and T_{sRx}. It can therefore be concluded that γ'-dissolution is not the general triggering mechanism for secondary recrystallization in high γ' ODS alloys (4). Independently, Mino and co-workers came to the same conclusion for the ODS superalloy TMO-2 (7). However, it is expected that the γ'-solvus forms a lower bound for the recrystallization because it is hard to imagine that grain boundaries can move with a high velocity in the presence of γ'-precipitates.

3.4 Dispersoid Coarsening

An alternative explanation of what triggers secondary recrystallization is that dispersoid coarsening is responsible for the observed increase of the grain boundary velocity (7). However, the results of a number of experiments make it unlikely that dispersoid coarsening is actually triggering secondary recrystallization:
1. DTA experiments on the alloys MA 760 and ALLOY 10 (which have the same composition, but differ in yttria content by a factor of 2) showed identical recrystallization behaviour. This shows that the triggering mechanism is rather insensitive to dispersoid size and volume fraction.
2. Dispersoid size distributions in MA 6000 were determined in three conditions: (i) fine grained (as-extruded) material, (ii) fine grained material quenched from just below T_{sRx} and (iii) coarse grained material quenched directly after exceeding T_{sRx} (4). Fig.7 shows that the dispersoid size distributions in the three conditions are virtually the same and that no dispersoid coarsening before or during recrystallization could be observed.

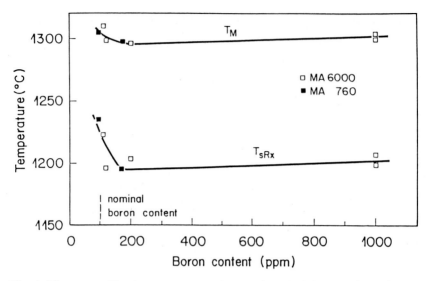

Fig. 4: The recrystallization temperature T_{sRx}, and the incipient melting point T_M, of MA 6000 and MA 760 with various boron contents.

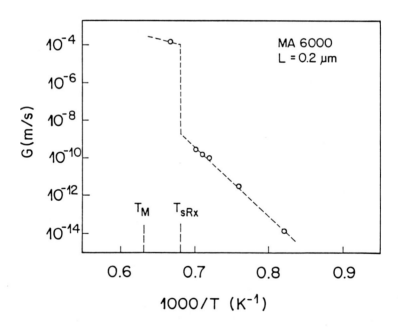

Fig. 5: The average grain boundary velocity of MA 6000 as a function of temperature. The abrupt increase of the grain boundary velocity with increasing temperature represents the transition from normal to abnormal grain growth.

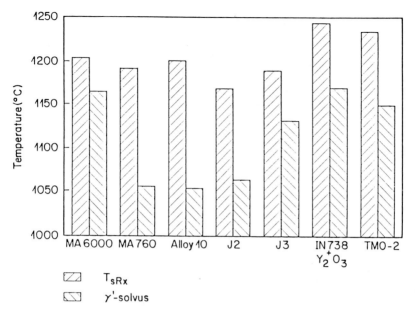

Fig. 6: γ'-solvus and recrystallization temperature for several ODS alloys.

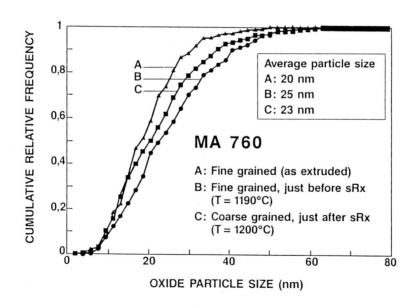

Fig. 7: Oxide particle size distribution in MA 760 measured in three different conditions. No significant dispersoid coarsening is observed.

3. It can be shown with Eqs.(1), (3) and (5) that dispersoid coarsening can at the most cause an increase of the average grain boundary migration rate by a factor of 2 to 3. It was observed, however, that the grain boundary migration rate increases by six orders of magnitude. Moreover, dispersoid coarsening is a very continuous process, which conflicts with the observation that the increase of the grain boundary velocity at the onset of secondary recrystallization is very abrupt.

3.5 Grain Boundary Break-Away

If neither γ'-dissolution nor dispersoid coarsening is triggering secondary recrystallization, then the question remains what is causing the onset of abnormal grain growth. In this section an alternative explanation will be given that is able to explain the phenomena observed during secondary recrystallization in high γ', ODS alloys.

One of the most characteristic features of the onset of abnormal grain growth is the sudden increase of the grain boundary migration rate by six orders of magnitude, (Fig.5). In principle, this can be explained by either an abrupt change in (i) pinning force P_P, (ii) driving force P or (iii) grain boundary mobility M, as is obvious by considering Eq.(1). The first possibility, a sudden decrease of the pinning force (for instance, γ'-dissolution or dispersoid coarsening), is unlikely in view of what has been discussed before. The second possibility, a sudden substantial increase of the driving force, can be ruled out, because it would require a major decrease in grain size. This leaves the third possibility which will form our basic hypothesis: the onset of abnormal grain growth is caused by a sudden change in the grain boundary mobility.

The second hypothesis is concerned with the nature of this sudden change. It is assumed that the sudden change of the grain boundary mobility is caused by an abrupt change in the dragging effect of solutes that are segregated at the grain boundaries, i.e. the so called break-away behaviour. The course of events at the onset of secondary recrystallization is envisaged to be as follows: At a specific temperature some grain boundaries of normally growing grains attain a critical velocity that enables them to break away from the cloud of segregated solute atoms that surrounds them. This increases the grain boundary mobility abruptly by some orders of magnitude, which allows the grain boundaries to migrate at a much higher rate than the grain boundaries of grains that are still undergoing normal grain growth and that have not yet reached the critical grain boundary velocity. Only the few fastest normally growing grain boundaries attain the critical velocity at the recrystallization temperature where they transform to segregation-free, high velocity grain boundaries. This and the fact that the primary grains (i.e. the potential nuclei) are consumed very fast explain the relatively large secondary grain size.

Early theoretical work on the grain boundary dragging effect of solutes in high purity metals was carried out by Lücke and Detert (24), Cahn (25) and followed by others (26,27). The theories predict that under certain conditions a travelling grain boundary is able to tear itself free from the accompanying solute cloud. This process is referred to as break-away of a grain boundary. A stationary grain boundary is laden with atoms that have a negative interaction energy with the grain boundary. If the grain boundary starts to migrate with a low velocity, the cloud of segregated atoms will move along and at the same time start to exert a dragging force on the grain boundary. The atoms are envisaged as moving behind the boundary and making their diffusion jumps toward the boundary. If the grain boundary velocity increases, a point may be reached where the diffusion speed of the atoms is no longer sufficient to keep pace with the boundary and the boundary is able to escape or break away from the segregated solute cloud. Consequently, the dragging force disappears abruptly which will cause an abrupt increase in the grain boundary migration rate G. Break-away is, in fact, the transformation from solute drag controlled migration to the intrinsic drag mode, in which the "chemical composition of the grain boundary" is identical to the bulk composition, i.e. no segregation at the fast moving grain boundaries. The corresponding grain boundary velocity is called the intrinsic grain boundary velocity.

Fig.8 shows results of the Cahn model, which in principle is valid for high purity metals with small additions of an impurity element. The figure shows that for small and medium driving forces there is, with increasing temperature, a gradual transition from solute drag controlled grain boundary migration to solute independent migration. For microstructures with a relatively high driving force, as encountered in the ODS alloys currently under investigation, an abrupt increase of several orders of magnitude is predicted for the migration rate when a certain temperature is exceeded. This is exactly what has been observed in MA 6000, Fig.5.

Break-away is not a process unique to grain boundaries. The occurrence of a yield point in, for instance, low carbon steels and serrated grain boundaries in iron and other alloys (strain ageing, Portevin - Le Chatelier effect) is explained by break-away of dislocations. Sessile dislocations are surrounded by a cloud of segregated nitrogen and carbon atoms. If the dislocations are forced to move they will be able to break away from the solute atmosphere which is reflected in a drop of the stress (yield point, strain ageing) or in the serrations of a tensile curve (dynamic strain ageing) (28).

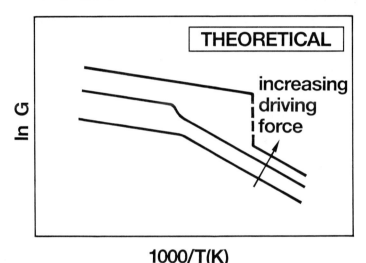

Fig. 8: Results of theoretical models: the grain boundary velocity G as a function of the temperature. With increasing temperature grain boundary migration changes from the solute drag controlled to the intrinsic drag mode. For microstructures with a high driving force this transition is rather abrupt.

The grain boundary composition as a function of the temperature is not known. It is likely that with increasing temperature the grain boundary active elements evaporate from the grain boundaries into the grains. This effect might play a determining role with regard to the triggering of secondary recrystallization.

If grain boundary break-away is indeed the triggering mechanism then the grain boundaries of the abnormally growing grains should attain the intrinsic grain boundary velocity (the grain boundary velocity in the absence of segregated solute atoms). Literature shows that measured intrinsic grain boundary velocities generally lie in the range 1-100 μm/s (29-31).

Regarding the relatively high driving pressure for grain growth in ODS alloys (even after sub-traction of the pinning pressure exerted by the dispersoids), the measured 150 μm/s fits well with these observations and it can be concluded that the grain boundaries of abnormally growing grains indeed move with the intrinsic grain boundary velocity, i.e. the fast moving grain boundaries during secondary recrystallization are free of segregated atoms.

The temperature at which grain boundaries will break away will be influenced by the "chemi-cal composition of the grain boundary". This explains why boron, which is known to segre-gate at the grain boundaries, has such a strong effect on the secondary recrystallization tem-perature, T_{sRx}.

It is interesting to note that secondary recrystallization occurring in silicon-iron, Fe-3Si, is also strongly influenced by grain boundary active elements (boron, sulphur and nitrogen) (22,23). It is not known whether the occurrence of secondary recrystallization in this alloy can be explained by grain boundary break-away as well.

It should be considered that the dispersoids are not only exerting a pinning force on the grain boundaries. A grain boundary that is closing in on a dispersoid experiences an attractive force that causes an acceleration of the grain boundary. Especially at higher migration rates, this effect might become noticeable and explain why the influence of the dispersoids on abnormal grain growth is so surprisingly small.

4. Secondary grain size and morphology

4.1 Isothermal Recrystallization

Samples of a round as-extruded bar of MA 6000 were given a heat treatment, henceforth called pre-anneal, below TsRx and the primary grain size (mean intercept length) was measured in a transmission electron microscope. The samples were subsequently isothermally

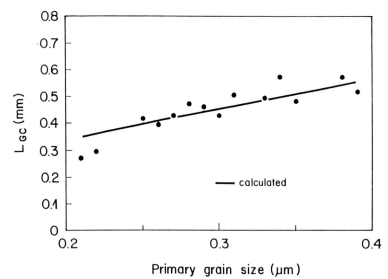

Fig. 9: The secondary grain size as a function of the primary grain size. The line repre-sents the calculated values according to eq.(13) with C = 915 m^{-1} and M_{in}/M_{sd} = 10^6.

recrystallized with the heat treatment 0.5h/1250°C/AC and the grain sizes in longitudinal and transverse direction were determined (L and l, respectively). The results are given in Table 2. The longitudinal grain size after pre-annealing and isothermal recrystallization varies between 0.4 and 1.4 mm, whereas the starting material has a secondary grain size (longitudinal) of 0.45 mm. Thus, one of the effects of pre-annealing is an increase of the secondary grain size. A correlation exists between the primary and secondary grain size in that a large primary grain size yields a large secondary grain size, Table 2 and Fig.9. In Fig.9 the profile diameter, L_{GC}, (the diameter of a round grain with a corresponding surface area) is used as the representative secondary grain size rather than the longitudinal grain size:

$$L_{GC} = \sqrt{\pi/4 \ L \ l}$$

This allows the comparison of grain structures with different grain aspect ratios.

The fact that the secondary grain size increases with increasing primary grain size (at the instant of secondary recrystallization) can be explained qualitatively with a rather simple model.

Heat Treatment		L_{prim} (μm) (calculated)	L_{prim} (μm) (measured)	L_{sec} (longitudinal) (mm)	l_{sec} (transverse) (mm)	GAR ($=L_{sec}/l_{sec}$)
T(°C)	t(min)					
1155	5	0.28	0.31	0.93	0.19	5.0
1155	15	0.34	0.25	1.28	0.20	6.4
1140	5	0.26	0.26	1.20	0.10	12.0
1140	15	0.30	0.30	1.10	0.13	8.15
1140	60	0.39	0.34	1.42	0.15	9.5
1110	15	0.25	0.26	0.74	0.18	4.1
1110	30	0.28	0.27	0.96	0.15	6.4
1110	60	0.31	0.34	1.06	0.19	5.6
1110	120	0.35	0.30	1.10	0.17	6.6
1110	180	0.38	0.34	1.18	0.22	5.4
1110	600	0.48	0.42	not recrystallized		-
1050	600	0.29	0.29	0.74	0.23	3.2
1050	1440	0.33	0.32	0.74	0.26	2.8
950	14400	0.23	0.23	0.41	0.17	2.4
As-received		0.215	0.215	0.45	0.13	3.5

* Calculated according to Eq.(1) in [4]

Table 2: The primary grain size of as-extruded MA 6000 after the indicated heat treatment and the secondary grain size after subsequent isothermal recrystallization (1/2 h/1250°C/AC).

The secondary grain size is a function of the grain boundary migration rate during abnormal growth, G, and the nucleation rate, N_v, which is defined as the number of secondary grains nucleating per unit time and unit volume of unrecrystallized material. It should be remarked here that nucleation is not the ideal term to describe the first phases in the development of a secondary grain. Generally, nucleation refers to the formation of a new phase, which is principally different from the surrounding matrix, like in solidification (the formation of a solid nucleus in liquid matrix) and primary recrystallization (the formation of an ordered nucleus in a heavily distorted matrix). At the onset of abnormal grain growth, however, it is believed that an already existing primary grain suddenly starts to grow to become a secondary grain. There is no principle difference between the constitution of the "nucleus" and that of the surrounding matrix, only the size is different. Nevertheless, we will use the term nucleation here in lack of a better alternative.

The classic work of Johnson and Mehl (32) shows that

$$n = 1.01 \ (N_v/G)^{1/2} \qquad (6)$$

with n being the total number of secondary grains per unit area on a plane of polish. If it is assumed that the grains are more or less spherical then:

$$n = 1 \ / \ (1/4 \ \pi D^2) \qquad (7)$$

and

$$L_{sec} = 2/3 \ D \qquad (8),$$

with D being the grain diameter and L_{sec} being the mean intercept length of the (secondary) grains. The combination of Eqs.(6), (7) and (8) gives L_{sec} as a function of the grain boundary velocity during abnormal grain growth, G_{agg}, and N_v;

$$L_{sec} = 3/4 \ \sqrt{\pi} \ (G_{agg}/N_v)^{1/4} \qquad (9)$$

Further, the combination of Eq.(1), Eq.(3) and Eq.(5) yields an expression that gives G as a function of the primary grain size:

$$G = M \ 2 \ \gamma \ (1/L_{prim} - 1/L_Z) \qquad (10)$$

where L_{prim} is the actual primary grain size and L_Z is the Zener grain size (mean intercept length, section 2.2). Eq.(10) is valid for both normal and abnormal grain growth. For abnormal grain growth the intrinsic grain boundary mobility, M_{in}, should be used and for normal grain growth the solute drag controlled mobility, M_{sd}, which is about six orders of magnitude lower than M_{in}, as was mentioned before.

The next step is to find an expression for the nucleation rate. Although nucleation is a complicated process with a strong statistical character, the understanding of the triggering mechanism, as described in Section 3, might lead us in the right direction. It was concluded that a potential nucleus actually develops into a secondary grain if one of its boundaries is fast enough to break away from its cloud of segregated solute atoms. For this reason we assume in a first approximation that the nucleation rate is proportional to the (average) grain boundary velocity during normal grain growth, G_{ngg}. Further, we may assume that each single primary grain is a potential nucleus. The number of nuclei that will actually develop is proportional to the number of potential nuclei, i.e. the number of primary grains per unit volume. This gives (assuming spherical grains with diameter D):

$$N_v \sim G_{ngg} \qquad \text{and} \qquad N_v \sim 1/(\pi/6 \ D^3)$$

or,

$$N_v = C \, G_{ngg} \, 6/\pi \, D^{-3} \tag{11}$$

or, in terms of the mean intercept length L_{prim} (Eq.8)

$$N_v = C \, G_{ngg} \, 6/\pi \, 8/27 \, L_{prim}^{-3} \tag{12}$$

with C being the proportionality constant.

Combining the equations (9), (10) and (12) gives the relation between the primary grain size L_{prim} and the secondary grain size L_{sec} for the case that the primary grain size, L_{prim}, is smaller than the critical grain size: $L_{prim} < L_{crit}$:

$$L_{sec} = 3/4 \sqrt{\pi} \; (C \, M_{in}/M_{sd} \, 6/\pi \, 8/27 \, L_{prim}^3)^{1/4} \tag{13}$$

$$= \sim 1.15 \; (C \, M_{in}/M_{sd})^{1/4} \, L_{prim}^{3/4}$$

For $L_{prim} < L_{crit}$ secondary recrystallization will not occur, Section 2.3. For $C=915$ m^{-1} and $M_{in}/M_{sd}=10^6$, Eq.(13) is in excellent agreement with experimental values, Fig.9.

Investigations of other batches of MA 6000 showed that apart from the primary grain size at the instant of secondary recrystallization the cleanliness and homogeneity of the material determines to a large extent the secondary grain size. The largest grains are obtained in very clean and homogeneous material.

4.2 Directional Recrystallization

Experiments

Directional recrystallization (or zone annealing) experiments were performed in which a hot zone was moved through 2 mm thin MA 6000 strips. Heating took place by means of a high frequency induction coil. In the experiments the speed of the strips and the temperature gradient were varied. High temperature gradients were obtained by withdrawing the strip from a water bath that was placed just below the induction coil. Low temperature gradients were obtained by placing an insulation jacket around the workpiece before it entered the induction coil, Fig.10. Actual temperature gradients obtained range from 8 to 40°C/mm. The speed was varied between 1.5 and 12 mm/min. The maximum temperature was measured with a ratio pyrometer and kept constant at 1280°C. A thermocouple attached to the strip determined the thermal history of the workpiece as it travelled through the induction zone. Transverse gradients were avoided by using only 2 mm thick specimens.

Results

After zone annealing the specimens were ground, polished, etched, and the longitudinal and tranverse grain sizes (mean intercept length) were determined. The temperature gradient at the recrystallization front was determined from the temperature - time curves. The results are presented in Table 3 and Fig.11. The process parameter that has the strongest influences on the resulting grain structure is clearly the speed. Both the longitudinal and transverse grain size, as well as their ratio (the GAR) show a decreasing tendency as the speed increases. The temperature gradient in the range between 8 and 40°C/mm shows no direct correlation with the grain structure which, at first sight, is rather surprising.

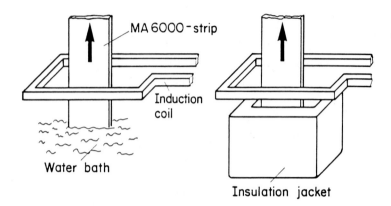

Fig. 10: Zone annealing experiments. Variation of the temperature gradient by cooling and insulation of the MA 6000 strip.

The longitudinal grain size

Fig. 11 shows that three different speed ranges can be observed. The occurrence of these three speed ranges can be explained as follows. During steady state zone annealing the secondary grains at the recrystallization front grow with the same velocity as the speed of the specimen, i.e. a certain grain growth rate is forced upon the secondary grains. At low speeds, the growing (secondary) grains have no problem of keeping up with the (moving) T_{sRx}-isotherm, because the grains grow at a rate slower than their "natural" rate of grain growth, as encountered in isothermal recrystallization experiments, Fig. 12a. The temperature in front of

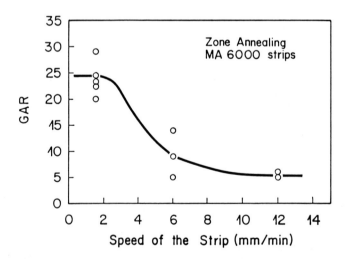

Fig. 11: The grain aspect ratio (GAR) as a function of the speed of the strip during zone annealing.

the recrystallization front is lower than T_{sRx} and nucleation of new grains is not expected. In principle, the grains will grow through the whole specimen (unless the adjacent grains interfere) and the longitudinal grain size is only limited by the size of the specimen. Indeed, it can be observed that most grains extend from one end of the specimen to the other.

If the speed is increased, the secondary grains are forced to grow faster and faster until the point is reached where the rate of grain growth attains its "natural" (maximum) value. Above this maximum value the grains cannot keep up with the T_{sRx}-isotherm any longer and the recrystallization front will be shifted to a higher temperature region, Fig.12b. This will increase the grain growth rate until the grains at the recrystallization front are growing at the same rate again as the speed of the specimen but at a temperature higher than the recrystallization temperature. The region between the recrystallization front and the T_{sRx}-isotherm is fine grained and since the temperature is above T_{sRx}, nucleation of new grains may take place. The secondary grains in the recrystallization front will impinge upon freshly nucleated grains and not be able to grow any further. Therefore, the longitudinal grain size will decrease. At still higher speeds, there is no way that the recrystallization front can keep up with the speed, not even at temperatures above T_{sRx}. In this case, the recrystallization front as such will disappear and everywhere in the hot zone nucleation will take place. The longitudinal grain size will be comparable to grain sizes after isothermal recrystallization, Fig.12c.

v (mm/min)	G (°C/mm)	L (µm)	l (µm)	GAR (=L/l)
1.5	31	5285	182	29
12.0	37	707	117	6
6.0	40	1769	130	14
1.5	39	4023	171	24
1.5	18	4030	205	20
12.0	23	670	130	5
6.0	15	1136	128	9
1.5	14	6400	266	24
6.0	14	587	132	5

Table 3: Results of zone annealing experiments.

Thus, the three different speed ranges can be summarized as follows (Fig.12):
- **Range A:** (for MA 6000: approx. 0-4 mm/min) The speed of the bar is lower than the "natural" grain growth rate at the recrystallization temperature, T_{sRx}. Nucleation of new grains is not expected and the longitudinal grain size is limited by the specimen size. The highest grain aspect ratios are reached in this range.
- **Range B:** (for MA 6000: approx. 4-9 mm/min) The speed is sligthly higher than the "natural" grain growth rate at T_{sRx}. The recrystallization front will be shifted to a higher temperature. In the region between the recrystallization front and the T_{sRx}-isotherm nucleation may occur. The longitudinal grain size will be smaller than in "Range A".

– **Range C**: (for MA 6000: approx. >9 mm/min) The speed is much higher than the attainable grain growth rates. There is no distinct recrystallization front which makes the situation very similar to isothermal recrystallization. This is also reflected in the average longitudinal grain size which is comparable with those obtained in isothermal recrystallization experiments.

The practically important range of speeds is Range A. In this range the highest grain aspect ratios are expected. The transition from this range to Range B is determined by the natural rate of grain growth which, according to isothermal recrystallization experiments, for MA 6000 is approx. 150 μm/s (9 mm/min), section 2.4. Interpretation of the zone annealing results give a somewhat lower value, namely between 1.5 and 6 mm/min. The longitudinal grain sizes measured after zone annealing at 12 mm/min are similar to the grain sizes observed after isothermal recrystallization (< 1 mm).

The transverse grain size

The transverse grain size, l, is mainly determined by the material itself and not so much by the process parameters during zone annealing. This cannot be deducted from the results

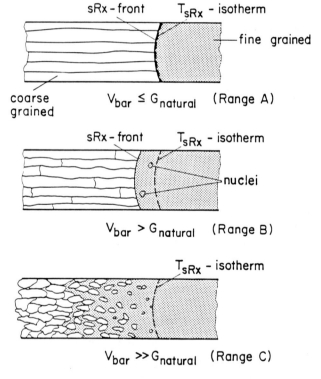

Fig. 12: The formation of secondary grains with increasing speed of the strip during zone annealing.

presented here, but becomes clear if various batches of MA 6000 are compared with each other or if MA 6000 is compared with MA 760. The transverse grain sizes yielded by the zone annealing experiments of MA 6000 vary between 100 and 300 μm. In earlier batches transverse grain sizes in excess of 3 mm were not unusual (see for instance Fig.1 in (33)). In MA 760 the transverse grain size after zone annealing normally lies around 2 mm. The reason for these differences may be found in the cleanliness and homogeneity of the material. Fig.13 shows the recrystallization front in MA 6000 after quenching. The light etching reveals the elongated prior particle boundaries in the fine grained area. At the recrystallization front it can be seen that secondary grains tend to grow within the prior powder particles. The prior particle boundaries form obstacles for transverse grain growth and as such influence the transverse grain size.

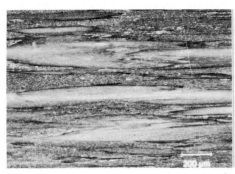 

a)   b)

Fig. 13: The microstructure of an MA 6000 strip after quenching during zone annealing. The light etching reveals the prior particle boundaries.
a) Fine grained just in front of the recrystallization front.
b) The recrystallization front (secondary grains at the right).

5. Conclusions

The following main conclusions can be drawn with regard to secondary recrystallization in high γ', ODS nickel-base superalloys:

1.   The onset of secondary recrystallization can be explained by grain boundaries breaking away from the atmosphere of segregated atoms that surround them.

2.   The secondary grain size after isothermal recrystallization increases with increasing primary grain size. This is explained by the fact that the number of potential nuclei decreases with increasing primary grain size. Consequently less nucleation will take place and the resulting secondary grain size will be larger.

3.   The longitudinal grain size and the grain aspect ratio after directional recrystallization (zone annealing) are strongest influenced by the speed at which the workpiece is moved through the hot zone. The temperature gradient in the range 8-40°C/mm has no significant influence. The transverse grain size is strongly influenced by the cleanliness and homogeneity of the material.

174

References

(1)   R.K.Hotzler and T.K.Glasgow: in J.K.Tien et al. (eds.); Conf.Proc. Superalloys 1980, American Society for Metals, Metals Park, Ohio, USA, 1980, 455.
(2)   R.K.Hotzler, T.K.Glasgow: Met.Trans.A 13A (1982) 1665.
(3)   R.F.Singer and G.H.Gessinger: Met.Trans.A 13A (1982) 1463.
(4)   C.P.Jongenburger and R.F.Singer: Conf.Proc. 1st ASM Europe Technical Conference, Sept.1987, Paris.
(5)   R.L.Cairns, L.R.Curwick and J.S.Benjamin: Met.Trans.A 6A (1975) 179.
(6)   G.H.Gessinger: Met.Trans.A 7A(1976)1203.
(7)   K.Mino, Y.G.Nakagawa and A.Ohtomo: Met.Trans.A 18A (1987) 777.
(8)   K.Mino, K.Asakawa, Y.G.Nakagawa and A.Ohtomo: Conf.Proc. PM Aerospace Materials, Bern, November 12-14, 1984.
(9)   M.S.Masteller and C.L.Bauer in; F.Haessner (ed.): Recrystallization of Metallic Materials, Riederer Verlag, Stuttgart, 1978, 254.
(10)   C.Zener: quoted by C.S.Smith; Trans.Met.Soc. AIME 175 (1948) 47.
(11)   M.Hillert: Acta Metall. 13 (1965) 227.
(12)   T.Gladman: Proc.Roy.Soc. (London), 294A (1966) 298.
(13)   T.Gladman: in N.Hansen et al. (eds.); Conf.Proc. 1st Riso Int. Symp. on Metallurgy and Materials Science, Roskilde, Denmark, Sept. 8-12, 1980, 183.
(14)   K.G.Wold and F.M.Chambers: J.Austr.Inst.Metals 13 (1986) 2,79.
(15)   C.J.Tweed, N.Hansen and B.Ralph: Met.Trans.A 14A (1983) 2235.
(16)   C.J.Tweed, B.Ralph and N.Hansen: Acta Metall. 32 (1984) 1407.
(17)   M.F.Ashby, J.Harper and J.Lewis: Trans.Met.Soc. AIME (1969) 413.
(18)   P.M.Hazzledine, P.B.Hirsch and N.Louat: in N.Hansen et al. (eds.); Conf.Proc. 1st Riso Int. Symp. on Metallurgy and Materials Science, Roskilde, Denmark1980, 159.
(19)   E.Nes, N.Ryum and Hunderi: Acta Metall. 33 (1985) 11.
(20)   R.V.Miner: Report Lewis Research Center, Report No. NASA TM X-2545, April 1972.
(21)   T.J.Garosshen, T.D.Tillman and G.P.McCarthy: Met.Trans.A 18A (1987) 69.
(22)   H.E.Grenoble: IEEE Trans.Mag. MAG-13 (1977) 5,1427.
(23)   H.C.Fiedler: Met.Trans.A 8A (1977) 1307.
(24)   K.Lucke and K.Detert: Acta Metall. 5 (1957) 628.
(25)   J.W.Cahn: Acta Metall. 10 (1962) 789.
(26)   P.Gordon and R.A.Vandermeer: in Recrystallization, Grain Growth and Textures, Seminar Proc. ASM, 16-17 Oct.1965, Metals Park, Ohio.
(27)   M.Hillert and B.Sundman: Acta Metall. 24 (1976) 731.
(28)   G.E.Dieter: Mechanical Metallurgy, McGraw-Hill, New York, 1986, 201.
(29)   K.T.Aust and J.W.Rutter: Trans. AIME 215 (1959) 119.
(30)   J.W.Rutter and K.W.Aust: Acta Metall. 13 (1965) 181.
(31)   P.Gordon and R.A.Vandermeer: Trans. AIME 224 (1962) 917.
(32)   W.A.Johnson and R.F.Mehl: Trans.AIME 135 (1939) 416.
(33)   W.Hoffelner and R.F.Singer: Met.Trans.A 16A (1985) 393.
(34)   I.A.Bucklow: Interim report 3785/3/85, COST 501, Project UK5 - The Welding Institute, Cambridge.

Problems Related to Processing of Ferritic ODS Superalloys

G. Korb, Metallwerk Plansee, Reutte, Austria

Introduction

The well known iron-base ODS alloys are primarily used for applications in the field of high-temperature stress together with hot-gas corrosion e.g. for fabrication of burner chambers, burner tubes, nozzles etc. Requirements, with respect to quality and long-term thermo-shock resistivity, are increasing dramatically. Therefore, improvement of properties mentioned above and minimization of respective failures are an urgent requirement. Some of the "deseases" of ODS materials are well known and accepted but at temperatures applied today (1350°C and above), faults will develop very quickly and may shorten the life time of respective parts in an inadmissable manner. This paper deals mainly with one interesting phenomenon, the formation of pores within fully-dense material at high annealing temperatures (or during application) for long times of exposure.

Experimental and observed phenomena

Iron-based ODS superalloys produced via the PM-route (mechanical alloying, encapsulation, thermomechanical treatment) demonstrate a significant feature when deformed and recrystallized at high temperatures: correct secondary recrystallization to the required elongated coarse grain structure is only achieved when the thermomechanical treatment is carried out in a special limited range of temperatures and reduction steps (Figs. 1,2,3).

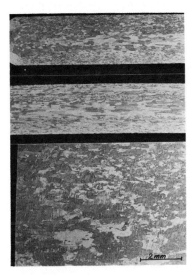

Fig. 1: Microstructure of ferritic ODS alloy processed under too high a temperature

Fig. 2: Microstructure of ferritic ODS alloy processed under lower temperatures and more specific conditions of deformation

Fig. 3 shows an "optimum microstructure" of a profile-rolled bar with a grain aspect-ratio from about 50:1 to 100:1 and only a few grains over the entire cross section. Excellent stress-rupture properties in the temperature range of 900 to 1200°C were measured.

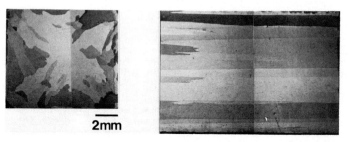

Fig. 3: Microstructure of ferritic ODS alloy processed under optimum conditions

When specimens are kept at the recrystallization temperature of e.g. 1350°C for a longer period (2 to 200 hrs), pores appear and significant growth of these cavities takes place. This phenomenon can be observed, by optical microscopy, only in coarse-grained specimens. Pore growth in its earlier stages depends also on prior forming operations and the final grain size within the examined section.

It must be assumed that either pores form solely during grain coarsening or develop from micropores, initially present within the fine grains, agglomerate and grow during the heat treatment. Fig. 4 gives an SEM fracture micrograph which demonstrates that small pores are already present in the primary microstructure.

Fig. 4: SEM micrograph of fracture surface in fine grain ODS Alloy

Figs. 5 to 7 demonstrate how pore growth depends on annealing time and material. One can clearly see that all materials made via the PM route contain porosity. Contrary to this, molten, deformed and recrystallized FeCrAl does not contain any porosity at all, even when heat-treated for an extremely long time. In the case of alloy MA 956 earlier investigations (1) claimed that porosity results from the argon content of this material originating from mechanical alloying under argon as a protective atmosphere. In fact the argon content of

MA 956 was found to be about 30 ppm (table 1) but all other specimens did not contain any inert gas at all. Since they all develop the pores likewise, argon cannot be the unique reason for pore formation.

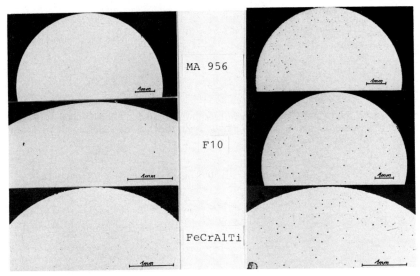

MA 956

F10

FeCrAlTi

Fig. 5: Porosity after 2 hrs heat treatment (MA 956, F10 and FeCrAlTi)

Fig. 6: Porosity after 16 hrs heat treatment

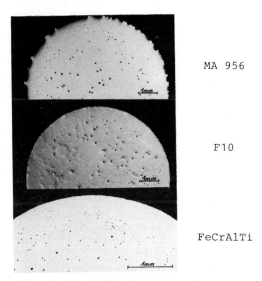

MA 956

F10

FeCrAlTi

Fig. 7: Porosity after 128 hrs heat treatment

Table 1:   Argon contents of materials

Material	MA 956 PM	F10 PM	FeCrAl PM	FeCrAl cast mat.
ppm Ar	30	0	0	0

Interstitial impurities

Trace analysis of C, N, O and H gave the results shown in table 2. Oxygen content is different, but the values for hydrogen, which have be considered more closely later on, are very similar for all materials.

Table 2.   Content of interstitial elements in different FeCrAl-materials

	MA 956 PM	F10R PM	F10L PM	FeCrAlTi* PM	FeCrAl* PM	FeCrAl* Aluchrom (cast)
O_2 (%)	0.20	0.42	0.18	0.016	0.02	0.01
N_2 (ppm)	280	300	300	190	150	95
H_2 (ppm)	8	8	5	7	6	4
C (ppm)	150	180	250	105	450	125

*   without Y_2O_3, not milled

After long exposure times (128 hrs) at 1350°C in air, the hydrogen content decreases to below 1 ppm. Additional variation of conditions, as e.g. the heat treatment atmosphere (hydrogen, inert gas, air or vacuum) has no significant influence on the formation of pores.

An early technological experiment was carried out with the aim to prevent pore formation – assuming they originate from chemical impurities and/or chemical reactions at 1350°C. For this purpose powder compacts with a density of appr. 50% (i.e. with totally open porosity) were sintered at 1350°C under a vacuum of 10^{-5} mbar. They were then processed to final, fully densified bodies by extrusion.

Heat treatment of these recrystallized bars gave a reduction in porosity only in the earlier stages of the treatment (2 hrs, 16 hrs, Fig. 8, Fig. 9). After 128 hrs a similar picture can be observed, irrespective of whether the specimens were vacuum sintered or not. The same results were also found when the bars were sintered under hydrogen.

Changes of density

Exact measurements of the density of heat treated specimens are given in Fig. 10. Cyclic annealing causes a much faster and significant density decrease of up to 1 or 2% After a larger number of cycles, the curve seems to lead to a more or less constant value, whereas isothermal treatment causes a smaller loss of density. This behaviour contradicts the hypothesis that a gas pressure within the pores leads to pore growth assuming a constant gas volume.

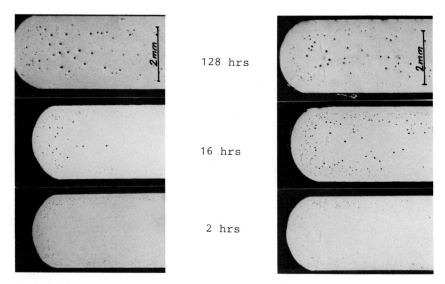

Fig. 8: Pore growth after preceding
1350°C vacuum sintering step

Fig. 9: Pore growth without preceding
sintering step

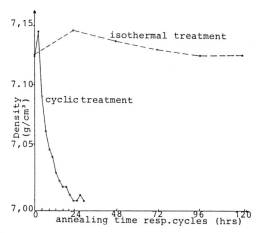

Fig. 10: Density of specimens versus time of annealing, resp. cycles

Micrographs, SEM investigations

SEM observations of the pores on fracture surfaces yielded the following results: in the early stages of their development the pores exhibit different geometries, i.e. more or less with irregular or polyhedral surfaces. In the final stage all pores have an approximately spherical shape.

Characteristic "rocks" of relatively coarse particles with a typical size of appr. 0.2-2 μm can be found on such inner surfaces. These particles were identified as Al_2O_3, $Al_2O_3 \cdot Y_2O_3$ or TiN respectively (Fig. 11).

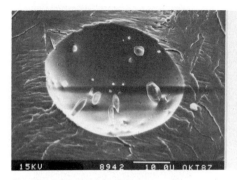

Fig. 11: Inner surface of a pore with nonmetallic particles

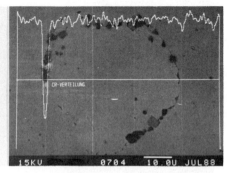

Fig. 12: Circularily arranged oxides

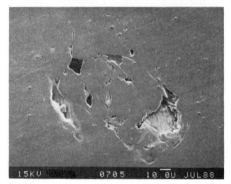

Fig. 13: Partially filled pores

Further interesting structures can be observed relatively often in the longtime treated specimens by optical and SEM methods: the oxides are arranged in circles (Fig. 12) and sometimes structures are found where either a pore is probably at the beginning of formation or near "closing" by material diffusion (Fig. 13). In any case the chemical composition within such "rings" was found to be identical to the matrix material.

Considerations and hypotheses for possible mechanisms of pore formation

The following facts have to be considered as certain:

- All PM FeCrAl specimens exhibit the described porosity, even those not containing yttria and/or titanium and those not mechanically alloyed. In the latter cases atomized FeCrAl powder was only densified, deformed and recrystallized.

- FeCrAl alloys fabricated via the melting route are not prone to pore formation under the conditions described.

- After remelting, PM products do not exhibit pore formation.

- Hence the porosity cannot be a phenomenon of the FeCrAl alloy system itself.

- The chemical composition of both the PM and molten materials is within the same range also with respect to interstitial impurities.

- Previous vacuum degassing of the powder at 1350°C and 10^{-5} mbar does not prevent porosity or influence the final appearance of the pores.

One hypothesis related to the small increase of density during the initial treatment at 1350°C (Fig. 10) was that pore formation is closely associated with the secondary recrystallization process. To check this, a specimen was recrystallized under HIP conditions, i.e. 1350°C and 2000 bars. This initially lead to a pore-free material, but a subsequent additional heat treatment under normal conditions again caused pores. Hence, this hypothesis does not hold. Another hypothesis was that the pores could be caused by an internal gas pressure, but thermodynamic calculations showed (2) that no reaction can be found within the given system for developing a sufficiently high pressure. Only molecular hydrogen could possibly cause such pressures, but it was found by calculation that the permeation of hydrogen through the alloy to the surface should lead to equalization of pressures within seconds.

Outgassing experiments under ultra high vacuum conditions gave a significant increase of hydrogen pressure in the range of 1100°C to 1250°C (3) but at present it is not yet clear whether this is an effect of the surface or whether it is due to hydrogen release from the inner parts of the specimen. Mass spectroscopy studies of fracture experiments under ultra high vacuum conditions are in preparation.

One of the most conspicuous facts is the close connection between pores and relatively coarse oxides. Together with the above mentioned fact that only PM products exhibit the described behaviour, the considerations must lead to the conclusion that these non-metallic particles can be the cause of pore formation. In order to check this, the following experiment was carried out: two slightly oxidized surfaces of "Aluchrom" (FeCrAl produced via the melting route) were hot rolled together so that a "cold welding seam" was the result which naturally included an oxide layer.

A subsequent heat treatment at 1350°C produced the well-known pores exactly on the line of the seam (Fig. 14 and Fig. 15). Fig. 16 gives an SEM photograph which shows that the oxides on the inner side of the pore can be found exactly along the cold welding seam.

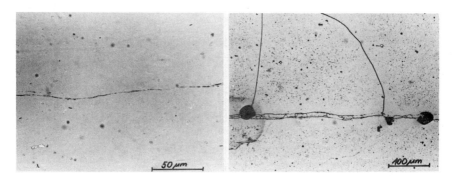

Fig. 14: Cold welding seam between two pieces of "Aluchrom"

Fig. 15: Pores along the cold welding seam in "Aluchrom"

Hence it is proved that oxides on or within the PM particles are at least one initial cause for the pores in the FeCrAl system. It can be suspected that is is mainly Al_2O_3 which causes the described phenomena.

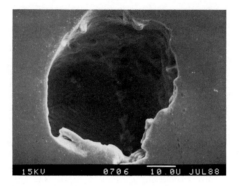

Fig. 16: Oxides along the cold welding seam within the pore. SEM photograph

The growth mechanism responsible for expanding the initial cavities is not yet totally clear, but the experiments described allow only two possibilities:

1. The initial stage may be due to a pure thermo-physical process like agglomeration of stacking-faults.
2. Chemical processes related to the mentioned hydrogen and/or hydrogen-water equilibrium in connection with the oxides might then be the reason for further pore formation. This is not yet certain and additional experiments have been started.

Conclusions

The influence of the described porosity on long-term properties like cyclic thermal shock resistance was found to be detrimental in MA 956 (1). Argon-free materials give much better results in spite of porosity. Although exact data are not yet available up to now a negative influence on stress-rupture properties at high temperatures must also be expected. On the other hand it is well known from PM parts technology that a controlled level of porosity does not necessarily have a detrimental effect on crack properties because crack propagation can be even arrested by pores.

In order to totally prevent porosity, no definite and reliable solution can be given at present. In any case, inert gas should be avoided. Pore formation could be reduced considerably if coarser oxides (0.1 to 1 μm, especially in the form of agglomerates) can be prevented and if the oxygen content can be kept at a low level.

Use of - of course, more expensive - ultra pure master-alloys and more intensive mechanical alloying processes (i.e. extremely intensive high-energy ball mills) could help to attain the two latter aims.

References

(1) D. Hedrich, Internal report (Daimler-Benz), not published
(2) P. Ettmayer, D. Sporer, Internal report, Technical University of Vienna, not published
(3) B. Heine, Internal Reports, MATFO-Programm, not published

C:
MECHANICALLY ALLOYED
MATERIALS AND THEIR PROPERTIES
C. 1:
ODS SUPERALLOYS

185

185

185

High Temperature Properties of Dispersion Strengthened Materials Produced by Mechanical Alloying: Current Theoretical Understanding and Some Practical Implications

Eduard Arzt, Max-Planck-Institut für Metallforschung, Stuttgart

Abstract

The typical effects of dispersoid particles and grain structure on creep of dispersion-strengthened materials are described. The current theoretical understanding of dislocation processes in these materials is reviewed extensively. It is argued that recent models based on dislocation detachment from dispersoid particles can have important practical implications for future alloy design and optimization. For creep data extrapolation, threshold stress concepts should be used with caution. Achievements and shortcomings of current models for grain boundary effects are also discussed. It is concluded that the creep behaviour of polycrystalline dispersion-strengthened materials is not fully understood and necessitates further work.

1. Introduction

Dispersion-strengthened materials for high-temperature applications were the main objective behind the development of "Mechanical Alloying" (1,2). The process as invented by Benjamin (3) involves high-energy ball milling of the starting powders together with the dispersoid particles, followed by consolidation and thermomechanical processing (for a recent review see Singer and Arzt (4)). In a similar process, termed "Reaction Milling", Jangg et al. (5-7) succeeded in forming well-distributed dispersoid particles by chemical reaction of milling additions with the powder. Typical dispersoid microstructures obtained by these two methods are shown in fig. 1.

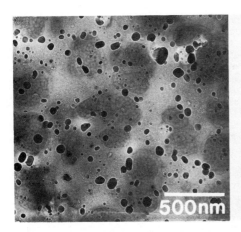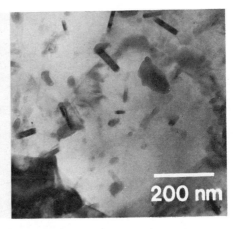

Fig. 1: TEM micrographs of two modern dispersion-strengthened engineering materials; left: the nickel base superalloy Inconel MA 6000 with Y_2O_3 dispersoids (small dark particles) and γ' precipitates (larger shadows in the background), produced by Mechanical Alloying (from Ref. 4); right: the aluminium alloy Dispal (AlC2) with Al_4C_3 (dark platelets) and Al_2O_3 dispersoids (light equiaxed particles), produced by Reaction Milling (from Refs. 14, 25).

The advantage of dispersion-strengthened alloys lies in the retention of useful strength up to a relatively high fraction of their melting points (~ 90 %) where other strengthening mechanisms, e.g. precipitation hardening or solid solution strengthening, rapidly lose their effectiveness. Compared with other processes, e.g. internal oxidation or co-precipitation, mechanical alloying techniques offer great versatility with regard to dispersoid-matrix combinations because they circumvent thermodynamic limitations imposed by the (desired) insolubility of the dispersoid material. Also dispersoids produced by mechanical alloying can be combined with other hardening phases: the prime example are oxide-dispersion strengthened (ODS) superalloys, which combine excellent high temperature strength (due to the dispersoids) with good intermediate temperature strength (due to ordered precipitates) (1-4).

The flexibility of mechanical alloying techniques has great potential for producing new dispersoid-matrix material combinations. Up to recently, the choice of dispersoid materials has been largely empirical, with much more attention being paid to thermal stability than to the efficiency of dispersoids as dislocation obstacles. As new alloy systems (e.g. aluminium, intermetallics) are now being developed for high-temperature applications, guidelines for achieving optimum dispersion strengthening would be particularly welcome. Recent attempts to model high temperature strength in dispersion strengthened materials appear to have approached a stage at which new conclusions, many of which are still tentative, for obtaining dispersoids with high efficiency and for grain structure optimization can be drawn.

The aim of this paper is therefore to review and critically discuss the current theoretical understanding of dispersion strengthening at high temperatures and to suggest conclusions for alloy design, optimization and use. Dispersoid effects are treated in section 2, while the influence of the grain structure is discussed in section 3. An extended version of this paper will be published elsewhere (8).

2. Effects of Dispersoid Particles on Creep Strength

2.1 Experimental Facts and "Threshold Stresses"

The presence of dispersoid particles causes a pronounced retardation of creep at high temperatures and comparatively low stresses; this effect is now well documented, e.g. (9-14). Two typical examples of the resulting strain rate vs. stress behaviour are shown in fig. 2. Unlike the dispersoid-free counterparts, the dispersion-strengthened materials exhibit creep rates which, on decreasing applied stress, fall extremely rapidly to (eventually immeasurable) low values. The "stress exponents", i.e. the slopes in fig. 2, which measure the stress sensitivity of the creep rate, can assume values in excess of 100. Such a behaviour is inconsistent with creep laws for conventional materials which usually exhibit stress exponents between 3 and 10. Also the temperature dependence of creep is often observed to be "abnormal" in the sense that temperature-compensated plots as in fig. 2 do not produce a common curve for different temperatures. This implies that the activation energy for creep appears to be higher than that for volume diffusion, which again is contrary to classical dislocation creep.

Because of these shortcomings a semi-empirical constitutive equation containing a "threshold stress" σ_0 has been proposed for dispersion-strengthened materials (15-17):

$$\dot{\epsilon} = A' \frac{D_V G b}{k_B T} \left[\frac{\sigma - \sigma_0}{G} \right]^{n'} \tag{1}$$

where $\dot{\epsilon}$ is the creep rate, σ the applied stress, G the shear modulus, n' the stress exponent, D_V the lattice diffusivity, b the Burgers vector, k_B Boltzmann's constant, T the absolute temperature and A' a dimensionless constant.

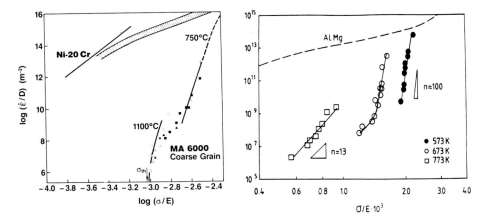

Fig 2: Typical creep behaviour of dispersion-strengthened materials; the creep rate $\dot{\epsilon}$ is normalized by the lattice diffusivity D, the applied stress σ by Young's modulus E: left: Inconel MA 6000, comparison of coarse-grain with fine-grain and dispersoid-free (Ni-20 Cr) material (4), and right: dispersion strengthened Al-4 Mg in the fine-grained condition (apparent grain boundary weakening occurs at 500 °C) (14,25).

This expression gives an arbitrarily high "apparent" stress exponent $n = d(\log \dot{\epsilon})/d(\log \sigma)$ at stresses just above σ_0. By allowing σ_0 to decrease with increasing temperature, a high "apparent" activation energy can also in principle be formally justified.

If eq. 1 is accepted as a formal representation of creep data for dispersion-strengthened materials, then an understanding of the micromechanisms of creep will hinge on a theoretical justification for σ_0. It must be emphasized here already that the "threshold stress" σ_0 cannot be a true threshold for creep deformation, in the sense that at stresses below it the strain rate is identical to zero - as is suggested by the form of eq. 1. With further mechanistic background, this modified constitutive equation will be seen merely as an approximation. In fact, it will be shown in section 2.2, that recent models for the mechanism of dispersion strengthening at high temperatures lead to serious conceptual reservations concerning uncritical use of a "threshold stress" concept.

Bearing these qualifications in mind, one can still attribute some practical significance to the value of the apparent "threshold stress" σ_0 as it characterizes the stress level at which dispersion-strengthened materials become far superior to their dispersoid-free counterparts. Further, for practical applications the "threshold stress" will be a design-limiting strength property: service stresses must always be kept below it in order to avoid occurrence of unacceptable creep deformation.

Threshold stress values have been determined by various authors for several dispersion strengthened alloy systems, including the mechanically alloyed iron-base alloy MA 956, and the nickel-base alloys MA 754, and MA 6000. Unfortunately, it is not always possible to compare these thresholds because they have been extracted from the creep data in different ways. For the sake of comparison, an attempt will be made here to evaluate "threshold stresses" from published creep data in a consistent way. In order to preclude grain boundary effects, only data for single crystals or coarse elongated grains (tested in the longitudinal direction) are included. The data were replotted as $(\dot{\epsilon}/D_V)^{1/n}$ vs. σ/E (where E is Young's modulus) and extrapolated linearly to $\dot{\epsilon} = 0$. The diffusivity normalization produced in most cases parallel lines for different temperatures, which generally improved the reliability of the extrapolations. Rather than force-fitting the data to stress exponents suggested by theoretical considerations, values typical of the dispersoid-free material were chosen: for example $n = 4.6$ for Ni-20Cr based ODS alloys.

The results are plotted in a normalized form in fig. 3. Absolute temperatures have been divided by the melting point (or the approximate solidus temperature). The "threshold stresses" have been normalized with respect to the Orowan stress given by

$$\sigma_{Or} = \frac{0.84\,M}{2\pi(1-\nu)^{1/2}} \frac{Gb}{\ell} \ln\left(\frac{r}{b}\right) \tag{2}$$

where M is the Taylor factor (or reciprocal Schmid factor for single crystals), ν is the Poisson number ($\nu = 0.3$), ℓ the mean planar particle spacing, and r the mean particle radius.

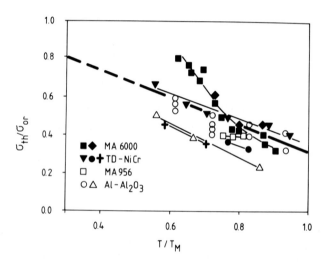

<u>Fig. 3:</u> Plot of experimental "threshold stresses", normalized with respect to the calculated Orowan stress, as a function of homologous temperature, for coarse-grain or single crystals materials: MA 6000 (Refs. 11,4), MA 956 (Ref. 63), Al-Al$_2$O$_3$ (Refs. 12,13), TD-NiCr (Refs. 9,64,65).

Despite some numerical uncertainties the following observations can be made in fig. 3:

1) The Orowan stress is a useful normalization parameter for "threshold stresses", as it brings the data from very different materials, whose absolute "threshold stresses" differ by more than an order of magnitude, to within a factor of about 2. It appears plausible that similar mechanisms are responsible for the creep strength in these alloys.

2) All the threshold stresses are below the theoretical Orowan stress, and decrease with increasing temperature. These deviations from the Orowan stress are important and will need to be explained theoretically.

In summary, the "threshold stress" concept appears to have some merit for formal rationalization of the creep behaviour in dispersion strengthened alloys. Without further mechanistic background, the applicability of this concept must however remain suspect. We therefore turn to the relevant theories of dispersion strengthening at high temperatures, which will be reviewed in the following section.

2.2 Theoretical Understanding of Dispersoid Effects on Creep Strength

This section is entirely devoted to theoretical attempts to explain, in terms of dislocation theory, the high stress sensitivity of the creep rate ("threshold stress" behaviour) in materials with low volume fractions of dispersoids. The problem has received great attention over the past fifteen years. Early papers attempted an explanation in terms of the Orowan process and its thermal activation (18). But once the "threshold stresses" were found to be significantly lower than the Orowan stress of the particle dispersion, subsequent models focussed on the process by which dislocations can circumvent hard particles, especially by climbing around them. Following detailed transmission electron microscopy of this process, more attention has been paid recently to the interaction between dispersoid particles and dislocations. New models have been proposed which abandon the idea of climb being the rate-limiting event. The development, including latest results, is described below.

Dislocation Climb Models

When dispersoid particles are considered as impenetrable obstacles which force gliding dislocations to climb a certain distance until they can continue to glide, a retardation of creep is predicted, but a high stress sensitivity cannot be explained in this way (19,20). The maximum stress exponent which is obtained theoretically is below $n=4$ (21).

Only by realizing that the dislocation has to increase its line length in order to surmount the dispersoid, can a climb-related "threshold stress" be justified. This process has been modelled by Brown and Ham (22) and Shewfelt and Brown (23), who assumed that climb is "local"; this means that only the portion of the dislocation which is in close proximity with the particle-matrix interface undergoes climb while the remaining segment stays in the glide plane (fig. 4 - broken dislocation line running along particle contour).

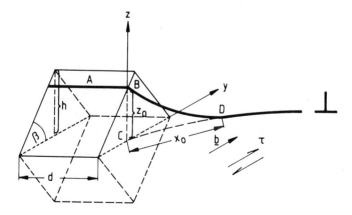

Fig. 4: Geometric assumptions for modelling climb of dislocations over dispersoids: "equilibrium climb" (28) - full line - and "local climb" (22) - broken line.

For climb over the dispersoid to occur, internal energy must be gained everywhere along the dislocation path, which requires the stress to exceed a "threshold stress" σ_{th} for local climb:

$$\sigma \geq \sigma_{th} = \frac{1}{2} \left[\frac{dL}{dx} \right]_{max} \sigma_{Or} \qquad (3)$$

This threshold stress is proportional to the Orowan stress. The other important parameter is the differential of the dislocation line length L with respect to advance distance x, at the point of its maximum. This quantity is called "climb resistance". During "local climb" the dislocation profiles the particle contour and therefore the climb resistance is related to particle shape. A detailed analysis shows that the threshold stress amounts to about 70 % of the Orowan stress for cubes oriented as in fig. 4 (22), and to about 40 % for spheres (23). We note that these threshold considerations lead to values which appear to be in the same range as experimental values (cf. fig. 3). The temperature dependence however remains unexplained.

As anticipated by Shewfelt and Brown (23), there are serious objections that can be raised against the basic assumption of "local" climb. The postulate concerning dislocation shape in the vicinity of the dispersoid particle must be regarded as unduly restrictive. Lagneborg (24) has argued that local climb would be an extreme non-equilibrium process: the sharp bend in the dislocation at the point where it leaves the particle-matrix interface will in reality be rapidly relaxed by diffusion, leading to more "general" climb. Because the additional line length then required depends on the kinetics, a "back stress" is predicted which scales as the applied stress. This would result only in a retardation of creep, with the same stress exponent as for the particle-free material; such a behaviour would clearly be in contradiction with experimental data on dispersion-strengthened materials. In addition, as discussed by Rösler (25), there appears to be a flaw in the calculations, leading to gross overestimation of the diffusive fluxes.

In the meantime it has become clear that a finite, but small threshold stress must always exist: in order to thread over and under dispersoid particles in a random arrangement, a small elongation of the dislocation line is inevitable. The resulting threshold stress for "general climb" over spherical particles (23,26) amounts to negligible numerical values (less than 10 % of the Orowan stress) for low volume fractions up to 10 %. When a random particle distribution is assumed (26) and when Friedel statistics is applied (27) these values are further reduced to less than about 2 %. In conclusion, it can be said that climb-related threshold stresses are sensitive to the details of the climb process especially in one regard: the degree to which climb is localized at the particles.

In order to solve the long-standing problem of climb localization, Rösler and Arzt (28) have considered the kinetics of dislocation climb over cuboidal particles. The only assumption concerning the dislocation geometry was that of local equilibrium (fig. 4): the chemical potential for vacancies along segment BD (where it is lowered by the curvature) is set equal to the chemical potential along AB (where it is negatively biased by the applied stress). Local equilibrium can be assumed to form rapidly by short-range diffusion, while the supply of new vacancies for climb and dislocation advance typically requires diffusion over the distance of one particle spacing. This leads to a more natural dislocation configuration for which the diffusion kinetics is then evaluated.

The resulting constitutive equation for "equilibrium climb" does not contain any adjustable parameters:

$$\dot{\epsilon} = \frac{\rho G b^4 B}{k_B T \, d^3} \left[a_p D_p + \pi D_V \ell d \right] \left[\frac{\sigma - \sigma_{th}^r}{\sigma_{Or}} \right]^n \qquad (4)$$

where the stress exponent n lies between 3 and 4, depending slightly on particle shape, B contains the particle shape, ρ is the density of mobile dislocations, d is the particle width according to fig. 4, ℓ is the particle spacing, $a_p D_p$ is the cross section of a dislocation core times its diffusivity, D_V is the volume diffusivity, σ the applied stress, σ^r_{th} a small threshold stress for "general" climb and σ_{0r} the Orowan stress.

An important result which emerges from this treatment of "equilibrium climb" is that truly "local" climb is always unstable, but the extent to which climb is localized in the vicinity of the dispersoid particle increases with applied stress. This implies that the number of vacancies required for climb (which is proportional to the length of the climbing segment) decreases as the stress increases, and therefore the dislocation velocity depends non-linearly on stress. The maximum climb-related stress exponents predicted by this model ($n \approx 6$) are however far below the values typical of dispersion-strengthened materials.

In summary, it seems that the assumption of dislocation climb alone cannot explain the high stress sensitivity for creep in dispersion-strengthened materials. In other words, if the only effect of dispersoid particles consisted in forcing the dislocations to climb over them, such particles would be weak barriers to slow creep deformation at high temperatures.

Dislocation Detachment Models

In the search for a genuine threshold stress mechanism, attention has only recently turned to the details of the possible interactions between dispersoids and hard particles. Much insight has been gained by transmission electron microscopy of the dispersoid-dislocation configurations in creep-exposed specimens of dispersion strengthened alloys (29-31,14,25). The most detailed study to date has been conducted on the ODS superalloy MA 6000, which was creep-deformed at temperatures near 1000 °C and under stresses well below the Orowan stress. A typical micrograph, taken subsequently under "weak-beam" conditions by Schröder (30,31), is shown in fig. 5. The dislocation is seen in a situation where it has already surmounted a dispersoid by climb and adheres to the "departure side" of the dispersoid. Under "weak-beam" conditions, the contrast of the dislocation segment which resides in or near the dispersoid-matrix interface is often well visible.

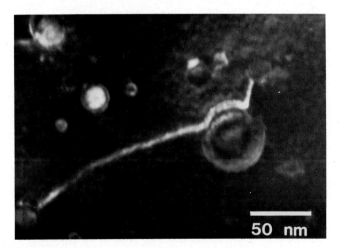

Fig. 5: TEM evidence for an attractive dislocation-dispersoid interaction during creep of Inconel MA 6000 (Refs. 30,31).

Similar, but less detailed observations were reported in the dispersion strengthened NiCr alloy MA 754 by Nardone and Tien (29), who were the first to point out the possibility of pinning on the "departure-side", and in dispersion strengthened Al by Rösler (25). While these TEM studies do by no means constitute unambiguous proof, they strongly suggest a new dislocation mechanism: particle bypassing by climbing dislocations may not be controlled by the climb process (as postulated in the early threshold stress models), but by a resistance to dislocation detachment from the particle. In other words, "threshold stresses" for creep would have to be attributed to an attractive particle-dislocation interaction.

Recent theoretical analysis confirms indeed that the assumption of an attractive particle-dislocation interaction at high temperatures is plausible. Srolovitz et al. (32) have solved the elastic problem of an edge dislocation interacting with a (cylindrical) particle, subject to the boundary conditions that both tangential tractions and gradients in the normal tractions at the particle-matrix interface are instantaneously relaxed. Such relaxation processes can be accomplished at elevated temperatures by boundary and volume diffusion, and by sliding of the phase boundary. Calculation of the characteristic relaxation times suggests that relaxation, at least of deviatoric stresses, will indeed occur quickly; therefore the dispersoid can be treated as a void (with an internal pressure corresponding to the hydrostatic component of the applied stress). The Srolovitz solutions show that at high temperatures the elastic interactions between dispersoids and dislocations become unimportant: even an infinitely stiff particle (which repels dislocations at low temperatures) can produce a strong attractive force, provided it is not coherent with the matrix. Srolovitz et al. (32) further suggest that core delocalization will occur in the interface. Therefore, the unpinning stress should be similar to that for voids, i.e. approximately equal to the Orowan stress (33).

There appears to be, however, an important discrepancy of these predictions with some TEM observations. The "weak-beam" micrograph in fig. 5 shows clearly that the identity of the dislocation is preserved in the vicinity of the particle-matrix interface and the local strain is not noticeably reduced by core spreading. Contrary to the behaviour of a void, the dispersoid evidently allows only a modest relaxation of the dislocation. That the detachment threshold can act as the decisive dislocation barrier even at small degrees of relaxation has been established by Arzt and Wilkinson (34). They treat the effects of an attractive interaction on the energetics of dislocation climb. In their model a variable attractive interaction is introduced by assigning a line tension to the dislocation segment at the particle which is lowered by a factor k (k<1). This parameter describes the extent to which the dislocation relaxes its energy by interaction with the particle-matrix interface. For k=1, no relaxation and, thus, no attractive interaction occurs; k=0 signifies maximum interaction which is approximately realized by the presence of a void.

The authors have calculated force-distance profiles for local climb under the action of different attractive forces. The main result is that an attractive interaction causes a significant threshold stress which must be exceeded in order to detach the dislocation from the back of the particle and which is given by:

$$\sigma_d = \sqrt{1 - k^2} \; \sigma_{Or} \qquad (5)$$

This "detachment threshold" is independent of the shape of the particle and its position with respect to the glide plane. It applies regardless of the details of the climb process which precedes detachment.

Because of the reduced line tension of the dislocation segment at the dispersoid, the "threshold stress" for local climb is lowered. This leads to an important conclusion: only a very modest attractive interaction, corresponding to a relaxation of about 6% (k ≈ 0.94), is required in order for dislocation detachment to become the event which controls the threshold stress. Because, in view of Srolovitz' calculations, some relaxation will always occur at the interface, these results lend strong support to the detachment process as a serious candidate for a "threshold stress" mechanism.

Recently, a kinetic model for dislocation climb which allows for the effects of an attractive interaction has been developed by Arzt and Rösler (35). In order to complete the model, thermally activated dislocation detachment from dispersoids needs to be considered also. This aspect has been analyzed in detail by Rösler (25) and Rösler and Arzt (36,37). The resulting constitutive equation for "detachment-controlled creep" is given by:

$$\dot{\epsilon} = \dot{\epsilon}_0 \exp\left[-\frac{Gb^2r}{k_BT}(1-k)^{3/2}\left(1-\frac{\sigma}{\sigma_d}\right)^{3/2}\right] \qquad (6)$$

where $\dot{\epsilon}_0 = 3 D_V \ell \rho / b$ and σ_d is given by eq. 5.

Eq. 6 is plotted in fig. 6 for different values of k. Several features which are in qualitative agreement with experimental results can be identified: Dispersoids with a strong attractive interaction (k<0.85) do indeed produce a region of high stress exponent which extends over many orders of magnitude in strain rate; this gives the appearance of a "threshold stress", which falls with increasing temperature. As the interaction gets weaker (or, equivalently, if the temperature is raised), a concave curvature at low stresses becomes apparent; this corresponds to a loss of strength which would not be expected by linear extrapolation of the data from higher strain rates. It is also clear from fig. 6 that there is no "universal" stress exponent for dispersion-strengthened materials; rather the stress sensitivity depends on temperature, stress and dislocation-dispersoid interaction.

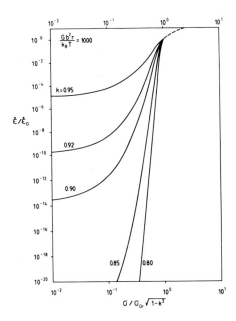

Fig. 6: Theoretical prediction of the creep rate (normalized) as a function of stress (normalized) on the basis of thermally activated dislocation detachment from attractive dispersoids (eq. 6), as a function of interaction parameter k (from Ref. 37).

When the interaction parameter k and the pre-exponential factor are adjusted, eq. 6 can be made to describe the creep behaviour of several dispersion strengthened materials extremely well. Extensive data evaluation shows that the value of k, obtained by fitting the theory to the creep data, varies characteristically with the type of dispersoid (table 1): carbides in aluminium seem to be most efficient in attracting dislocations, while oxides appear less suitable (25,8,37). Also the rapidly solidified Al–Fe–Ce alloy studied by Yaney et al. (38,39) fits well into the theoretical scheme. Its "anomalous" creep behaviour can be fully described by assigning a weak interaction effect (k=0.95), again independent of temperature, to the intermetallic precipitates. This interpretation is also consistent with Yaney et al. (38), who have suggested that the loss of strength at higher temperatures cannot be attributed to thermal instability of the microstructure.

Material	Reference	Dispersoid	k
AlC2	25, 14	$Al_2O_3 + Al_4C_3$	0.75
AlC0	25, 14	Al_2O_3	0.84
AlMg4C1	25, 14	$MgO + Al_4C_3$	0.85
Al–Fe–Ce	38	intermetallic	0.95
MA 6000	4	Y_2O_3	0.92

Table 1: Interaction parameter k for different materials, determined by fitting eq. 6 to experimental creep data (25).

2.3 Practical Implications

On the basis of the detachment model described above several conclusions can be drawn, some of which must necessarily remain speculative at the present state. Above all, the "threshold stress" concept, when applied uncritically, can be seriously misleading. Contrary to the implications of the local climb models there exists no significant minimum "threshold stress" (apart from the small threshold for general climb) that must be exceeded in order to sustain creep deformation. Laboratory creep rates may indeed give the impression of a "threshold stress" behaviour but eventually, at sufficiently low strain rates, an upward curvature of the curves is inevitable (fig. 6). Until now such a loss in strength has been attributed alternatively to the dissolution, coarsening or deformability of dispersoid particles (in TD–NiCr single crystals (9), and Al (13)) or to the onset of grain boundary sliding or damage formation (MA 754, Ref. 10, and Al, Ref. 14).

For the purpose of life–time prediction, it must be realized that extrapolations from laboratory data to low creep rates can be seriously in error. Only materials with highly attractive dispersoids (k<0.8) would permit linear extrapolation. The value of k, which is therefore important, can be estimated, by a method proposed by Rösler (25), from the values of the apparent stress exponent and the apparent activation energy for creep. In conclusion, even in cases where near–"threshold stress" behaviour is observed, it seems warranted, in the interest of clarity, to use the term "pseudo–threshold" instead.

The practical perspectives in terms of alloy design are enticing but still vague. Given a certain usable volume fraction of dispersoid (which is limited by minimum ductility requirements), the "detachment model" predicts that there should be an optimum particle size (and spacing). Physically, this comes about because (i) the probability of thermally activated detachment is raised for small dispersoids, and (ii) large particles are associated, at a given volume fraction, with a low Orowan stress and hence a small athermal detachment stress σ_d. This further implies that the creep of strength of alloys with initial particle size below this optimum value would be insensitive to moderate particle coarsening.

Furthermore, "detachment" discriminates different dispersoids by the degree of dislocation relaxation, suggesting that the properties of the dispersoid–matrix interface should be of critical importance. Interestingly, table 1 suggests tentatively that particles dispersed by mechanical alloying tend to be more efficient than those produced by (rapid) solidification. It remains to be seen whether concepts such as interface modification by segregation alloying can be put to use in order to further improve the efficiency of dispersoids.

3.   Grain Boundary Effects on Creep Deformation and Fracture

The previous considerations, which concerned the role of dispersoid particles in impeding the motion of lattice dislocations, can be strictly valid only for single crystals, in which grain boundaries do not contribute to creep deformation. Grain boundary weakening is of particular importance for polycrystalline dispersion-strengthened materials which as engineering materials are subjected to moderate stresses at high homologous temperatures. Because of processing limitations, single crystals of dispersion strengthened materials are difficult, if not impossible, to produce in useful dimensions. Therefore the effects of grain boundaries on deformation and fracture have received considerable attention in the literature; the understanding is however still incomplete as will be shown below.

3.1 Some Experimental Observations

In general, grain boundaries can lower the high-temperature strength in at least three ways: i) stress-induced vacancy accumulation leads to cavity formation on grain boundaries transverse to an applied tensile stress and, subsequently, to premature intergranular failure; ii) grain boundary sliding results in stress concentrations which accelerate dislocation creep; and iii) by acting as vacancy sinks and sources, grain boundaries promote additional grain deformation due to diffusional creep ("Nabarro-Herring creep").

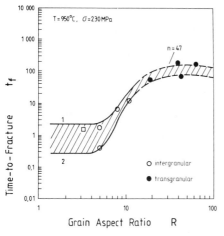

Fig. 7: Creep rupture time of coarse grained Inconel MA 6000 as a function of grain aspect ratio after recrystallization by zone annealing. Shaded area shows range of theoretical predictions based on eq. 7 (from Ref. 42).

In order to suppress premature fracture, nickel-base ODS alloys are commonly recrystallized by zone annealing to obtain an extremely elongated, coarse grain structure (40). Fig. 7 shows that the decisive parameter is the grain aspect ratio (GAR), see also (41). Fine grains, which constitute recrystallization defects and do not necessarily differ in chemical composition from the otherwise coarse-grained matrix, are particularly susceptible to the formation of creep damage. They reduce creep life by acting as starting points for transgranular cracks (42). The property degradation which results from this mechanism has been found to be even more aggravated under combined creep-fatigue loading (43,44).

An important corollary of the severe vulnerability of grain boundaries in ODS materials is a pronounced anisotropy of tensile creep properties. While (short term) tensile strength (45) and short-term compressive creep strength (11) are relatively insensitive to loading direction with respect to the extrusion axis (grain elongation), tensile creep strength is substantially reduced in the transverse directions (46) (fig. 8). The reason is again the occurence of premature intergranular fracture at the longitudinal grain boundaries, with a concomitant severe reduction in ductility. At low stresses creep rates are affected even under compressive transverse loading, which leads to considerable deformation below the "threshold stress" (46). This type of loading produces γ'-free diffusion zones on grain boundaries parallel to the compression axis (fig. 8), which has been interpreted as being indicative of a contribution by diffusion creep. What is atypical of diffusion creep, however, is the high stress exponent that applies in this regime. Similar evidence for diffusion creep has been found earlier by Whittenberger (47).

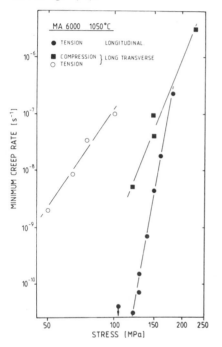
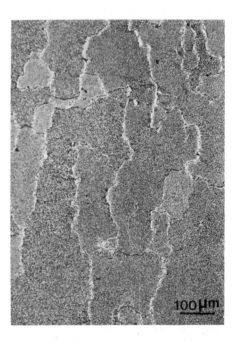

Fig. 8: Results of creep tests with Inconel MA 6000 in the long transverse direction (from Ref. 46): left: comparison of creep rates, under tension and compression with longitudinal data (tension); right: evidence for diffusion creep in the form of precipitate-free zones on a transverse section after compressive loading in the long transverse direction (stress axis vertical).

Grain size per se is very important for the rate of creep deformation in nickel-base ODS alloys: fine-grain (as-extruded) material exhibits creep strengths at least an order of magnitude lower than recrystallized material as is seen in fig. 9. By contrast, the difference in creep strength between fine-grained and recrystallized $Al-Al_2O_3$ appears to be much smaller ($\approx 30 \%$) at a similar homologous temperature. A grain coarsening treatment may thus be less advantageous in the latter case.

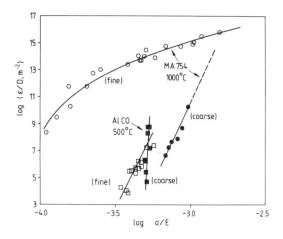

<u>Fig. 9:</u> The effect of grain size on the creep strength of Ni-base ODS alloy MA 754 (data from Ref. 48) and of Al-Al$_2$O$_3$ (data from Refs. 25,49,50).

3.2 Theoretical Understanding of Grain Boundary Effects

Models for the growth of creep cavities in elongated grain structures typical of some modern dispersion-strengthened materials have been developed by Arzt (51), Arzt and Singer (52), Stephens and Nix (53), and Zeizinger and Arzt (42). In view of the high resistance to dislocation creep, all models assume pore growth to occur by diffusion; this is supported by observations of particle-free zones in the ligaments between creep cavities (42). The theories differ in the type of accommodation process necessary for assuring compatibility between the deformation of damaged and undamaged grains. When sliding along longitudinal grain boundaries provides sufficient accommodation, a dependence on the grain aspect ratio arises naturally from the coupling of grain boundary sliding and cavity growth, between which the applied stress is distributed. A shortcoming of this approach is the linear stress dependence of the creep rate. Also, sliding could not be detected in model experiments (42), which casts some doubt on the applicability of the model to alloys like MA 6000.

Another possibility for damage accommodation is by dislocation creep, enhanced by the shedding of load from cavitating grain boundaries to the adjacent intact grains. A numerical solution for this case by Stephens and Nix (53) gives a strong dependence of the rupture time on GAR. The experimental strain-rate/stress dependence of MA 754 with uniform fibre grain morphology is well described by this model, except at low stresses where creep strength is consistently overestimated. Zeizinger and Arzt (42) take a similar modelling approach which yields the following analytical equation for the rupture time as a function of grain aspect ratio R:

$$t_f = \frac{0.085 \ k_B T \lambda^3}{\delta D_b \Omega \sigma} \left[1 - (1 - \frac{1}{R})^n \right] + \frac{\epsilon_B}{\dot{\epsilon}_u} (1 - \frac{1}{R})^n \qquad (7)$$

where λ is the cavity spacing, ϵ_B the total strain due to void growth and $\dot{\epsilon}_u$ the creep rate of uncavitated material. This equation, which was derived on the basis of a model by Cocks and Ashby (54) for equiaxed grains, is found to describe the GAR-dependence of the rupture time in MA 6000 well. Also the poor rupture properties in the transverse direction can be understood because a low GAR applies in this case.

In the light of these investigations it appears that the GAR-dependent "threshold stresses for diffusional creep" reported by Whittenberger (55) could also be a consequence of the coupling between grain boundary damage processes and accommodation by dislocation creep with a pseudo-threshold stress. As the GAR increases, the accommodation process becomes increasingly more important and the material exhibits a larger fraction of the pseudo-threshold for matrix creep.

We now turn to the case in which adherence of the grain boundaries is maintained during creep, but shear stresses are relaxed by grain boundary sliding. This mechanism has originally been invoked to explain the inferior creep strength of low-GAR grain structures (56). Detailed modelling of the coupling between grain boundary sliding and dislocation creep (57) shows, however, that the stress concentrations resulting from sliding are quite small: the loss in creep strength due to grain boundary sliding is predicted to be only about 20 %. Apart from a narrow transition region, the stress exponent of the dislocation creep process is retained down to low strain rates. The only experimental results which do not immediately contradict this prediction seem to be the creep data for dispersion-strengthened aluminium measured by Rösler et al. (50) and Joos et al. (49) (fig. 9). The grain boundaries in this material apparently do not contribute to premature creep fracture, which is in contrast to MA 6000. In MA 754 the strength difference between fine and coarse grain material (fig. 9) is much higher and cannot be interpreted in terms of grain boundary sliding alone (as discussed by Nix (48)). A possible explanation for this apparent difference may be connected with the lack of texture in fine-grained MA 754 as opposed to the Al alloy which has a strong <111> texture (49).

Finally, diffusion creep can occur in fine-grained materials, which requires grain boundaries both, to act as vacancy sinks and sources, and to slide. Several models have been put forward to predict the effect of grain boundary dispersoids on diffusional creep. One type of model, pioneered by Ashby (58) and detailed by Arzt et al. (59), considers the emission and absorption of vacancies at the grain boundaries (the "interface reaction") as the rate-limiting process; this idea is supported by atomistic studies of grain boundaries which suggest a fairly well-defined structure: vacancies can be produced or absorbed only at grain boundary dislocations which have to move non-conservatively in the grain boundary plane. Because these dislocations are pinned by grain boundary particles, a minimum "threshold stress" for diffusional creep is predicted which is equal to the Orowan stress or the "local climb" stress for boundary dislocations. A difficulty in applying this model lies in the assumption that must be made with regard to the Burgers vector of the boundary dislocation; it is only clear that it is smaller than for lattice dislocations. In analogy to dislocation creep, an attractive interaction between boundary dislocations and particles could also play a role, as suggested by TEM studies of grain boundary dislocations interacting with particles (60). But this effect has not been included in the models so far.

In another type of model it is assumed that the grain boundaries, but not the particle-matrix interfaces, act as perfect sources and sinks for vacancies. A stress concentration then builds up at the particles, which must be relaxed before creep can continue. Relaxation can occur by the nucleation of lattice dislocation loops (61), or alternatively, defect loops can be nucleated in the particle-matrix interface (62). Both processes give rise to a "threshold stress" for diffusion creep.

It is difficult to ascertain how important the role of diffusion creep is in the engineering alloys considered in this review. The experimental evidence for diffusion creep is limited to occasional observation of particle-free zones (e.g. 46,55). The stress dependence of the creep rate however is always found to be non-linear, but lower than in coarse-grained material. Such intermediate stress exponents (n = 12...16) are difficult to justify theoretically. It is clear from the absolute magnitude of the creep rates that diffusion creep is suppressed or strongly inhibited in the fine-grained dispersion-strengthened Al alloys.

3.3 Practical Conclusions

Grain boundaries can be extremely vulnerable microstructural elements in dispersion-strengthened high-temperature materials. The effect of GAR on creep rupture time of nickel-base alloys is well established and understood. It follows that, in order to minimize the detrimental influence of grain boundaries, a high GAR in the loading direction must be achieved. Also the generally poor properties in the transverse directions are due to the presence of grain boundaries and could possibly be improved in one transverse direction by increasing the GAR in that direction as well. Diffusion creep apparently places a lower bound on the creep rates in the transverse directions; this effect can be miminized by increasing the width of the elongated grains. Since this will tend to lower the overall GAR in the longitudinal direction, an optimum balance between longitudinal and transverse properties would need to be struck for a given application. Fine-grain regions should be eliminated as far as possible in order to exploit the full potential of dispersion-strengthening with regard to creep and high-temperature fatigue properties. Single crystals of course would in principle be an ideal solution.

The advantages of grain coarsening appear to be much smaller for dispersion-strengthened aluminium than for nickel alloys. Therefore the use of fine-grained materials for high-temperature applications may be feasible in some cases. The reasons for this difference in grain boundary behaviour are not understood. Clearly more scientific and technological work on the creep behaviour of extremely fine-grained dispersion-strengthened alloys is urgently necessary.

4. References

(1) J.S. Benjamin: this volume.
(2) R.C. Benn and P.K. Mirchandani: this volume.
(3) J.S. Benjamin: Met. Trans. 1 (1970) 2943.
(4) R.F. Singer and E. Arzt: in "High Temperature Alloys for Gas Turbines and Other Applications 1986", W. Betz, R. Brunetaud, D. Coutsouradis, H. Fischmeister, T.B. Gibbons, I. Kvernes, Y. Lindblom, J.B. Marriott, D.B. Meadowcroft (eds.), D. Reidel Publishing Company, Dordrecht, 1986, p. 97.
(5) G. Jangg and F. Kutner: Aluminium 51 (1975) 641.
(6) G. Jangg: Radex-Rundschau (1986) 169.
(7) G. Jangg: this volume.
(8) E. Arzt: Res Mechanica, in press.
(9) R.W. Lund and W.D. Nix: Acta Met. 24 (1976) 469.
(10) J.J. Stephens and W.D. Nix: Met. Trans. 16A (1985) 1307.
(11) J.D. Whittenberger: Met. Trans. 15A (1984) 1753.
(12) A.H. Clauer and N. Hansen: Acta Met. 32 (1984) 269.
(13) W.C. Oliver and W.D. Nix: Acta Met. 30 (1982) 1335.
(14) E. Arzt and J. Rösler: in "Proc. 117th TMS Annual Meeting", Phoenix (1988), in press.
(15) C.R. Barrett: Trans. AIME 239 (1967) 1726.
(16) J.H. Gittus: Proc. R. Soc. A342 (1975) 279.
(17) J.C. Gibeling and W.D. Nix: Mat. Sci. Eng. 45 (1980) 123.
(18) P. Guyot: Acta Met. 12 (1964) 941.
(19) G.S. Ansell: in "Oxide Dispersion Strenthening", G.S. Ansell, T.D. Cooper and F.V. Lenel (eds.), Gordon and Breach, New York, 1968, p.61.
(20) M.F. Ashby: in "Proc. Second Int. Conf. on Strength of Metals and Alloys", ASM, Metals Park, Ohio, 1970, p. 507.

(21) J.H. Holbrook and W.D. Nix: Met. Trans. 5 (1973) 1033.
(22) L.M. Brown and R.K. Ham: in "Strengthening Methods in Crystals", A. Kelly and R.B. Nicholson (eds.), Elsevier, Amsterdam, 1971, p. 9.
(23) R.S.W. Shewfelt and L.M. Brown: Phil. Mag. 30 (1974) 1135 and 35 (1977) 945.
(24) R. Lagneborg: Scripta Met. 7 (1973) 605.
(25) J. Rösler: Ph.D. Thesis, University of Stuttgart, 1988.
(26) E. Arzt and M.F. Ashby: Scripta Met. 16 (1982) 1285.
(27) W. Blum and B. Reppich: in "Creep Behaviour of Crystalline Solids", B. Wilshire and R.W. Evans (eds.), Pineridge Press, Swansea, 1985, p. 83.
(28) J. Rösler and E. Arzt: Acta Met. 36 (1988) 1043.
(29) V.C. Nardone and J.K. Tien, Scripta Met. 17 (1983) 467.
(30) J.H. Schröder and E. Arzt: Scripta Met. 19 (1985) 1129.
(31) J.H. Schröder: Ph.D. Thesis, University of Stuttgart, 1987
(32) D.J. Srolovitz, M.J. Luton, R. Petkovic-Luton, D.M. Barnett and W.D. Nix: Acta Met. 32 (1984) 1079.
(33) R.W. Weeks, S.R. Pati, M.F. Ashby and P. Barrand: Acta Met. 17 (1969) 1403.
(34) E. Arzt and D.S. Wilkinson: Acta Met. 34 (1986) 1893.
(35) E. Arzt and J. Rösler: Acta Met. 36 (1988) 1053.
(36) J. Rösler and E. Arzt: this volume.
(37) J. Rösler and E. Arzt: to be published.
(38) M.L. Öveçoglu, and W.D. Nix: in "Proc. 117th TMS Annual Meeting", Phoenix, AZ, January 1988, in press.
(39) D.L. Yaney, M.L. Öveçoglu and W.D. Nix: this volume
(40) R.F. Singer and G.H. Gessinger: in: G.H. Gessinger, Powder Metallurgy of Superalloys, Butterworth, London, 1984, p. 213.
(41) R.L. Cairns, L.R. Curwick and J.S. Benjamin: Met. Trans. 6A (1975) 179.
(42) H. Zeizinger and E. Arzt: Z. Metallkde. 79 (1988), in press.
(43) D.M. Elzey and E. Arzt: in "Superalloys 1988", S. Reichman et al. (eds.), TMS, 1988, p. 595.
(44) D.M. Elzey and E. Arzt: this volume.
(45) INCOMAP, Data Sheet, 1983.
(46) R. Timmins and E. Arzt: Scripta Met. 22 (1988) 1353; more to be published.
(47) J.D. Whittenberger: Met. Trans. 4 (1973) 1475.
(48) W.D. Nix: Proc. Superplastic Forming Symposium, Los Angeles, ASM, Metals Park, 1984.
(49) R. Joos: Diploma Thesis, University of Stuttgart, 1988.
(50) J. Rösler, R. Joos and E. Arzt: to be published.
(51) E. Arzt: Z. Metallkde. 75 (1984) 206.
(52) E. Arzt and R.F. Singer: in "Superalloys 84", M. Gell, C.S. Kortovich, R.H. Bricknell, W.B. Kent and J.F. Radavich (eds.), TMS-AIME, Warrendale, 1984, p. 367.
(53) J.J. Stephens and W.D. Nix: Met. Trans. 17A (1986) 281.
(54) A.C.F. Cocks and M.F. Ashby: Progr. Mat. Sci. 27 (1982) 189.
(55) J.D. Whittenberger: Met. Trans. 8A (1977) 1155.
(56) B.A. Wilcox and A.H. Clauer: Acta Met. 20 (1972) 743.
(57) F.W. Crossman and M.F. Ashby: Acta Met. 23 (1975) 425.
(58) M.F. Ashby: Scripta Met. 3 (1969) 837.
(59) E. Arzt, M.F. Ashby and R.A. Verall: Acta Met. 31 (1983) 1977.
(60) G.L. Dunlop, J.-O. Nilsson and P.R. Howell: J. Microscopy 116 (1979) 115.
(61) J.E. Harris: Met. Sci. J. 7 (1973) 1.
(62) B. Burton: Mat. Sci. Eng. 11 (1973) 337.
(63) R. Petkovic-Luton, D.J. Srolovitz and M.J. Luton: in "Frontiers of High Temperature Materials II, J.S. Benjamin and R.C. Benn (eds.), INCO Alloys International, New York, p. 73.
(64) J. Lin and O.D. Sherby, Res Mech. 2 (1981) 251.
(65) J.H. Hausselt and W.D. Nix, Acta Met. 25 (1977) 1491.

Properties of Oxide Dispersion Strengthened Alloys

J. Daniel Whittenberger, NASA-Lewis Research Center, Cleveland, OH

Introduction

Dispersion strengthened alloys contain a low volume fraction of extremely fine, uniformly distributed, inert second phase particles in a metallic matrix. Through appropriate processing, materials containing less than 5 percent of a strenghtening phase in the form of nominally 30 nm diameter dispersoids approximately 0.1 μm apart can be fabricated, where the effect(s) of the particle strenghtening are far greater that that predicted by simple reasoning: for example the Rule of Mixtures. The enhanced mechanical properties found in dispersion strengthened alloys are due to both direct and indirect effects of the particles. The interference of the dispersoids with the ability of dislocations to freely glide or climb is an example of direct effect, while the grain morphology and preferred crystallographic orientation of the alloy are typical indirect effects. In many cases, thermal fatigue resistance and creep strength for instance, the secondary influences of the dispersoid are extremely important, and they are regarded to be as significant as the distribution of particles for explanation of the observed behavior of dispersion strengthened alloys. The following paper describes the general characteristics of several commercial or semicommercial oxide dispersion strengthened (ODS) alloys with emphasis on those mechanical properties where these materials display superior behavior or behaviors not generally found in other high temperature alloys.

Representative ODS Alloys

Table I lists typical characteristics (composition, grain structure, texture, and dispersoid statistics) for several types of ODS alloys. The Ni-base alloys fall into several natural groups which depend on composition and tend to predict maximum use temperature and oxidation behavior. The first group consists of the alloys TD-Ni and DS-Ni which conceivably are useful to 1730 K, but because they are NiO formers, they possess limited oxidation resistance. Historically these Ni-2ThO$_2$ materials were surpassed initially by the chromia forming, thoriated alloys, TD-NiCr and DS-NiCr, and more recently by the yttria strengthened alloy MA 754. Maximum use temperature for these Ni-20Cr materials approaches 1675 K. The next group with the nominal matrix composition of Ni-16Cr-4Al, useable to about 1650 K, represented an attempt to develop extremely oxidation resistant alloys via formation of stable Al$_2$O$_3$ scales. The final Ni-base alloy, MA 6000, is the end product of considerable effort (29,30) to produce a combined γ' + dispersion hardened alloy which possesses properties similar to the best Ni-base superalloys at low and intermediate temperatures as well as the strength of the simpler ODS materials at elevated temperatures. Due to its significant alloying content, MA 6000 has the lowest maximum use temperature, ~ 1570 K, of the family of Ni-base ODS alloys. In spite of a great deal of research (31,32) directed toward the development of bcc iron based ODS alloys, only MA 956 has been brought to commercialization. This alloy has a higher maximum use temperature than TD-Ni, ~ 1750 K, and also superior oxidation behavior derived from its basically resistance heater wire composition of Fe-20Cr-4.5Al. Recent efforts in ODS alloys have been directed at Al-based materials, and this work has yielded both pure, solid solution and potentially heat treatable alloy matrix systems (33). As opposed to the work on ODS Fe- and Ni-base materials, such ODS Al alloys are generally being developed for ambient temperature properties.

In terms of long time, elevated temperature particle stability, thoria appears to be relatively inert in Ni-base matrices and only slowly undergoes Ostwald ripening (34), although the possible dissolution and subsequent reprecipitation of small ThO$_2$ particles in Ni and Ni-Cr has been reported (35). Investigators at INCO have shown that the dispersoid in yttria containing mechanically alloyed Fe- and Ni-base products was actually $3Y_2O_3 \cdot 5Al_2O_3$ (21,36), and some questions remain concerning long term, elevated temperature stability of such particles (37). While Al_2O_3 dispersoids in MA Al-base materials are quite stable, even after 823 K annealing, the Al_4C_3 particles start to measurably coarsen above 673 K (33). Although TD-Ni, DS-Ni and DS-NiCr are essentially free of undesirable second phases, the processing (5) involved in the production of TD-NiCr resulted in relatively high fraction of ~ 1 μm diameter Cr_2O_3 particles. Fe- and Ni base ODS alloys produced by mechanical alloying (MA) have second phases such as carbonitrides, oxides and metal-rich particles; however such particles are of a much smaller size than those found in conventional wrought or cast superalloys. Additionally alloys produced by mechanical alloying can have processing defects which are strung out in the major working directions. With the exception of the Al materials, all the ODS alloys listed in Table I possess preferred orientations in, at the very least, the major direction of working. Such texture can affect the mechanical behavior; additionally preferred orientation affects the definition of grain structure where the term "grains" refers to crystals separated by low angle boundaries rather than the highly misoriented grain boundaries found in normal polycrystalline materials.

Although the general goal for most ODS alloys is high temperature strength, until recently no one microstructure was typical of ODS alloys. Data in Table I and Fig. 1 illustrate several of the grain structures found in commercially available alloys. Several early alloys {Ni-2ThO$_2$ (Fig. 1(a)) and DS-NiCr (Table I) were able to combine small grain dimensions (on the order of a few μm) and adequate strength at high temperatures. However work on Fe- and Ni-base alloy matrices has generally demonstrated that larger, elongated grain structures are necessary for strength. Typical examples are TD-NiCr (Table I) which has pancake shaped grains and MA 754 in Fig. 1 (b). This latter figure illustrates the grain aspect ratio (GAR) nature of many ODS alloys, where GAR is equal to the grain dimension in the testing direction divided by the average diameter in the orthogonal directions; thus the longitudinal direction has a high GAR (> 1) while the long transverse value is low (< 1). The belief that large grain structures are necessary for high temperature strength has lead to the development of recrystallization procedures which yield grains with dimensions ranging from mm's to tens of cm's in MA 6000 and MA 956 (Table I). As current MA Al alloys are being developed for ambient temperature use (33), they possess extremely small, thermally stable equiaxed grain structures (Fig. 1(c)) to take advantage of grain size strengthening effects.

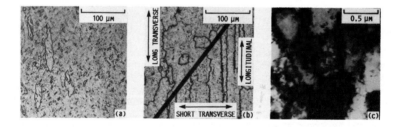

Fig. 1: Photomicrographs of representative ODS Alloys. (a) DS-Ni {Photograph courtesy Dr. M. Ruscoe, Sherritt Gordon Mines, Ltd.}, (b) MA 754 (38), (c) Transmission electron photomicrograph of an experimental MA Al-Li-Mg alloy after a 500 h anneal at 373 K {Photograph courtesy of Dr. J. Weber, IncoMAP}.

Table 1: Characteristics of Several ODS Alloys

Alloy	Composition Wt. Pct.	Form	Grain Dimensions — Major Working Direction (Perpendicular)	Grain Dimensions (Parallel)	Texture Short Trans	Texture Long Trans	Texture Longitudinal	Dispersoid Diameter nm	Dispersoid Spacing μm	Refs.
TD-Ni	Ni-2ThO_2	Bar	1.3μm Dia	25μm	[100] wire		[001]	37	.23	1,2
DS-Ni	Ni-2ThO_2	Sheet	Oriented Lamellae: <15μm Dia	<100μm	(100)			20	–	3
TD-NiCr	Ni-20Cr-2ThO_2	Sheet Bar	50-100μm Dia Single Crystals	200-700μm	(210)		[001]	10	.1	4,5 6
Ds-NiCr	Ni-20Cr-2ThO_2	Sheet	Bands of the ThO_2-rich and ThO_2-free material				[001]			7
MA 754	Ni-20Cr-0.3Al-0.5Ti-0.6Y_2O_3	Bar	~150μm Dia	~500μm	[1̄10]	[110]	[001]	14	.2	8,9
HDA-8077	Ni-16Cr-4Al-1.7Y_2O_3	Bar	~90μm Dia	~1400μm	[100] wire			~30		10,11
YD-NiCrAl	Ni-16Cr-4Al-1Y_2O_3	Bar	~.7mm Dia	~3.5mm	[1̄10]	[110]	[001]	25	.11	12,13
MA 757	Ni-16Cr-4Al-0.6Y_2O_3	Bar	~100μm Dia	~250μm	[1̄10]	[110]	[001]			13,14
MA 6000	Ni-15Cr-4.5Al-2.5Ti-2Mo-4W-2Ta-1.1Y_2O_3	Bar	~3mm Dia	~20mm	[111]	[1̄12]	[011]	30	.1	15,16,17,18
MA 956	Fe-20Cr-4.5Al-0.5Ti-0.6Y_2O_3	Bar	~2cm Dia	>10cm Dia	[3̄32] (126)[411]	[110]	[113]	25	.11	19,20,21,22
		Sheet	.1-1mm	~1cm Dia	[110] or		(307)[532]	25		
R-400	Al-0.2Fe-0.15Si-0.9Al_2O_3 (vol.pct.)		elongated grains, 0.2-10mm long 1.5 < GAR < 30 ~0.3μm grains		unknown			27	.18	23,24,25
MA Al	Al-0.5C-2O				minor, if any		minor, if any	30x100 Al_2O_3 Flakes ~100 Dia Al_4C_3		26,27,28
Al-9052	Al-4.Mg-1.1C-0.5O	Bar	Equiaxed, ~0.3μm		minor, if any		minor, if any	30x100 Al_2O_3 Flakes ~100 Dia Al_4C_3		26,27,28

Basic Mechanical Properties of ODS Alloys

Elastic Modulus - Due to the preferred crystallographic orientation of most Fe- and Ni-base ODS alloys, the elastic properties are dependent on testing direction. The variation of Young's Modulus with temperature is shown in Fig. 2(a) for several ODS alloys. Clearly the bcc alloy MA 956 is much stiffer than the fcc nickel materials; in fact its modulus is approximately twice that of the [100] orientated TD-Ni and MA 754. As the behavior for untextured MA 754 is similar to that of cast or wrought Ni-base superalloys as well as MA 754 in the long transverse direction (40), longitudinally oriented MA 754 is much less rigid than conventional alloys. On the other hand the near agreement among the sets of data for MA 6000 and untextured MA 754 indicates that the former alloy should perform elastically as γ'-strengthened materials.

Tensile Properties - The ultimate tensile strengths of ODS Ni alloys are compared to those for their nondispersion strengthened matrix in Fig. 2(b); clearly at elevated temperatures the strengths of the ODS versions are greater than the parent alloy. Much tensile data exists for ODS Al alloys including the experimental high temperature mechanically alloyed Al-Ti Alloy E5 (Al-11.6Ti- 1.9C-0.7O) (46), DISPAL 3 (a "reaction milled" pure Al matrix dispersion strengthened alloy containing Al_2O_3 and Al_4C_3 particles) (33), Al-9052, MA Al containing 2.5 to 3.5 percent Al_2O_3 (26), R 400 (a powder blended dispersion strengthened Al alloy) (24,25), and SAP 930 (Sinter Aluminium Powder containing 7 wt pct Al_2O_3 (24,33). With the exception of the weaker R 400 based alloys and SAP 930, all the materials possess similar tensile strengths to 450 K; however at higher temperatures only E5 and DISPAL 3 have promising properties. The lack of elevated temperature strength in the mechanically alloyed material Al-9052 is understandable since its intended use is high performance marine applications (47).

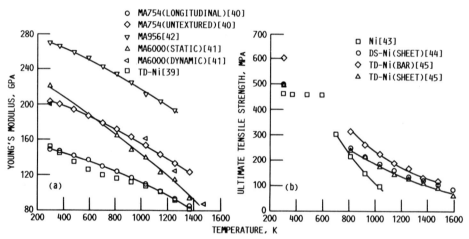

Fig. 2: Young's modulus (a) and ultimate tensile strength (b) as a function of temperature for several ODS alloys.

Fatigue properties - With the possible exception of MA 6000 relatively limited mechanical fatigue data exist for ODS alloys; however the work to date has indicated that ODS alloys, at least in the longitudinal direction, exhibit better properties than conventional materials, and examples of such behavior are shown in Fig. 3. Comparison of room temperature low cycle strain - life plots for MA, ingot and powder metallurgy versions of nominal Al-2.8Li-1.5Cu-

1Mg materials (Fig. 3(a)) demonstrate that the unaged mechanically alloy product is more fatigue resistant than the overaged ingot or powder metallurgy alloys. Similar behavior was also observed when underaged conventional materials were compared to the unaged MA alloy (48). Such improvements have been related to the presence of the dispersoid and differences in microstructure and texture. The dispersoid leads to more homogeneous deformation as the particles prevent planar slip both during loading and unloading portions of a cycle. Such homogeneity of the strain retards crack nucleation and directly yields improved fatigue life. Additionally the processing involved in the manufacture of ODS alloys is a factor as the size of inclusions can be much smaller in ODS alloys than their nondispersed conventional forms; hence these defects are less likely crack initiation sites. In fact the fatigue strength can be a factor of 2 greater in the longitudinal direction than that in the transverse direction (Fig 3(b)), and this difference has been attributed [17,18], in part, to the size of potential failure initiation sites (i.e., internal inclusions for longitudinal testing and surface connected processing defects for transverse testing). While weaker, it can be seen in Fig. 3(b) that the fatigue strength in the transverse direction is still approximately equal to that of conventional cast alloys.

Thermal fatigue testing of several ODS alloys (49-53) in the longitudinal direction has shown that Ni-base materials have excellent properties (Fig.3(c)). These studies which involved heating to about 1400 K in 30 to 240 s followed by an equally fast cooling illustrated that the thermal fatigue behavior of MA 754, HDA 8077 and MA 6000 was approximately equal to that of directionally solidified Ni-base superalloys. In general the thermal fatigue resistance of the ODS alloys has been ascribed to the low elastic modulus [100] texture (Table I) of the materials, which is identical to that for directionally solidified superalloys. However the [110] texture of MA 6000 indicates that orientation is not the only factor which leads to good thermal fatigue properties (53). In contrast to Ni-base materials, rapid thermal cycles lead to poor thermal fatigue resistance for the Fe-base alloy MA 956 (Fig. 3(c)). Several reasons (18,49-53) have been advanced for this behavior: (1) the bcc crystal structure; (2) argon entrapped during the mechanical alloying; (3) the higher elastic modulus, preferred orientation direction of this alloy; and (4) adherence of the oxide scale. Unfortunately it is not known which factor(s) is responsible for the apparent low thermal fatigue properties of MA 956. Two studies (10,51), however, do question the conclusion that MA 956 suffers from an inherent lack of thermal fatigue resistance. Prabhakar and Unnam (51) subjected MA 956 sheet to a more gentle heating/cooling schedule (780 s heat up to 1365 K, 420 s hold at temperature, followed by air cooling for 1200 s) which simulates an advanced aerospace vehicle reentry, and they did not find any evidence of cracking after 1200 cycles. Additionally Lowell, et al. (10) undertook cyclic high velocity oxidation and hot corrosion testing of a variety of superalloys where modified vane shaped specimens where subjected to Mach 0.3 combusted gas streams (1373 K for oxidation testing or 1173 K for hot corrosion studies) for 3600 s followed by 180 s in a similar velocity, ambient temperature air blast. For MA 956 the oxidation testing was discontinued after 3790 cycles, and the hot corrosion exposures stopped after ~ 2300 cycles. Neither of these two experiments yielded observable thermal fatigue cracking or environmental attack. These latter results are particularly significant in that (1) MA 956 can withstand fairly severe heating/cooling conditions without signs of failure and (2) the oxidation/hot corrosion resistance of this alloy far exceeds Ni-16Cr-4Al based ODS alloys as well as bare and coated superalloys.

Time to Rupture - The measurement which most easily and directly demonstrates the high temperature, high strength capability of ODS materials is the stress rupture test, and typical data are presented in Fig. 4. The thoriated Ni and TD-NiCr generally possess lives on the order of 100 to 1000 h for stresses ranging from ~ 70 to ~ 50 MPa. The strength of DS-NiCr, whose structure consists of bands of thoriated and nonthoriated Ni-20Cr (Table I), is somewhat weaker with stress levels between ~ 40 and ~ 30 MPa causing failure in about 100 to 1000 h respectively. MA 754 tested in the longitudinal direction is significantly stronger than the other Ni or Ni-20Cr alloys where the 100 to 1000 h range in life exists over a narrow band of stress centering about 100 MPa. in the long transverse direction, on the other hand,

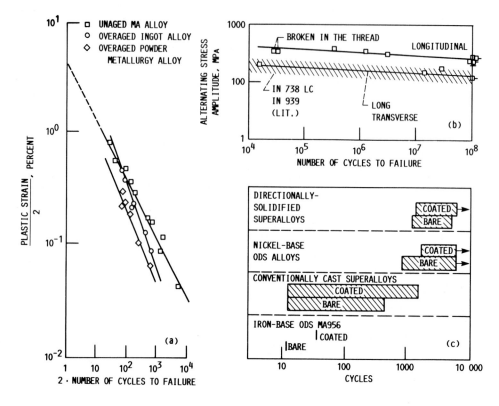

<u>Fig. 3:</u> Comparison of the fatigue behavior of ODS alloys to conventional materials (a) room temperature plastic strain - cycles to failure behavior for ODS, ingot and powder metallurgy versions of Al-2.8Li-1.5Cu-1Mg (48); (b) The 1123 K high cycle fatigue strength of MA 6000 tested in the longitudinal and long transverse directions compared to several conventional alloys (17); (c) Range in number of thermal fatigue cycles to crack initiation for several classes of high temperature alloys (52). Each cycle consisted of immersion in an air fluidized bed at 1400 K for 180 s followed by 180 s at 630K.

the strength of MA 754 is limited. Such differences in behavior are ascribed (55-57) to the GAR's (Table I) which are much greater in the longitudinal direction than the long transverse direction (Fig. 1(b)). Similar behavior, due to differences GAR (Table I), was also found in MA 757 (Fig. 4(b)) tested in the longitudinal and long transverse directions. The stress rupture curves for MA 757, MA 956 and YD-NiCrAl tested in the longitudinal direction are nearly parallel to the time axis; thus small changes in stress can greatly increase/decrease the time to failure. MA 6000 (Fig. 4(c) also shows "flat" 1366 K stress rupture curves along with a large difference in strength depending on testing direction. Additionally Fig. 4(c) illustrates the strength advantage of MA 6000 over conventional Co-base (MAR- M509) and Ni-base (B-1900 and MAR-M421) high temperature alloys. Similar comparisons for Ni/Ni-20Cr (Fig. 4(a) or Ni-16Cr-4Al/Fe-20Cr-4.5Al Fig. 4(b) can not be realistically made because of the weakness of these alloys at 1366 K. Time to rupture data for several Al alloys tested at 773 K are illustrated in Fig. 4(d), where the curve shown for R 400, which has a low Al_2O_3

content (~ 0.9 vol. pct.) was estimated from the creep data of Clauer and Hansen (23) and the assumption of an approximately 1 percent strain on rupture (24). In all cases the expected lives are strong functions of the applied stress. For Al alloys high dispersoid/particle content definitely improves strength, and the properties of DISPAL 3 are significantly better than the other materials.

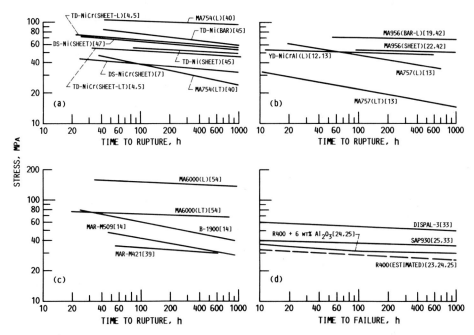

Fig. 4: Time to rupture behavior for representative ODS alloys. (a) Ni and Ni–20Cr, (b) Fe–20Cr–4.5Al and Ni–16Cr–4Al, (c) Superalloys, and (d) Al alloys. Parts (a–c) at 1366 K; part (d) at 773 K.

As first noted by Bourne, et al. (55), high grain aspect ratios in the testing direction lead to good high temperature strength for simple ODS alloys. In a similar vein, Arzt and Singer (58) found that increasing the GAR of MA 6000 from ~ 5 to >30 leads to a hundred-fold increase in the stress rupture life under the same test conditions (Fig. 5(a)). Accompanying this improvement, they found a change in the fracture mode from intergranular at low GAR to trans-granular at high GAR. It was reasoned that this transition was the result of cavity growth on transverse grain boundaries being constrained by sliding along boundaries parallel to the stress axis (Fig. 5(b)). At low GAR values the weak transverse boundaries can easily fail and separate with the necessary displacements along the longitudinal boundaries accommodated by sliding. However as the longitudinal boundaries are elongated (GAR increased), grain boundary sliding becomes more difficult.

Steady state creep – For many ODS alloys creep resistance is quite dependent on grain structure where fine or ultrafine equiaxed grain structures can yield superplastic or superplastic-like behavior (59). While such tendencies are of interest for fabrication techniques, the microstructure of the final product must possess adequate creep strength. In general this means a large grain size, high GAR materials. There are, however, exceptions: for example DS–Ni

(Fig 4(a)) and most dispersion strengthened Al alloys (Fig. 4(c)). The following will be restricted to those structures which yield reasonable creep properties.

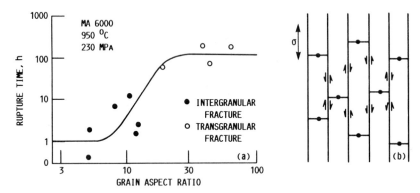

Fig. 5: Time to rupture as a function of GAR for MA 6000 tested at 1323 K and 230 MPa (a), and (b) schematic representation illustrating the constraint of grain boundary sliding along longitudinal boundaries on the ability of transverse boundaries to separate (58).

(a) Al- and Ni-base materials: Considerable elevated temperature creep testing of ODS alloys has been undertaken, and many materials appear to have a threshold stress σ_0 defined as the critical stress below which measurable creep does not occur. Evidence for such behavior can be seen in Fig. 6(a) for the ODS Al alloy R 400 tested between 573 and 873 K where the strain rate - flow stress curves in the higher temperatures and slower strain rates regimes are becoming vertical. Similar tendencies can be seen when normalized creep data are examined. Clearly both for MA Al (Fig. 6(b)) and single crystal TD-NiCr (Fig. 6(c)) extreme changes in strain rate can occur as stress is varied about stress/modulus ratios of 10^{-3}. A corresponding behavior is also shown in the stress rupture curves (Fig. 4) where "flatness" can be associated with a threshold stress.

If strain rate data $\dot{\epsilon}$ are analyzed in terms of the power-law equation

$$\dot{\epsilon} \propto \frac{Eb}{kT} \left[\frac{\sigma}{E} \right]^n \exp(-Q/RT) \tag{1}$$

(where E is an elastic modulus, b is the magnitude of the Burgers vector, σ is the applied stress, n is the creep stress exponent, Q is the activation energy for creep, R is the universal gas constant, T is the absolute temperature, and k is Boltzmann's constant), unusually high values of n and Q have been found. For instances Wilcox and Clauer (1) report n=40 and Q=795 kJ/mol for TD-Ni bar, and several investigations (15,16,29) present data for MA 6000 where 10<n<50 and 600<Q<800 kJ/mol. Due to the tendency for little deformation under relatively large stresses, such high values of n and Q have been rationalized by replacing σ in eqn (1) with an effective stress σ' where $\sigma'=\sigma-\sigma_0$. With this approach it is possible to reduce the values of n and Q to those of the pure matrix (16,61). The appearance of a threshold stress for creep in ODS alloys is of great practical importance, as it signifies that these materials can be used under some finite load at extreme temperatures without deformation. Therefore the factors controlling the magnitude of threshold stresses for creep are under active investigation, and results have been recently reviewed by Arzt (62) and Raj (63).

There appear to be at least two different threshold stresses in ODS Ni alloys (64): one associated with dislocation controlled creep, and the other related to diffusional creep / grain boundary sliding mechanisms. For the case of single crystalline or high GAR materials, it has been suggested that the threshold stress is the result of an attractive dislocation - dispersoid interaction due to relaxation of the dislocation core at the interface (65-69). Confirmation of this idea has been presented in the form of several transmission electron photomicrographs of "departure" side pinning by Nardone, et al. (68) as well as Arzt, Rösler and Schröder (70). In contrast to the core relaxation concept, where the dispersoid is assumed to be nondeformable, Raj (63) has examined the case where particles deform under the action of a fast approaching dislocation, and this model leads to an apparent threshold stress. The consequences of both models are illustrated in Fig. 7, where it is predicted that core relaxation k_r must exceed 24 and 38 percent respectively in Ni-Y_2O_5 and Ni-ThO_2 ODS alloys for the Arzt-Wilkinson model (69) to control the threshold stress. If the relaxation is less for the Ni alloys or for the Al-Al_2O_3 system, Raj's model indicates that deformation of the oxide will be the factor which controls the threshold stress. The most interesting aspect of these two models lies in the difference between predictions of the most efficient dispersoid. If core relaxation is controlling, then a stable array of voids (ie. no dispersoid) would be best; however, if particle yielding is governing, the dispersoid-matrix combinations would be important.

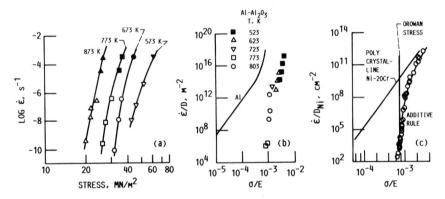

Fig. 6: Typical creep behavior in ODS alloys: (a) minimum creep rate - stress behavior for the Al - Al_2O_3 alloy R 400 (23); (b) normalized creep rate - stress behavior for MA Al and pure Al (60) and (c) single crystal TD-NiCr and polycrystalline Ni-20Cr (6). Creep rate normalized with respect of lattice diffusion coefficient D_1 and stress normalized by Young's modulus E or shear modulus G.

For dispersion strengthened alloys with a "bamboo" grain structure or tested in a low GAR direction, a second type of threshold stress for creep becomes important. It can be associated with grain boundary deformation mechanisms via (1) diffusional creep through the ability of boundaries in ODS alloys to act as sources/sinks for vacancies, and/or (2) resistance to grain boundary sliding along longitudinal boundaries. Such threshold stresses are dependent on GAR as demonstrated in Fig. 8(a) for a variety of Ni-base ODS alloys; however there does not appear to be a straightforward relationship between grain size of moderate dimensions (100 to 400 μm) and σ_0 (2). While not well understood, this threshold stress is important because the creep strain tends to be localized at the transverse grain boundaries, and it can severely affect post creep test mechanical properties (71).

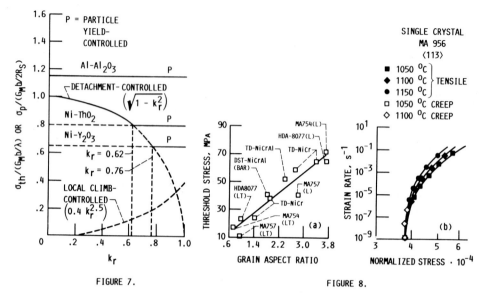

FIGURE 7.   FIGURE 8.

Fig. 7: Comparison of dislocation detachment model (69) and particle yield controlled model (63) for threshold stresses in ODS alloys.

Fig. 8: Threshold stress for creep in moderate grain size ($100 <$ average $< 400~\mu m$) ODS Ni-base alloys (2,13) as a function of the grain aspect ratio (a), and (b) minimum creep rate - normalized stress curves for MA 956 illustrating threshold stress behavior (73).

(b) MA 956: Although the bcc alloy MA 956 also exhibits threshold stress characteristics (Fig. 8(b)), it is not clear that this alloy can creep in the traditional sense. At high stresses or imposed fast strain rates, MA 956 can readily be deformed (19,22,72,73). However under tensile conditions designed to produce slow deformation, it appears that MA 956 "creeps" by a crack nucleation and growth process (19,22,72). Evidence cited for this conclusion include:
1. Strain to failure which decreases with strain rate (Fig. 9(a));
2. Extremely high dependency of strain rate on stress (Fig. 9(b)) {in terms of eqn (1) the stress exponent would be 100};
3. The observation of crack-like features on the fracture surfaces;
4. Cracks in the test section which appear to be extending by the joining of cavities which nucleate and grow ahead of the running crack (Fig. 9(c));
5. Identical fractures in constant velocity and creep/stress rupture tests.

Apparently the threshold stress exhibited by MA 956 bar tested in the longitudinal direction or sheet (tested in either the longitudinal or long transverse direction) is associated with the formation and/or growth of cracks; thus it is a third type of threshold stress and is quite different than those found in ODS Ni alloys.

Diffusional creep - Diffusional creep is extremely important in ODS alloys as it can affect the distribution of strengthening particles. An understanding of diffusional creep is best approached by first examining the effects of diffusion in dispersion strengthened materials. For instance the microstructure of DS-NiCr, which consists of alternating ThO_2-containing and ThO_2-free bands (Table I), is the result of the method to produce ODS Ni-20Cr sheet by pack chromizing Ni-2ThO$_2$ sheet (7). In this case an initially uniform distribution of ThO$_2$ within thoriated Ni alloys was partitioned into bands by the diffusion of Cr into the ODS Ni where the ThO$_2$-free regions are a direct consequence of substitutional diffusion with grain boundaries acting as vacancy sources (74).

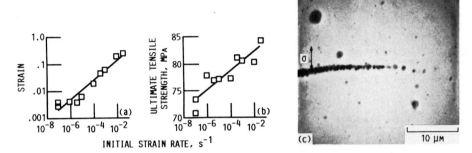

Fig. 9: Various aspects of crack formation in MA 956 bar tensile tested in the longitudinal direction at 1366 K (72): (a) strain - strain rate behavior,strain rate behavior, (c) SEM photomicrograph of crack formed in gage section at $\dot{\epsilon}$ = 8.3 x 10^{-6}s^{-1}.

Diffusional creep is the result of an applied tensile stress forcing transverse boundaries to act as vacancy sources which increases the local vacancy concentration at these regions. Owing to their higher local chemical potential, vacancies diffuse away through the lattice {Nabarro-Herring Creep (75,76)} or along grain boundaries {Coble Creep (77)} to areas of lower potential, and the counter-current mass flow produces creep in the direction of tension. If convenient sinks for vacancies are available, a steady state diffusional creep rate can be established as reflected by the Nabarro-Herring and the Coble creep models. However it is not necessary to have vacancy sinks in order for a Nabarro-Herring type creep to take place. Increasing the grain size is the usual method employed to reduce diffusional creep and, hence, the formation of dispersoid-free bands. Unfortunately this approach can not be completely effective unless the material is a single crystal, for a tensile stress acting on any transverse boundary will force it to act as a vacancy source and increase the local vacancy concentration over the equilibrium level. As the excess vacancies will diffuse into the grain interior and the vacancy concentration at the source is being held constant by the applied stress, a pseudo vapor - infinite solid diffusion couple geometry is established. Through this analogy the thickness of dispersoid-free bands w' can be calculated as a function of stress and time t for the unconstrained situation (78), where

$$w' = 2.26 \frac{\sigma \Omega (D t)^{0.5}}{kT} \qquad (2)$$

with D being the diffusion coefficient and Ω being the atomic volume. While the rate of increase in the width of the dispersiod-free band diminishes with time, quite wide regions compared to the normal interparticle spacing can be developed in relatively short periods. For example an ODS Fe or Ni alloy tested at 1400K (D=10^{-14}m/s^2) and 35 MPa, eqn (2) predicts that an approximately 2.5 μm thick dispersoid-free band would form in 100 h.

The formation of particle-free regions in an ODS alloy during creep can have a profound effect on subsequent behavior, because these areas will have different properties than the dispersion strengthened material. For instance, without the strengthening dispersoids these regions will be: (1) much weaker {no barriers to dislocations} and thus unable to support their share of the load; (2) prone to creep cavity formation; and (3) probably less oxidation/corrosion re-sistance. In short <u>diffusional creep can destroy the very microstructure that was designed to prevent dislocation creep</u>, and examples (71,79) of dispersoid-free regions formed during creep testing have been found in ODS alloys.

<u>Residual properties</u> – If diffusional creep (or any other active creep mechanism(s) which causes microstructural damage at the boundaries) occurs in ODS alloys, then subsequent test-ing could lead to premature failure due to concentration of deformation in the dispersoid-free/damaged regions. This indeed was the case for TD-NiCr, where it was observed (5,80) that minor creep deformation (< 0.2 pct) at 1365 K was capable of drastically reducing the tensile ductility and somewhat lowering the ultimate tensile strength. A number of studies in-volving tensile testing of ODS Ni-base alloys following creep exposures (2,4,13,71,79,81) have been undertaken, and it has been found that creep degradation is:

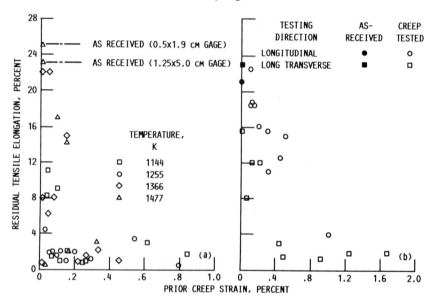

Fig. 10: Residual room temperature tensile elongation as a function of prior creep strain for (a) TD-NiCr creep tested between 1144 and 1477 K (4) and (b) MA 754 creep tested at 1365 K (79).

(1)   dependent on the amount of prior creep strain but not the actual creep temperature-stress conditions as is illustrated in Fig 10(a) by the room temperature tensile elonga-tions of TD-NiCr which had been creep tested between 1144 and 1477 K; and

(2)   somewhat dependent on the grain aspect ratio in the testing direction, as shown in Fig. 10(b) where MA 754 tested in the longitudinal direction (GAR = 3.7) is more likely to retain its as received ductility after creep than material tested in the long transverse directions (GAR = 0.7).

While it is clear from Fig. 10 that creep damage can occur in materials with GAR's < 4, it is probable that testing in high GAR directions would not result in degradation as creep would take place by dislocation motion with little or no grain boundary damage. This belief follows directly from the observations of Arzt and Singer (58) who noted that the fracture path in

MA 6000 switches from intergranular to transgranular when the GAR exceeds 10. Thus it would appear that dispersion strengthened materials with grain aspect ratios in excess of 10 would not be subject to losses in residual properties. Unfortunately a high GAR direction in an ODS alloy generally means that a very low GAR direction(s) also exists, and such boundaries would be quite susceptible to creep damage. Even though creep in a multigrained dispersion strengthened material can lead to microstructural damage, ODS Ni-base alloys subjected to high temperature, high stress exposures will not be damaged if creep does not occur. Therefore the threshold stress for diffusional creep/grain boundary sliding mechanisms represents the maximum safe stress which can be tolerated by the alloy without introducing microstructural defects such as dispersiod-free bands, grain boundary cavities and/or oxides. The only remaining dilemma is to either experimentally or theoretically determine this threshold stress; unfortunately this is not a trivial problem.

Summary

In this short review an attempt has been made to present and contrast the behavior of ODS alloys with their matrix compositions and more conventional alloys. In general the properties of the ODS materials are better than the comparative materials. However, in many cases, direct comparisons can not be made as ODS alloys are intended to be utilized under conditions where few conventional materials have useful lives. Enhanced mechanical properties can be traced to the presence of the dispersoid and, sometimes, to the lesser number and size of second phases. With regard to application of ODS alloys in severe temperature - stress regimes, most materials exhibit threshold stress behavior. In fact the data indicate the existence of at least three different threshold stresses due to: (1) dislocation - particle interactions; (2) diffusional creep/grain boundary sliding; and (3) for MA 956 crack nucleation/growth processes. It is extremely important to know which threshold stress is in effect as exposure of ODS alloys to stresses which exceed either of the latter two threshold stresses can lead to unexpected failure by cracking or degradation of residual properties.

References

(1)   B.A. Wilcox and A.H. Clauer: Trans. AIME 236 (1966) 570.
(2)   J. D. Whittenberger: Met. Trans. A 8A (1977) 1155.
(3)   R. W. Fraser, S. G. Berkley, and B. Hessler: "Dispersion Strengthened Nickel for Gas Turbine Applications," Metal Powder Report 40 No. 10. (1985).
(4)   J. D. Whittenberger: Met. Trans. A 7A (1976) 611.
(5)   L.J. Klingler, W.R. Weinberger, P.G. Bailey and S. Baranow: NASA CR-120796, 1971.
(6)   R.W. Lund and W.D. Nix: Acta Metall. 24 (1976) 469.
(7)   R. C. Cook and L. F. Norris: NASA CR-134484, 1973.
(8)   J. J. Stephens and W. D. Nix: "Superalloys, 84" (eds. Gell, M., Kortovich, C. S., Bricknell, R. H., Kent, W. B. and Radavich, J. F.) AIME, Pittsburgh, PA3
(9)   J. J. Stephens and W. D. Nix: Met. Trans. A 16A (1985) 1307.
(10) C.E. Lowell, D.L. Deadmore and J.D. Whittenberger: Ox. Metals 17 (1982) 205.
(11) D.L. Klarstrom and R. Grierson: NASA CR-134901, 1975.
(12) M.E. McAlarney, R.M. Arons, T.E. Howson, J.K. Tien and S. Baranow: Met. Trans. A 13A (1982) 1453.
(13) J. D. Whittenberger: Met. Trans. A 12A (1981) 193.
(14) J. D. Whittenberger: NASA TP 1781, 1981.
(15) T.E. Howson, D.A. Mervyn and J.K. Tien: Met. Trans. A 11A (1980) 1609.
(16) J. D. Whittenberger: Met. Trans. A 15A (1984) 1753.
(17) W. Hoffelner and R. F. Singer: Met. Trans. A 16A (1985) 393.

(18) R.F. Singer and E. Arzt: "High Temperature Alloys for Gas Turbines and Other Applications" (eds. W. Betz, R. Brunetand, D. Coutsouradis, H. Fischmeister, T.G. Gibbons, I. Kvernes, Y. Lindblom, J.B. Marriott and D.B. Meadowcroft) Reidel, 1986, p. 97.
(19) J. D. Whittenberger: Met. Trans. A 12A (1981) 845.
(20) R. Petkovic-Luton and M.J. Luton: "Strength of Metals and Alloys, ICSMA 7" Pergamon Press ,1986, p. 743.
(21) J.J. Fischer, I. Astley and P. Morse: "Superalloys: Metallurgy and Manufacture" (eds. B.H. Kear, D.R. Muzyka, J.K. Tien and S.T. Wlodek) Claitor's Publishing Div., 1976, p. 361.
(22) J. D. Whittenberger: Met. Trans. A 9A (1978) 101.
(23) A.H. Clauer and N. Hansen: Acta Metall. 32 (1984) 269.
(24) N. Hansen: Powder Metallurgy 10 (1967) 94.
(25) N. Hansen: Powder Metallurgy 12 (1969) 23.
(26) J.S. Benjamin and M.J. Bomford: Met. Trans. A 8A (1977) 1301.
(27) R.F. Singer, W.C. Oliver and W.D. Nix: Met. Trans. A 11A (1980) 1895.
(28) P.S. Gilman: "Aluminum Lithium Alloys II" (eds. T.H. Starke and E.A. Starke), 1983, p. 485.
(29) H.F. Merrick, L.R. Curwick and Y.G. Kim: NASA CR-135150, 1977.
(30) Y.G. Kim and H.F. Merrick: NASA CR-159493, 1979.
(31) R.E. Allen and R.J. Perkins: Final Report Naval Air Systems Command Contract No. N00019-71-C-0100, 1972.
(32) I.G. Wright and B.A. Wilcox: Met. Trans. 5 (1974) 957.
(33) J.H. Weber: "Dispersion Strengthened Aluminum Alloys," Handbook of Composite Materials, (eds. S. Ochiai), Marcel Dekker, to be published.
(34) P.K. Footner and C.B. Alcock: Met. Trans. 3 (1972) 2633.
(35) R.W. Lund and W.D. Nix: Acta Metall. 24 (1976) 469.
(36) J.S. Benjamin, T.E. Volin and J.H. Weber: High Temp. High Press. 6 (1974) 443.
(37) J.K. Tien: "Frontiers of High Temperature Materials II" (eds. J.S. Benjamin and R.C. Benn) INCO Alloys International, 1983, p. 119.
(38) L.J. Ebert, M.L. Telich, M. Holly and B. Roopchand: US Army Tank Automotive Command R&D Report No. 12896, 1983.
(39) "High Temperature, High Strength Nickel Alloys," INCO Alloys International.
(40) IncoMAP INCONEL alloy MA 754 Data Sheet.
(41) P.P Millan Jr. and J.C. Mays: NASA CR 179537, 1986.
(42) IncoMAP INCONEL alloy MA 956 Data Sheet.
(43) Alloy Digest Data Sheet.
(44) Sherritt Gordon Data Sheet, DS-Ni.
(45) Aerospace Structural Materials Handbook.
(46) Mirchandani, P.K. and Benn, R.C.: "Experimental High Modulus Elevated Temperature Al-Ti Base Alloys by Mechanical Alloying," Proceedings of SAMPE Metals and Metals Processing Conference, 1988 (2-6 August88, Dayton, OH)],
(47) IncoMAP INCONEL alloy Al 9052 Data Sheet.
(48) R.T. Chen and E.A. Starke Jr: Mat. Sci. and Eng. 67 (1984) 229.
(49) P.G. Bailey: NASA CR-135269, 1977.
(50) K.E. Hofer, V.L. Hill, and V. E. Humphreys: NASA CR-159842, 1980.
(51) O. Prabhakar and J. Unnam: "Thermal Fatigue Resistance of Iron Base Oxide Dispersion Strengthened Alloy Sheet Material", TMS-AIME Paper Selection, Paper No. F82-12, 1982.
(52) J. D. Whittenberger and P. T. Bizon: Int. J. Fatigue 4 (1981) 173.
(53) Y. G. Kim and H. F. Merrick: "Superalloys 1980" (eds. J.K. Tien, S.T. Wlodek, H. Morrow III, M. Gell and G.E. Maurer) ASM, 1980, p. 551.
(54) IncoMAP INCONEL alloy MA 6000 Data Sheet.
(55) A.A. Bourne, J.C. Chaston, A.S. Darling and G.C. Bond: British Patent 1,134,492 (1962).

(56) R.W. Fraser and D.J.I. Evans: "Oxide Dispersion Strengthening" Gordon and Breach, 1968, p. 375.
(57) B.A. Wilcox and A.H. Clauer: Acta. Metall. 20 (1972) 743.
(58) Arzt, E. and Singer R. F. "Superalloys 1984", (eds. M. Gell, M., Kortovich, C. S., Bricknell, R. H., Kent, W. B. and Radavich, J. F.)) AIME, 1984, p. 369.
(59) J. K. Gregory, J. C. Giebling and W. D. Nix: Met. Trans. A 16A (1985) 777.
(60) Oliver, W.C. and Nix, W.D.: Acta Metall. 30 (1982) 1335.
(61) S. Purushothaman and J.K. Tien: Acta Metall. 26 (1978) 519.
(62) Arzt, E.: "Threshold Stresses for Creep of Dispersion Strengthened Alloys," Handbook of Composite Materials (eds. S. Ochiai) Marcel Dekker, NY, to be published.
(63) Raj, S.V.: "Creep and Creep Fracture of Dispersion-Strengthened Alloys," Handbook of Composite Materials (eds. S. Ochiai) Marcel Dekker, NY, to be published.
(64) C. M. Sellers and R. Petkovic-Luton: Mat. Sci. Eng. 46 (1980) 75.
(65) D. Srolovitz, R. Petkovic-Luton and M.J. Luton: Scripta Metall. 16 (1982) 1401.
(66) D. Srolovitz, R. Petkovic-luton and M.J. Luton: Acta Metall. 31 (1983) 2151.
(67) D. Srolovitz, M.J. Luton, R. Petkovic-luton, D.M. Barnett and W.D. Nix: Acta Metall. 32 (1984) 1079.
(68) V.C. Nardone, D.E. Matejczyk, and J.K Tien: Acta Metall. 32 (1984) 1509.
(69) E. Arzt and D.S. Wilkinson: Acta Metall. 34 (1986) 1893.
(70) E. Arzt, J. Rösler and J.H. Schröder: "Creep and Fracture of Engineering Materials and Structures" THE INSTITUTE OF METALS, London, 1987, p. 217.
(71) J.D. Whittenberger: Met. Trans. 4 (1973) 1475.
(72) J.D. Whittenberger: Met. Trans. 10A (1979) 1285.
(73) R. Petkovic-Luton, D. J. Srolovitz and M. J. Luton: "Frontiers of High Temperature Materials, II", International Nickel Company, 1983, 73.
(74) J.D. Whittenberger: Met. Trans. 4 (1973) 715.
(75) F. R. N. Nabarro: "Report of a Conference on Strength of Solids", London, 1948, p. 75.
(76) C. Herring: JAP 21 (1950) 437-445.
(77) R. L. Coble: JAP 34 (1963) 1679.
(78) J.D. Whittenberger: "Diffusion and Diffusional Creep in Dispersion Strengthened Alloys," Handbook of Composite Materials (eds. S. Ochiai), Marcel Dekker, to be published.
(79) J.D. Whittenberger: Met. Trans. 8A (1977) 1863.
(80) J.W. Davis: Proc. Natl. Tech. Con. on Space Shuttle Materials SAMPE, 1971, p. 335.
(81) J.D. Whittenberger: Met. Trans. 9A (1978) 1327.

Properties and Applications of Iron-Base ODS Alloys

H.D. Hedrich, Daimler-Benz AG, Stuttgart West Germany - FRG-

1. Introduction

In up-to-date stationary gas turbines, future-orientated aircraft engines and particularly in new small gas turbines and uprated diesel engines for vehicle propulsion, increasing process temperatures are needed in order to reduce the performance-related fuel consumption and to arrive at smaller unit volumes. This concept for new combustion engines requires an increased degree of the use of highly temperature-resistant materials.

The Mercedes-Benz research gas turbine is used as an example to show the requirements and the potential for an application of new high-temperature materials. For this up-to-date small gas turbine hot gas temperatures of 1250°C are implemented, and in order to further increase efficiency temperatures of 1350°C are envisaged. The basic concept of this two-shaft gas turbine is shown in a cutaway view in Figure 1. The high-temperature gas turbine has the following operating conditions under full load:

The fresh air taken in leaves the compressor at a pressure of 4 bar and a temperature of 200°C. The compressed fresh air is then heated up to 800 to 900°C in a rotating, regenerative heat exchanger. At this temperature the air reaches the combustion chamber and produces a hot gas with temperatures varying between 1250° to a maximum of 1350°C during the combustion of the fuel. This hot gas is guided through two mechanically independent turbine wheels, first through the compressor turbine and then through the power turbine. The hot gas flow passes the combustion chamber, then it flows through the inlet plenum and the stationary compressor turbine nozzle before it reaches the compressor turbine. Then the hot gas is guided to the power turbine through an inter-diffusor and a variable power turbine nozzle under optimum flow conditions. Then the hot gas which has now cooled down to 1000°C and whose pressure has been reduced to 1 bar is guided into the heat exchanger. In the regenerative heat exchanger the hot gas is then cooled down to 200° to 300°C, and the fresh air is preheated for the combustion process. Thus the thermal energy of the hot gas is exploited as completely as possible. To attain a high thermo-mechanical degree of efficiency intensive cooling of the components which are under the influence of hot gas is not permitted. Under these extreme thermal operating conditions without cooling of the components, component temperatures of much more than 1000°C are reached. The conventional high-temperature materials nickel- and cobalt-base superalloys very easily lose their strength and their resistance to corrosion or oxidation above a temperature of 1000°C. The use of new further advanced high-temperature materials is required for such operating conditions.

2. Choice of Materials for Operating Temperatures above 1000°C

The ceramic high-temperature materials on the basis of silicon carbide and silicon nitride are promising materials. Owing to the lack of ductility the use of ceramic materials requires special design and production methods. The specific structure of the ceramic materials makes them particularly suitable for use at high temperatures. Owing to the strong covalent and ionic bonding in ceramic materials, losses in strength occur at much higher temperatures than in metallic superalloys. Figure 2 shows the typical strength characteristics of silicon carbide and silicon nitride as well as some superalloys. The temperature range up to 1000°C or max. 1050°C can be covered with conventional nickel and cobalt superalloys. For operating temperatures in excess of 1000°C special

Fig. 1   The Mercedes-Benz experimental vehicular gas turbine

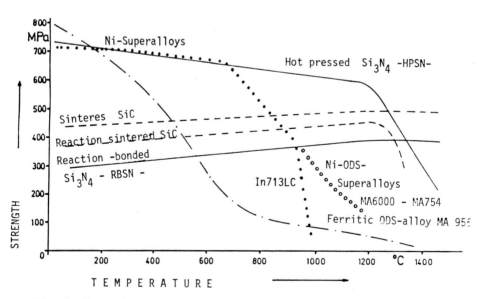

Fig. 2   Short time strength of ceramic materials, nickel-superalloys,
nickel-ODS-superalloys and a ferritic ODS-alloy

production methods for grain-structure optimization have to be applied in
addition to the purely alloy-related measures; these are: "directional solidification", "single crystal growth" or "mechanical alloying". Figure 3 shows
several possibilities to raise the temperature limits and increase strength.
The oxidation and corrosion resistance as well as the low melting temperature
range restrict the use of nickel and cobalt superalloys. The development of
efficient protection layer systems and oxide dispersion strengthening have
enabled higher operational temperatures for these superalloy systems to be
reached. However, the low melting temperatures of 1260°C to 1340°C restrict the
range of application of nickel- and cobalt-base alloys to 1100°C to 1150°C.

A very interesting ODS material is the iron-base alloy Incoloy MA 956. This ODS
material was derived from the conventional heating conductor alloys Kanthal or
Fecralloy. In addition to the excellent oxidation resistance the matrix alloy
of heating conductor material has a high melting point of 1480°C, which permits
higher component temperatures than with the nickel-base superalloys. This
ferritic ODS material is interesting and particularly suitable for stationary
and high-temperature components in turbine and diesel engine construction.

Part of the above-mentioned investigations made for the purpose of material
characterization and processing of the ODS ferrite was promoted and carried out
in the course of the European COST 501 research program.

3. Characteristics of the High-Temperature ODS-Ferrite

3.1 Oxidation and Corrosion Resistance

The chemical composition of the commercially available ODS ferrite Incoloy
MA 956 by INCO or Wiggin is based on the conventional Kanthal and Fecralloy
heating conductor alloys. In a mechanical alloying process the ODS ferrite is
produced in ball mills from prealloyed and elementary metal powders and yttria
powder and has the following basic chemical composition:
74 % b.w. Fe, 20 % b.w. Cr, 4.5 % b.w. Al, 0.5 % b.w. Ti and 0.5 % b.w. Y_2O_3.
As a result of the mechanical alloying process the yttria particles are distributed very finely in the iron-base matrix. In combustion and hot-gas atmospheres this iron-base ODS ferrite forms excellent, protecting alumina or
alumina-chromium(III) oxide spinel layers down to very low oxygen partial
pressures. These oxide layers which are highly resistant to thermodynamic
influences provide protection against oxidation and corrosion up to temperatures of 1350° to 1400°C. In addition to the chemical and thermal stability
these alumina layers have extremely good adhesive strength on the ODS ferrite.
As shown by the measurement results in Figure 4 these positive characteristics
become particularly apparent during cyclic oxidation. During a cyclic oxidation
process lasting one hour a firmly adhering Al_2O_3 top layer forms on the ODS
ferrite, which involves a small area-specific weight increase. By contrast,
nickel-base superalloys feature considerable losses in weight as a result of
spalling oxide layers already after a few oxidation cycles. As shown in Fig.5,
the excellent oxidation resistance of the ODS ferrite is proved also during
isothermal heat treatment when compared with nickel superalloys and even with
SiC and Si_3N_4 ceramic materials. With the ODS ferrite, oxidation takes place up
to temperatures of 1350°C in accordance with Tamman's scaling law, and a
protective Al_2O_3 layer is formed.

The oxidation resistance of the ODS ferrite is mainly determined by the aluminium content of the alloy. The total aluminium contents of the ODS alloy
MA 956 usually vary between 4 and 5 percent by weight. Much more important than
the integral Al content, however, is the active aluminium portion dissolved in

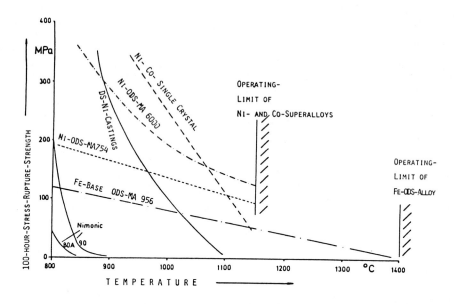

Fig. 3 Material strength and operating limit of nickel,
nickel-cobalt- and iron-base-superalloys

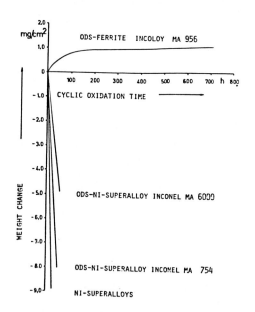

Fig. 4 Weight change of nickel- and iron-base-superalloys
during cyclic oxidation at 1100°C in air

221

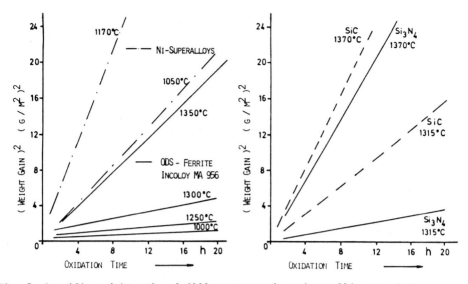

Fig. 5   Specific weight gain of different ceramic and metallic materials
during high temperature oxidation in air

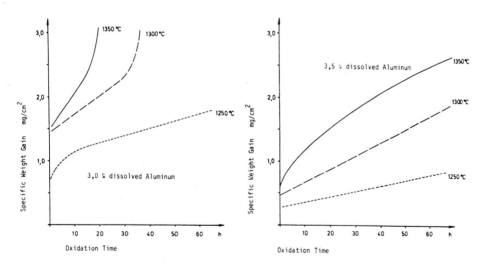

Fig. 6   Thermogravimetric specific weight change of ODS-ferrite sheet material
(o,3mm thick) with 3,0 and 3,5% dissolved Al-content during high temperature oxidation in air

the matrix. In commercial material batches this active dissolved Al content is between 3.0 and 3.5 percent by weight. Minor fluctuations in the Al content cause significant differences in oxidation characteristics, in particular with small material thicknesses. For example, when subjected to heat treatments at 1300°C, 0.3 mm thick sheets with only 3 percent by weight of active aluminium feature drastic weight increases as a result of increased oxidation already after 30 hours. Since the aluminium diffuses only to an insufficient degree to the surface, iron-chromium mixed oxides are formed which cannot provide protection against further oxidation. This cannot occur on ODS ferrites with 3.5 % b.w. dissolved aluminium, because under oxidation conditions up to 1350°C and even 1400°C the formation of a protective aluminia coating is ensured at all times - see Figure 6. Owing to the formation of stable alumina coatings, the active aluminium contents of the ODS alloy with more than 3.5 % b.w. ensure excellent corrosion restistance. The corrosion characteristics of such ODS ferrites can be compared with the resistance of MeCrAlY coatings; temperatures of up to 1400°C are permissible for short periods.

3.2 Mechanical Properties of the ODS-Ferrite Incoloy MA 956

With the ODS ferrite the iron-chromium-aluminium heating conductor base alloy is produced as the matrix material by means of mechanical alloying. To increase hot strength yttria particles are mixed in to obtain an oxide dispersion strengthened alloy. Finally, a hot-strength and coarse grain structure is built in a thermo-mechanical process with isothermal recrystallization heat treatment.

Mechanical strength of the ferrite matrix is considerably increased by finely distributed yttria dispersoids with a size of approx. 10 to 100 nm. The typical strength and formability characteristics of the various ODS ferrite structures are shown in Figs. 7 and 8. The strength of the fine-grained, dispersion-strengthened ferrite is consistently reduced until a temperature of 700°C is reached; at the same time, however, fracture elongation and thus the ductility of the material rises considerably. At temperatures in excess of 700°C ductility of the fine-grained ODS ferrite and its strength are consistently reduced. In tensile tests the highest elongations of approx. 100% and more are attained in the temperature range around 700°C. This favourable formability behaviour of the fine-grained ODS ferrite permits the production of the most diverse semi-finished products, like sheets with thicknesses between 0.3 and 6 mm, bars, rounds and forgings. When the thermo-mechanical treatment is properly conducted, i.e. with accurately matched forming and heat treatment steps, it is possible to add finally a recrystallization heat treatment with coarse-grain structure. Although this coarse-grain formation reduces the strength of the ODS ferrite in the temperature range up to 600°C, it considerably increases its high-temperature strength at temperatures in excess of 800°C. The effective grain aspect ratio and the grain size determine the hot strength as well as the creep and deformation resistance of the ODS material in the head direction.

3.3 High-Temperature Strength of the ODS-Ferrite

The high-temperature strength of the ODS ferrite can be characterized by creep and stress-rupture properties as well as hot tensile strength. Since hot tensile tests provide time and cost-saving quality control, they are suitable for industrial applications; variable strain rates have to be used in these tests. The strength characteristics of the ODS ferrites obtained from hot tensile tests are clearly a function of the strain rate.

$$\sigma = B \cdot \ln \dot{\epsilon} + C$$

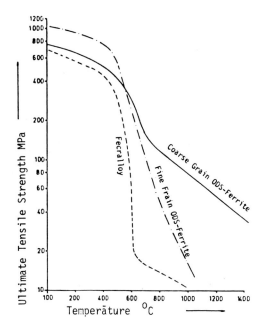

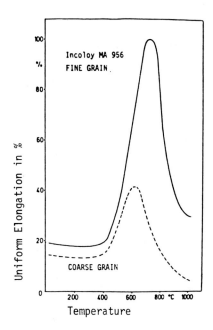

Fig. 7 Improvement in high temperature strength
of the ferritic "Fecralloy" by oxide dispersion strengthening and coarse grain
recrystallization

Fig. 8 Ductility of fine and
coarse grain ferritic ODS-alloy
as a function of temperature

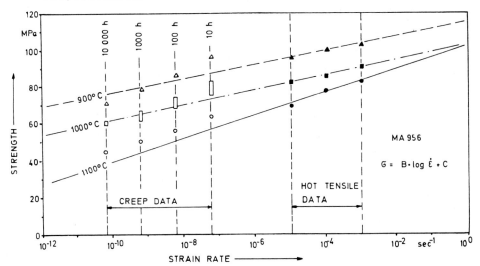

Fig. 9 High temperature tensile strength as a function of strain rate in relation to the creep strength of Incoloy MA 956

Hot tensile tests of material samples with known creep and stress rupture parameters show - see Figure 9 - that hot tensile data and stress rupture parameters can well be correlated in the case of the coarse-grain ODS ferrite. In the temperature range between 900 and 1100°C and in the strain rate range between $\dot{\epsilon} = 10^{-2}$ to $10^{-8} sec^{-1}$ the strength characteristics of bar and sheet materials made of the commercial ODS ferrite Incoloy MA 956 can be described with sufficient accuracy as a function of this deformation rate.

These laws and relationships apply to quite a large extent to the ODS ferrite, since only minor plastic deformations with relatively small dislocation move-ments occur during the relevant high-temperature tests. In the temperature range up to approx. 1200°C the grain structure of the ODS ferrite is largely stable and does not cause any additional effects which would be relevant in terms of strength.

3.4 Long-term High-Temperature Characteristics at Temperatures in Excess of 1200°C

Long exposure times at temperatures in excess of 1200°C result in a deterio-ration of the material characteristics of the ODS ferrite. Isothermal annealing at 1300°C is a relatively good and quick method of demonstrating the loss in strength of the ODS ferrite. This typical loss in strength can be ascertained in room and high-temperature tensile tests after isothermal annealing, as is shown by test results in Fig. 10. After just 5 hours' exposure at 1300°C, the initial strength of the coarse-grain recrystallized ODS ferrite drops from 700 N/mm² by approx. 10% to 630 - 620 N/mm². After an incubation period, this strength decrease is proportional to the logarithm of the ageing time. This loss in strength was initially interpreted as a potential Ostwald ripening process of the dispersoids; however, more thorough investigations revealed thermally activated formation of pores within the material being the reason. Ductility of the ODS ferrite is considerably decreased in the temperature range between 1000° and 1100°C due to this formation of pores; room temperature formability, however, is only slightly affected. The loss in strength of the ODS ferrite due to these high-temperature strains must be taken into considera-tion since especially the long-time strengths of the material are reduced due to the pore formation, as is demonstrated in Fig. 11.

3.5. Tendency to form Pores during High-Temperature Annealing

The distinct tendency to form pores gave rise to the assumption that gas con-tents of the material at high temperatures cause this phenomenon. The mechani-cal alloying process of the commercial ODS ferrite takes places in an argon inert gas atmosphere and thus the incorporation of adsorbed argon atoms is possible in mechanical alloying. Hot-extraction gas analyses carried out by Dr. Grallath at the Max Planck Institute for Material Technology, department of pure materials analytical chemistry, revealed characteristic argon contents of 20 to 30 µg Ar/g ODS ferrite MA 956 in all commercially available alloys. The argon gas content is entirely insoluble in the ferrite matrix and thus leads to pore formation at high temperatures in excess of 1200°C. Changes in density of the material are a further proof of the formation of pores. After extremely long annealing, the 20 to 30 ppm of argon cause pore volumes of 4 to 6 mm³/cm³ MA 956. The gas pores result from plastic yielding of the material at the pore edges so that the material is strained and undergoes a loss in ductility. Moreover, the internal notch effect in the material at the pore edges results in a high stress increase by the factor 2 as compared to the mean stress applied. These stress increases at the pore edges cause premature failure of the ODS ferrite. Figure 12 shows typical pore formation in ODS ferrites.

225

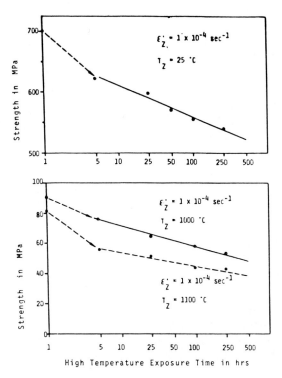

Fig. 10   Degradation of room- und high-temperature-strength as a
function of high temperature tempering time at 1300°C

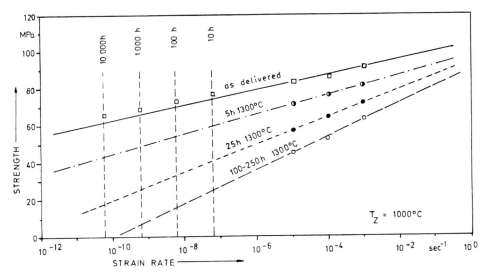

Fig. 11   Change of high temperature strength induced by isothermal
tempering at 1300°C - Testing Temperature 1000°C

226

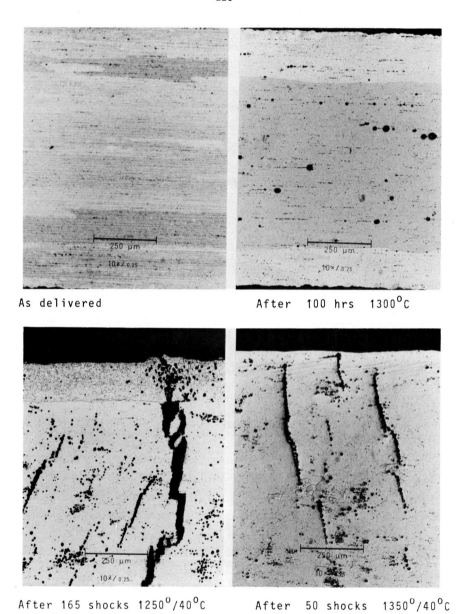

As delivered   After 100 hrs 1300°C

After 165 shocks 1250°/40°C   After 50 shocks 1350°/40°C

Fig. 12   Porosity formation in 2,0 mm sheet material of Incoloy
          MA 956 induced by isothermal tempering at 1300°C and
          cyclic thermal shock testing at 1250° and 1350°C

3.6 Thermal Fatigue and Thermal Shock Resistance

Thermal fatigue and thermal shock resistance are highly significant for
application in internal combustion engines. The temperature range between 1000
and 1350°C is of special interest as regards gas turbines. The cyclic temperature attack is produced by subjecting the test samples to hot and cold gas flow
paths at corresponding temperatures. At hot gas temperatures in excess of
1200°C the thermal shock resistance of the ODS ferrite is essentially controlled by the argon pore formation. Under these extreme strains, material
failure is not primarily induced by the temperature change-related elastic-
plastic thermal fatigue deformation of the ferritic matrix but by the argon
pore formation. First of all, argon pore formation facilitates crack initiation
and subsequently favours crack growth. Figure 12 shows typical thermal shock
cracks whose course is determined by the arrangement of the pores in the material.

4. Joining Techniques

In order to produce complex and high-temperature-resistant components, suitable
joining methods are technically and economically significant.

4.1 Welding Techniques

Owing to the materials used, fusion welding processes provide joining zones
with excellent corrosion and oxidation resistance. However, the hot tensile
strengths of fusion welded joints, like electron beam or TIG welds, are only 20
to 30% of the original ODS material values. The coagulation of the finely
distributed Y_2O_3 dispersoid triggered during melting causes the loss in hot
strength.

Welding techniques in which molten material is neither formed nor can remain in
the joining zone, like diffusion or friction welding processes, are particularly suitable for joining ODS ferrite materials. High-temperature strengths of
up to 50 to 60% of the base materials can be attained using these two methods.
The hot strength of the joint is mainly determined by the microstructure and
the grain structure of the joining zone. Figure 13 shows the strengths of
several welded joints.

4.2 Brazing Techniques

Vacuum brazing techniques are also suitable for the production of hot-strength
joints between several ODS ferrites on the one hand and between ODS ferrites
and other hot-strength alloys, like nickel superalloys on the other hand. Using
various high-temperature brazing fillers it is possible to make joints which
have approx. 50 to 60% of the base material's strength.

In the case of brazing joints oxidation and corrosion resistance is a problem,
because very often the brazing material has insufficient oxidation resistance.
Oxygen penetration into the material interior must definitely be avoided in the
area around the brazing joint, as otherwise corrosive damage may occur in the
joining zone very soon.

For the production of brazing joints with high bond strength excellent high-
vacuum conditions have to be maintained during high-temperature vacuum brazing,
to prevent the formation of alumina coatings on the joining surfaces of the ODS
ferrite.

228

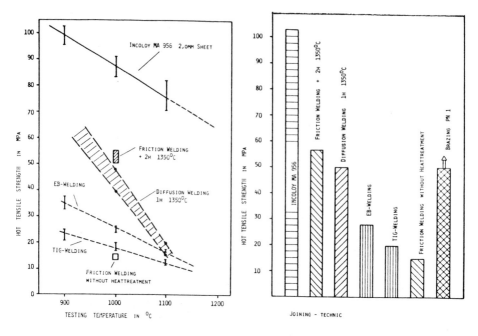

Fig. 13   Hot tensile strength of different joining technics in
ODS-ferrite Incoloy   MA 956

Fig. 14   Turbine inlet plenum   Fig. 15   Compressor turbine nozzle

5. Components in Gas Turbines and Diesel Engines made of the ODS-ferrite

For the time being, several complex stationary components which are subjected to the hot gas flow in the gas turbine without cooling and protection can be made exclusively of the ODS ferrite Incoloy MA 956. In the following, the application potential and the formability of the ODS ferrite is demonstrated by means of a few examples:

a) Combustion chamber components, like flame tubes, flame holders and hot-gas baffle plates can primarily be made of the ODS ferrite. The required structural components can be produced thanks to the good formability of the sheet metal.
b) The inlet plenum of the PWT 110 turbine is made exclusively of the ODS ferrite Incoloy MA 956. The component consists of two deep-drawn shells, several hot-spun parts and forged rings. The individual parts are joined by TIG welding - see Figure 14.

c) The compressor turbine nozzle is made either of forged vanes or machined vanes. The component is joined by electron-beam welds with a length of approx. 1.6 mm - see Figure 15.

d) The interdiffusor between compressor and power turbine with integrated adjustable power turbine nozzle is also produced of ODS material in the area where it is subjected to hot gas. The interdiffusor is hot-spun of 4 mmm thick Incoloy MA 956 sheet material. The finished interdiffusor component with welded-in ODS bearing bushings is shown in Figure 15. The variable vanes are made of the ODS ferrite or the nickel-base ODS material Inconel MA 754, see Figure 17.

e) As a result of increased combustion temperatures in the prechamber the ODS ferrite material can also be used in uprated and emission-optimized prechamber diesel engines. With the use of ODS-ferrite pins and precombuster nozzle cups combustion parameters can be varied within a wide range without having to expect fusion as a result of sulphur and carbon reactions or high-temperature corrosion attacks. A large-volume application of the ODS ferrite is expected for the diesel engine prechamber. Figure 18 shows the components of the prechamber which will be made of ODS ferrite.

Fig. 16 Outer cone of interdiffusor   Fig. 17 Power turbine vane

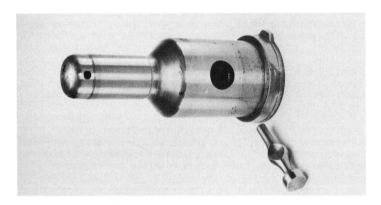

Fig. 18 Diesel-engine prechamber components manufactured from
         ODS-ferrite Incoloy MA 956

6. Conclusion

The various turbine- and diesel-engine components manufactured from the ODS-ferrite Incoloy MA 956 have successfully undergone several hundred hours running in
test rigs and vehicles. An urgent necessity would be to eliminate the formation
of argon pores. This, in our present opinion, would produce significant improvement in hot tensile strength, creep strength and resistence to thermal shock and
thermal fatigue. We wish the ODS-alloys every success in their progress to indu-
strialization.

Analysis and Modelling of the Creep-Fatigue Behaviour of Oxide Dispersion Strengthened Superalloys

D. M. Elzey and E. Arzt, Max-Planck Institut für Metallforschung, Stuttgart

Abstract

The creep-fatigue behaviour of two representative, commercial ODS (Oxide Dispersion-Strengthened) superalloys, Inconel MA 754 and MA 6000, both of which are produced by mechanical alloying, has been investigated. The mechanisms leading to failure are identified and discussed with particular emphasis on creep-fatigue interaction phenomena. Fine-grain regions are found to be the primary source of crack-initiation by the formation of creep damage. A finite difference approach to modelling of these damage processes is introduced. Results of the model are compared with experimentally determined crack initiation life. In order to explain waveform effects it is suggested that, in analogy with prestraining experiments, rapid plastic deformation can introduce residual stresses in the neighbourhood of hard particles such as oxide dispersoids during asymmetric creep-fatigue cycling. Subsequent creep damage rates are thus strongly influenced by the magnitude, sign and rate of prior plastic straining.

Introduction

The exceptional creep properties of dispersion strengthened alloys produced by mechanical alloying have resulted in extensive study of their creep behaviour. As a consequence, our understanding of creep-related phenomena in these materials is now comparable with that of currently-used, non-dispersion strengthened alloys (1). However, the high temperature fatigue behaviour of dispersion strengthened alloys has received relatively little attention although this aspect may well turn out to be design-limiting for potential applications. The current study concerns creep-fatigue of two commercially available ODS superalloys. Some results have already been presented and discussed in a previous paper (2) and are therefore only briefly reviewed here.

With increasing temperature and decreasing cyclic frequency, fatigue life will be determined by some combination of cycle-dependent and time-dependent damage processes. Complete evaluation of the mechanisms leading to failure requires consideration of the influences of one type of damage on another. These creep-fatigue interactions are frequently reflected in the effect of intergranular cavities on fatigue crack initiation and growth rates or the effect of cyclic plastic deformation on cavity nucleation and growth rates. An important consideration which is often overlooked when analyzing creep-fatigue behaviour is the state of stress produced by rapid plastic deformation and its influence on subsequent creep damage. It will be demonstrated in this paper that such considerations are necessary in order to explain the strong influence of cyclic waveform on creep-fatigue life.

Experimental

Present experimental work has been concentrated on the two yttrium-oxide dispersion-strengthened superalloys Inconel MA 754 and MA 6000. Both alloys are produced by INCO Alloys International, by the mechanical alloying process (3). Compositions and microstructural data are summarized in table 1.

MA 754 is strengthened by solid solution and by the dispersion. In addition, an elongated grain structure is obtained during heat treatment (see ref. 4 for a more detailed description of the microstructure). MA 6000 is strengthened at intermediate temperatures by γ'-$Ni_3(Al,Ti)$ precipitates while at temperatures above 900 °C the dispersion becomes the dominant source of strength. A bi-modal distribution of grain sizes exists in MA 6000; large grains with lengths on the order of several centimeters and grain aspect ratios (GAR) > 20 account for some 95% of the volume, while the remainder comprises a distribution of much smaller grains with lengths ranging from 100 μm to several millimeters and GAR between 1 and 20. For a more detailed description of the microstructure of MA 6000 the reader is referred to (5,6).

Creep–fatigue tests using total strain control have been carried out at 850, 950 and 1050 °C. Cycle forms included tensile hold–times (H–T), slow tensile loading followed by fast compression, commonly referred to as 'slow–fast' (S–F) and slow symmetric (S–S). Samples from tests run to physical separation of the specimen as well as interrupted experiments were then subjected to optical, scanning electron and transmission electron microscopy. Further experimental details will be given elsewhere (7).

Table 1. Elemental Composition of Materials and Microstructural Data

Alloy	Ni	Cr	Fe	Al	Ti	W	Mo	Co	Ta	C	B	Zr	Nb	Y_2O_3
MA 6000	Bal	15.5	–	4.5	2.5	3.8	2.0	–	1.9	.06	.01	.16	–	1.08
MA 754	Bal	20.5	.13	.30	.35	–	–	–	–	.06	–	–	–	0.5

	Grain Size [mm]	GAR	Particle Dia [nm]	Particle Spac [nm]	Texture
MA 6000	4–10	20	30	150	<110>
MA 754	0.5–3.0	10	15	125	<100>

Results

Examination of the creep–fatigue fracture behaviour of MA 6000 and MA 754 has provided ample evidence that crack initiation is caused by the presence of grain structure inhomogeneities in the form of fine grains (see also ref. 2). This is illustrated by figure 1, which shows a polished and etched axial section of an MA 6000 specimen tested to failure in 735 cycles. A crack has initiated at the transverse boundary between two fine grains and has propagated radially outward into the surrounding macrograin. Figure 2 represents a fracture surface in which more than 25 internal crack initiation sites could be identified.

Interrupted tests have shown that the transverse boundaries of fine grains are subject to cavitation damage. As typified by figure 3, cavities have an elliptical shape and range in size from 1-2 μm just prior to coalescence.

233

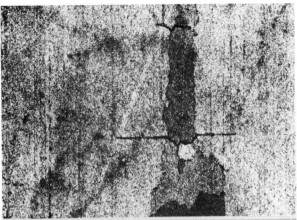

Fig.1.
Crack initiate internally at
fine grains during creep-
fatigue.(MA 6000, etched and
polished axial section, 850°C,
slow–fast, $\dot{\epsilon}_S=10^{-5}[s^{-1}]$,
$\dot{\epsilon}_F=10^{-2}[s^{-1}]$, failure after 242
cycles)

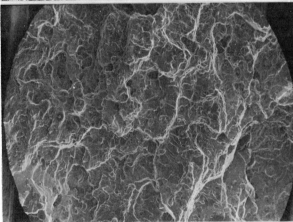

Fig.2.
Fracture surface of a sample
tested under 'slow–fast' conditions at 850°C. More than 25
individual crack initiation
sites may be identified.
(MA 6000, $\dot{\epsilon}_S=10^{-5}[s^{-1}]$,
$\dot{\epsilon}_F=10^{-2}[s^{-1}]$, $\Delta\epsilon/2=0.3\%$,
failure after 242 cycles)

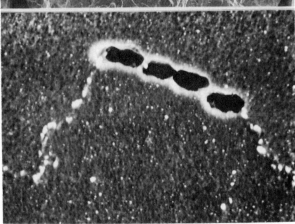

Fig 3.
Cavitation of the transverse
boundary of a fine, included
grain. Pores are typically
1–2 μm wide at coalescence
and are elliptical in shape.
(MA 6000, slow–fast, 850 °C,
$\dot{\epsilon}_S=10^{-5}[s^{-1}]$, $\dot{\epsilon}_F=10^{-2}[s^{-1}]$,
interrupted after 420 cycles)

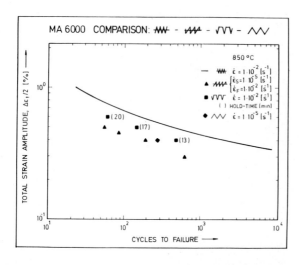

Fig.4.
Results of creep-fatigue tests of MA 6000 at 850°C illustrate the influence of waveform on cyclic life.

Figure 4 illustrates the effect of waveform on cyclic life of MA 6000 at 850 °C. It may be seen that the S-F cycle results in the shortest life, followed by S-S and H-T cycles. Results of symmetric fatigue cycling at a strain rate of 10^{-2} [s^{-1}] are also shown for comparison.

Although cracks were found to initiate internally at grain defects under all creep-fatigue conditions tested, metallographic examination often revealed features distinguishing one waveform from another. It was found for example that during tensile hold-time tests, spherical pores occur frequently on grain boundaries parallel to the loading axis and occasionally, also within the matrix. Pores not lying on a grain boundary appear to be associated with larger yttrium-aluminate particles. During S-F and S-S cycling, pores were found only on transverse grain boundaries. It was also observed that cracks initiating during tensile hold-time tests display a greater tendency to propagate intergranularly along longitudinal fine grain

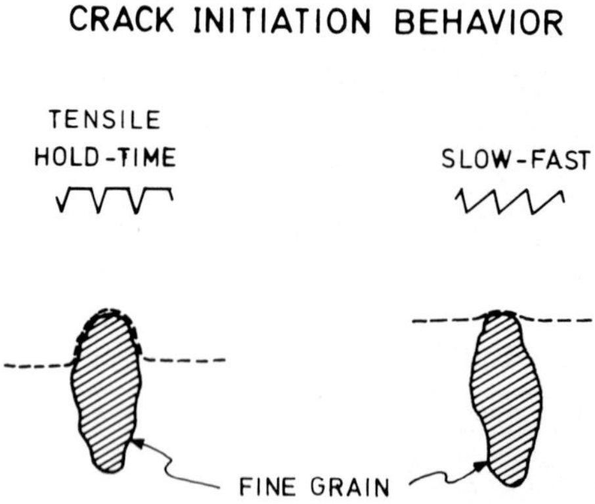

CRACK INITIATION BEHAVIOR

TENSILE
HOLD-TIME

SLOW-FAST

FINE GRAIN

Fig.5.
During tensile hold-time cycling, cracks tend to propagate along boundaries parallel with the stress axis before extending transgranularly into the surrounding matrix.

provides further confirmation that creep-fatigue life and the influence of waveform are being largely determined by local residual stresses.

Enhanced recovery at high temperatures will prevent the buildup of residual stresses as large as those produced during room temperature deformation, but it seems very likely that their effect may be significant. Residual stresses produced during RT prestraining have been found to persist over a significant fraction of creep life (9). Alloys containing oxide particles and/or grain boundary carbides and which fail or initiate cracks by intergranular cavitation, may be expected to exhibit a similar influence of waveform on creep-fatigue life.

Creep-Fatigue Crack Initiation Model

The following is a brief description of a model which has been developed in an attempt to quantify the mechanism leading to creep-fatigue crack initiation. The development has been tailored to the physical situation encountered in ODS superalloys: cracks initiate by cavity growth and coalescence on isolated, transverse grain boundaries. For a more detailed description, the reader is referred to an upcoming publication (7).

Cavity growth models have been successfully applied to a number of alloys evaluated under conditions of constant stress, but little work has been done to apply these concepts to situations wherein the applied stress continually varies. Analytical approaches based on creep models where a time-averaged stress during a creep-fatigue cycle is used, cannot be expected to reflect the influence of waveform; S-F cycles generally have a lower time-averaged tensile stress than an H-T cycle of identical stress amplitude, yet the S-F cycle is considerably more damaging. Instead, a finite difference approach is found to be more suitable since the damage can be integrated over time as a function of stress.

The stresses which act across the grain boundary in question are generally not the same as the externally applied stress because of load shedding to the surrounding grains and local relaxation of elastic stresses. Load shedding is accounted for by a compatibility condition requiring grain boundary displacements not to exceed displacements in the surrounding matrix, unless displacements can be accommodated by elastic stress relaxation. In order to better visualize the process of accommodation by local stress relaxation, consider a penny-shaped crack in an elastically loaded body; the crack opening displacement (which may be calculated by linear elasticity theory) is made possible by the relaxation of elastic stresses near the crack faces. Analogously, the grain boundary 'opens' as displacements due to cavitation occur and this opening may be accommodated by stress relaxation. The calculation of the grain boundary stress σ_{gb} is then determined as a balance between an increment in the applied stress and an increment of relaxed stress, i.e., $\sigma_{gb} = \sigma_{gb} + \Delta\sigma_a - \Delta\sigma_{rel}$.

Once the grain boundary stress is known for a given time increment, the change in pore radius ΔR under the influence of this stress may be calculated. The mechanisms of pore growth which are considered include grain boundary diffusion, for which ΔR may be written as (13,14)

$$\Delta R = \frac{2}{kT} \frac{\Omega}{\lambda} \frac{\delta D_b}{Q(\omega)} \frac{1}{R} \cdot \sigma_{gb} \cdot \Delta t \qquad (1)$$

and plastic hole growth, for which ΔR may be written as (15)

$$\Delta R = 0.6 \, B \, \frac{\lambda^2}{B} \cdot \frac{1}{R} \left[\frac{1}{(1-\omega)^n} - (1-\omega) \right] (\sigma_{gb} - \sigma_{th})^n \cdot \Delta t \qquad (2)$$

where the effect of a pseudo-threshold stress σ_{th} for creep has been included in the latter expression. The symbols used are defined as follows:

236

Ω:	atomic volume
δD_b:	grain boundary diffusion coef.
k:	Boltzmann's constant
T:	absolute temperature
ω:	cavitated area fraction

R: pore radius
B: Dorn's constant
n: stress exponent
λ: cavity spacing
$Q(\omega)$: shape factor

Crack initiation is assumed to be completed when the cavitated area fraction ω reaches a critical value of $\omega = 0.7$. The number of cycles to crack initiation predicted by the model is compared with experimental data for S–F tests in figure 6 and with data for H–T tests in figure 7. The model predictions are shown as a scatter band in both figures, which results from using a wide range of material constants and parameters. The scatter band can be shifted laterally by varying the critical area fraction of cavitated boundary.

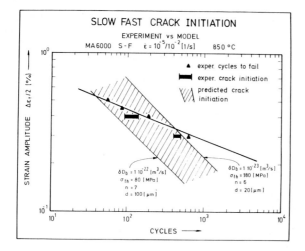

Fig.6.
Comparison of crack initiation life as predicted by the model with experimental data for slow–fast tests.

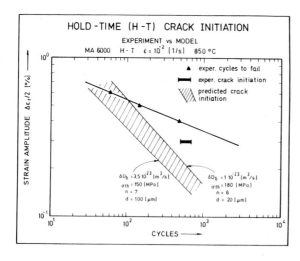

Fig.7.
Comparison of crack initiation life as predicted by the model with experimental data for tensile hold–time tests.

boundaries before extending transgranularly into the surrounding grains. This tendency is illustrated schematically in figure 5 and is compared with behaviour for S-F tests where the intergranular initiation site and transgranular crack growth surface are practically coplanar. Slow-slow tests are characterized by a generally lower density of crack initiation sites in comparison with S-F and H-T tests. Additionally, interrupted S-S tests demonstrated that cavities which nucleate during slow tensile loading may frequently be sintered during slow compression. Sintering did not occur during S-F and H-T tests.

Discussion

The nucleation and diffusional growth of cavities leading to intergranular fracture is a well known phenomenon in creep-resistant, low-alloy steels and superalloys exposed to low stress, high temperature conditions for extended periods. Although cavitation may occur in some materials during high frequency cyclic loading, the nucleation and growth of pores during high temperature fatigue is normally associated with diffusional or dislocation creep. For the case of MA 6000, the presence of precipitate-free zones between cavities and the spherical caps shape of the pores are taken to be indications that creep cavitation is occurring. The particular susceptibility of the boundaries of fine grains to fatigue and cavitation damage has been discussed previously (2). Briefly, fine grains have been found to have crystallographic orientations which may be very different than the highly textured <110> orientation shared by normally recrystallized macrograins. Thus, some fine grains will exhibit a significantly higher elastic modulus in the loading direction with the result that transverse boundaries are subjected to correspondingly higher stresses.

As for a variety of other alloys, including Al-5Mg (10), stainless steels (e.g. 11) and superalloys (e.g. 12), the S-F cycleform has proven to be the most damaging for MA 6000 (see fig. 4). The initiation of cracks also occurs most quickly during slow-fast cycling, as has been demonstrated by comparison of interrupted tests. This behaviour may be explained by considering the effect of local residual stresses. The rate of creep damage accumulation during the slow tensile portion of the cycle will depend on the local stress state existing at the start of creep loading. The local stress state depends, in turn, on the amplitude, sign and rate of the preceding plastic deformation. Studies of the effect of prestraining on subsequent creep damage rates demonstrates this point well: It has been shown that creep life is greatly reduced if prestraining is done in compression, while tensile prestraining is seen to exert little influence on subsequent creep life (8,9). This phenomenon is due to local residual stresses resulting from the plastic deformation of matrix material in the neighborhood of very hard particles such as oxides or carbides. Compressive prestraining results in residual stresses which complement the applied tensile stress. Thus, cavities which nucleate on grain boundaries transverse to the applied stress experience accelerated growth and coalescence. On the other hand, tensile prestraining induces a residual stress state which tends to oppose a tensile (creep) stress and so cavities on transverse boundaries may experience inhibited growth. It has been observed in fact, that tensile prestraining enhances growth of pores on grain boundaries parallel to the stress axis. It is to be expected that residual stresses created by rapid plastic straining exert a similar influence on creep damage rates during creep-fatigue loading.

Thus, two preconditions must be satisfied for residual stresses to be a significant influence during creep-fatigue; (1) the material must contain non-deforming inclusions (on grain boundaries to affect damage rate due to cavitation) and, (2) sufficient plastic deformation must occur athermally in the matrix surrounding the hard particle. In MA 6000 and MA 754, larger oxide dispersoids and yttrium-aluminate particles lying on or near grain boundaries are available to act as stress concentrations and as cavity nuclei. During H-T cycling, creep follows rapid tensile deformation which, in analogy with tensile prestraining tests, would be expected to promote cavitation on boundaries parallel with the stress axis while exerting little or no influence on the growth of pores on transverse boundaries. That this is indeed observed

The comparison of prediction with experiment in figures 6 and 7 is encouraging as order-of-magnitude agreement is found; this makes it plausible that crack initiation does indeed occur by a creep damage mechanism. However, two inadequacies are apparent in the model; firstly, the predicted curves are too insensitive to the strain amplitude, (i.e., curves are too steep), and secondly, although the prediction is acceptable for the S-F case, the influence of waveform is not properly accounted for in going to the H-T cycle.

The insensitivity of the prediction to cyclic strain amplitude is a consequence of the stress-based approach taken for modelling the damage rate. The cyclic plastic deformation, which is conventionally expressed as a function of the inelastic strain range $\Delta \epsilon_{pl}$, must be incorporated into the damage rate equation. The inability of the model to accurately predict the influence of waveform could be improved by including the influence of prior plastic deformation on creep damage rate. This aspect is currently being investigated by considering a residual stress term which is added to the grain boundary stress. The residual stress term depends on the loading prior to creep and is a function of strain amplitude, loading direction and strain rate. It must also be a function of time since recovery processes due to creep are occurring.

Summary

The creep-fatigue behaviour of the two ODS alloys MA 6000 and MA 754 has been investigated. Failure is found to occur in both alloys as a result of cracks which initiate at grain structure defects. These defects, which occur primarily in the form of fine, included grains, are susceptible to intergranular cavitation. Cracks thus initiated grow transgranularly and eventually coalesce, leading to fracture. Since this initiation process is found to take up a significant portion of the number of cycles to failure, its understanding and eventual suppression may be of great practical importance.

Investigation of the influence of waveform on the creep-fatigue life has shown the slow-fast waveform to be most damaging, followed by slow-slow and tensile hold-time cycles. The principal effect of waveform on cyclic life is ascribed to the influence of local residual stresses produced during rapid plastic deformation on subsequent creep damage rates. Matrix plastic deformation in the vicinity of non-shearable particles such as oxide dispersoids results in a localized stress state which may either accelerate or inhibit cavity growth, depending on the direction (compressive → accelerates; tensile → inhibits) of prior plastic straining. The occurrence of cavitation on grain boundaries parallel with the loading axis during H-T tests could also be rationalized by the presence of residual stresses.

A model for the simulation of the damage processes leading to crack initiation during creep-fatigue of ODS superalloys has been introduced. While able to predict crack initiation during S-F cycling at lower strain amplitudes, the current model fails to properly reflect the influence of waveform. Further work is being concentrated on the introduction of a means to include the influence of residual stresses on creep damage rates.

Acknowledgements

The authors are especially grateful to C. Weis for assistance with SEM studies and special metallography. Parts of this work have been carried out within the frame of the European Collaborative Programme COST 501. We would like to acknowledge the financial support of the Bundesministerium für Forschung und Technologie in the Federal Republic of Germany under project number O3ZYK1228.

References

(1)   E. Arzt, this volume.
(2)   D.M. Elzey, E. Arzt in: Superalloys 1988, S. Reichman et al, eds., The Metallurgical Society, 1988, pp.595-604.
(3)   G.A.J. Hack in: Frontiers of High Temp Materials II, J.S. Benjamin and R.C. Benn, eds., (USA: INCO Alloys International, 1983), pp.3-18.
(4)   J.J. Stephens, W.D. Nix: Met Trans 16 (1985) 1307.
(5)   J. Schröder in: Fortschr.-Ber. VDI, Reihe 5 (1987), Nr.131.
(6)   H. Zeizinger in: Fortschr.-Ber. VDI, Reihe 5 (1987), Nr.121.
(7)   D.M. Elzey: Ph.D. Thesis, Univ of Stuttgart, 1988, to be published.
(8)   B.F. Dyson, M.F. Loveday, and M.J. Rodgers: Proc. R. Soc. A 349 (1976) 245.
(9)   K. Shiozawa, J.R. Weertman: Acta Met 31 (1982) 993.
(10)   S. Baik, R. Raj: Met Trans 13A (1982) 1215.
(11)   S. Majumdar, P.S. Maiya: Canad. Metall. Quart. 18 (1979) 57.
(12)   M.Y. Nazmy: Met Trans 14A (1983) 449.
(13)   D. Hull, D.E. Rimmer: Phil. Mag. 4 (1959) 673.
(14)   M.V. Speight, J.E. Harris: Metal Sci. J. 1 (1967) 83.
(15)   A.C.F. Cocks, M.F. Ashby: Prog. Mat. Sci. 27 (1982) 189.

C. 2:
ALUMINIUM ALLOYS

Mechanical Alloying of Light Metals

R. Sundaresan and F. H. Froes, Air Force Wright Aeronautical Laboratories, Wright-Patterson AFB, Ohio, USA.

Abstract

The status of the mechanical alloying of the light metals aluminum, magnesium and titanium is presented. For the aluminum system the technology is already at the early commercial plant production stage for unalloyed, high strength, low density, elevated temperature and metal matrix composite materials. Magnesium alloys are less well developed but have been applied to supercorroding needs and for hydrogen storage. Because of its reactivity little has been done on MA of titanium alloys. However, experimental studies have indicated the potential for extended solubility and for the use of amorphous phase formation as a precursor to novel crystalline materials. Some projections on future advances that are likely to be made in the mechanical alloying of the light metals are made.

Introduction

In recent years major inroads have been made into the traditional materials of choice for structural applications in aerospace systems. For instance, organic composite materials and ceramics can compete successfully with metals in many applications. To combat this trend the metallurgist must be innovative in designing advanced metal systems. This covers both improved alloy chemistries, including the elimination of detrimental "tramp" elements, and careful control of the microstructure during subsequent processing.[1-3] One such technique now receiving increased attention is mechanical alloying (MA).

It is the purpose of the present paper to review the current status of the MA of light metals, which we as the authors have chosen to define as aluminum, magnesium and titanium. The reader will see that much has already been accomplished, particularly in the aluminum system. However, the surface has only just been scratched, and it is therefore also the pupose of this paper to make some projections on useful areas for further exploration. The MA approach offers one of a number of opportunities for the metallurgist to increase performance capabilities of light metals, and should receive considerable attention over the next few years.

Since the process of MA is being adequately covered in other papers at this meeting, we shall directly address the three alloy systems without a discussion of MA itself.

Aluminum System

For aluminum alloys the adherent surface oxide film on the powder
particles is incorporated into the interior of the processed
powders as a result of the repeated fracture and cold welding
which occur during MA. For control of the balance between
fracture and welding in the MA processing of aluminum alloys, it
was found that process control additives (PCA) were necessary.[4]
The PCA developed were typically organic compounds, which
conferred additional benefits with the formation of Al_2O_3 and
Al_4C_3 which were also incorporated into the interior of the powder
particles as dispersoids.

Unalloyed Aluminum

The first application of MA to Al system was naturally on
unalloyed Al. When MA was being developed, oxide dispersion
strengthened Al was already fully developed as SAP. However, the
oxide dispersion in SAP is inefficient compared to that of MA Al,
as seen from a comparison of the mechanical properties (Fig. 1).
[4,5] The temperature capability at a given strength level was
better than other dispersion strengthened alloys (Fig. 2) as was
the electrical conductivity (Table 1). The improved properties
result directly from the unique structure developed by MA
processing. The structure consists of very fine (0.2 to 0.5 μm)
equiaxed grains, and well distributed dispersoids 30 to 40 nm in
size (Fig. 3). The dispersoids have been clearly identified as
Al_2O_3 and Al_4C_3, that are formed from amorphous precursors during
elevated temperature exposure, such as the vacuum degassing
subsequent to the milling.[6] The structure developed in MA has
substantially greater thermal stability than conventional cast and
wrought aluminum alloys. While the tensile strength of commercial
alloys deteriorates sharply on elevated temperature exposure, MA Al
with no alloying additions shows no change in the structure or in
strength and ductility on thermal exposures as high as 490°C for
100 h.[7,8] Applications as high temperature, high strength
electrical conductor wires are foreseen for the alloy.[9]

Material	Tensile Strength (MPa)	Electrical Resistivity (μ.ohm.cm)
MA Aluminum	332-350	3.25 - 3.31
EC Al (H19 temper)	186	2.76
5005	186	3.22
SAP-930	241	3.22
SAP-895	324	3.46
SAP-865	371	4.64

Table 1: Electrical properties of MA aluminum compared with other
aluminum base conductors

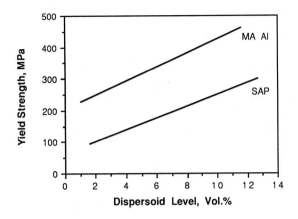

<u>Fig. 1(a)</u>: Yield strength of MA aluminum
and SAP as a function of dispersoid content

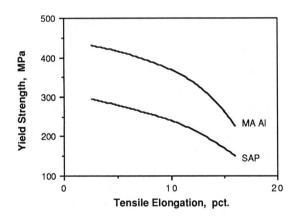

<u>Fig. 1(b)</u>: Yield strength of MA aluminum
and SAP as a function of ductility level

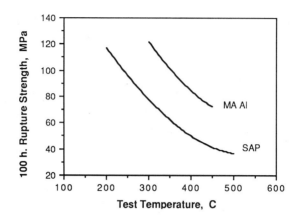

Fig. 2: Elevated temperature strength of MA aluminum and SAP

Fig. 3: Transmission electron micrograph of MA aluminum

High Strength Aluminum Alloys

Initial attempts at MA compositions corresponding to 2618 and 7075 showed that conventional alloying additions for grain refinement and precipitation hardening were deleterious.[5] With substantial strengthening from the inherent fine grain structure and the stable dispersoids, different alloying considerations apply to MA alloys. Subsequent alloy development sought to combine ambient temperature strength from limited solid solution strengthening and precipitation hardening, with elevated temperature strength from dispersoid content and composition. [5,10]

Based on such considerations, Mg was fully evaluated as an alloying addition.[11] The alloys IncoMAP AL-9052 and AL-9021 based on magnesium as an alloying addition have been well characterized for tensile strength, fatigue strength, toughness, corrosion resistance etc.[6,12-15] The properties of these alloys are listed in Table 2, and their fatigue characteristics are shown in Fig.4. AL-9052 is currently being considered for application as torpedo hulls and other marine applications.[16] AL-9021 was developed for aircraft structural applications, but is currently being projected for application as a matrix for SiC particle reinforced composites. AL-9021 also exhibits extended ductility, over 500%, at high strain rates of 1 - 10 /sec at 475°C.[17,18] The effect and its applications are yet to be assessed.

	IncoMAP AL-9052		IncoMAP AL-9021	IncoMAP AL-905XL
Composition	Al-4Mg-1.1C -0.8O		Al-1.5Mg-4Cu -1.1C-0.8O	Al-4Mg-1.5Li -1.2C-0.4O
Density	2.68 Mg/m^3			2.58 Mg/m^3
Mechanical Properties				
Longitudinal	YS	542 MPa	613 MPa	468 MPa
	UTS	577 MPa	634 MPa	525 MPa
	El	8 %	12 %	9 %
Transverse	YS	540 MPa	591 MPa	469 MPa
	UTS	570 MPa	612 MPa	497 MPa
	El	8 %	14 %	6 %
Short Trans.	YS	549 MPa	584 MPa	420 MPa
	UTS	568 MPa	597 MPa	481 MPa
	El	2 %	11 %	6 %
Fracture Toughness K_{Ic}	44 MPa.$\sqrt{m}$			32 MPa.$\sqrt{m}$
Elastic Modulus	76 GPa			80 GPa
Corrosion Rate	0.009 mm/year			0.003 mm/year
SCC Threshold	>380 MPa			345 MPa

Table 2: Mechanical properties of MA aluminum alloys

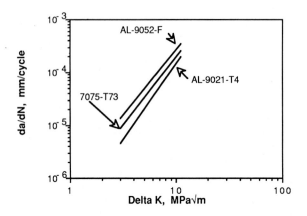

Fig. 4: Fatigue crack growth behavior of
MA aluminum alloys

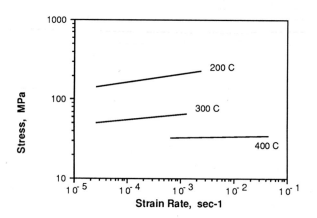

Fig. 5: Creep properties of MA aluminum-
lithium alloy AL-905XL

Low Density Alloys

IncoMAP alloy AL-905XL is an Al-Mg-Li MA alloy developed as a low density forging alloy for airframe applications.[19,20] One of the first among the Al-Li alloys to reach the commercial production stage, the alloy has been well characterized.[19-23] Some of the properties are shown in Table 2 and in Fig. 5. Extensive evaluation of the MA alloy have been carried out in comparison with Al-Li alloys processed through other routes [24]. With superior mechanical properties and corrosion resistance, and scale up already accomplished, the MA alloy seems to have an edge over other PM routes. However, alloying additions for precipitation hardening and for grain refinement have to be limited in the MA alloy. The Li content has also had to be limited to 1.5 wt% to avoid Al_3Li (δ') precipitates that have an adverse influence on fracture characteristics. Though AlLi (δ) and Al_2LiMg precipitates have not been detected, Li containing oxide precipitates are likely to be present and the precipitate content in the alloy cannot be estimated precisely. Thus alloying additions and their levels are still somewhat arbitrary. Further, the MA alloy as consolidated is considerably stronger than corresponding ingot metallurgy or conventional powder metallurgy alloys, making further fabrication steps more difficult.[24] Recently, it has been found necessary to further reduce the Li level in the alloy from 1.5 wt% to 1.3 wt% to counter thermal instability problems.[20] The effect of this change on other mechanical and chemical properties are not yet known. There have also been other process related problems still to be fully resolved, such as iron contamination from the processing vessels and media, and structural inhomogeneities resulting from inadequate powder processing.

	DISPAL 2	DISPAL 1 Si12	AlSi12CuMgNi
Dispersoid, %	10 - 12	6 - 10	-
Density, Mg/m^3	2.70	2.66	2.70
Hardness, HB	90 - 110	100 - 120	90 - 125
UTS, MPa	340 - 380	340 - 380	300 - 370
YS, MPa	300 - 340	300 - 340	280 - 340
El, %	8 - 12	4 - 8	1 - 3
Modulus, GPa	80	85	81
Fatigue strength, MPa	80 - 110	120 - 150	80 - 120
Thermal expansion coefff., $10-6/°K$	25	19	21
Thermal conductivity, $W/m°K$	159	157	155

Table 3: Properties of DISPAL alloys

High Temperature Alloys

DISPAL alloys developed in West Germany employ dispersion strengthening by Al_2O_3 and Al_4C_3 as in MA, but the alloy powders are processed directly with carbon black and controlled additions of oxygen from the milling atmosphere.[25-27] With this independent control of the dispersoid content, the alloys can be suitably tailored for the application. The properties of these alloys [27-31] are shown in Table 3, and Fig. 6. Applications as forged automotive pistons, as high temperature electrical conductors in Na-S battery, and as a structural material in aerospace cryogenic field are foreseen for the alloys.

Extensive research continues in development of new MA alloys. One direction in these developments is the combination of MA with rapid solidification, particularly in alloys designed for strengthening by intermetallic dispersions. In the Al-Fe-X (X=Ni, Ce) alloys there have been some reports indicating advantages in this route.[32-36] At intermediate temperatures, up to 250°C, the intermetallic particles provide the strengthening and the non-mechanically alloyed material is about as strong as the MA alloy. However, at higher temperatures, up to 500°C, MA material is considerably stronger. The creep rates of MA material are several orders of magnitude better (Table 4). The strength is retained in this case possibly because the oxide and carbide dispersions retain their strength and fineness adequately, while in the absence of these dispersions, the second phase particles cannot retain their strength. Further, the fine carbide and oxide dispersions on subgrain boundaries resist both recovery processes and coarsening of the intermetallic particles. Further extension of such alloying has also been reported, where by combining MA with RS in an Al-4Cu-1Mg-1.5Fe-0.75Ce alloy, the strength level could be increased to 435 MPa.[37]

Various Al-Ti alloys are also being explored by MA to realise the benefits of dispersion strengthening from both oxides/carbides and intermetallics. Both MA of elemental/master alloy blends [38,39], and MA of rapid solidified alloy powders [40,41] have been investigated. The results appear to be somewhat contradictory: excellent thermal stability of the structure has been reported in the elemental blend mechanically alloyed [38], but poor stability is reported in the RS+MA alloy [42]. However, it does appear that

Temperature, °K	Stress, MPa	Creep Rate in RS Alloy, /s	Creep Rate in RS+MA Alloy, /s
623	103	4.1×10^{-7}	6.3×10^{-10}
653	83	2.6×10^{-7}	1.6×10^{-9}
653	103	4.9×10^{-5}	1.4×10^{-9}

Table 4: Steady state creep rate of Al-8Fe-4Ce alloy

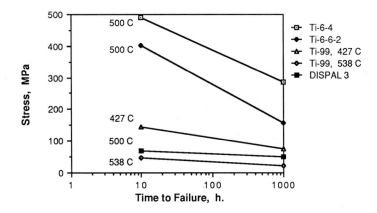

Fig. 6: Creep of a DISPAL alloy compared with titanium alloys

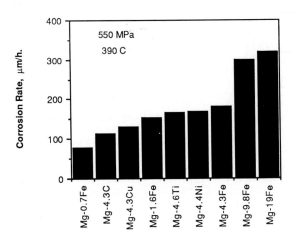

Fig. 7: Corrosion rates in "supercorroding" MA magnesium alloys

the good elevated temperature strength, high stiffness and encouraging levels of ductility [39] make this alloy system worthy of further exploration.

Metal Matrix Composites

Aluminum-base metal matrix composites are under development with MA as a processing route for SiC and BN particulate reinforcements.[43] Considerable development of the processing appears to have been realized, using both AL-9021 and AL-9052 compositions as the matrix material, with SiC particulate reinforcement up to 15 vol%. AL-905XL has also been evaluated as a matrix material for MMC.[22] While processing details are restricted, extensive property evaluation details are available.[44-46] Typical strength and stiffness values are shown in Table 5. Unlike other processing routes for similar composites, the particulate dispersions do not significantly enhance the strength of the matrix that has been already strengthened by the oxide and carbide dispersions. The matrix AL-9021 is thermally unstable and loses its strength at temperatures above 150°C due to dislocation annihilation, and the coarse SiC particles do not appear to result in any significant strengthening at high temperatures.[47] While the particle-matrix interfacial strength is substantial, well over 1000 MPa, coarse particles of SiC break during fast fracture, and the fracture toughness of the composite is low.[46] However, at high temperatures in the 425-450°C range, at high strain rates ($\dot{\epsilon} > 0.7$ /sec) the composite material with AL-9021 matrix shows extended ductility up to 300% total elongation.[47,48]

There have been several other exploratory studies on production of composites by MA. Al_2O_3 and spinel $MgAl_2O_4$ have been tried as the reinforcement by MA with a rapidly solidified Al-3Mg powder,

Temperature °C	YS MPa	UTS MPa	El %	E GPa
SiC/Al-9021				
25	420 (370)	546 (440)	6 (10)	81.8 (66.1)
100	424 (347)	483 (394)	4 (14)	
200	274 (124)	286 (137)	14 (54)	
300	77 (47)	81 (54)	12 (21)	
400	40 (30)	43 (32)	3 (10)	
SiC/Al-905XL				
25	- (562)	581 (584)	0 (7)	
150	334 (258)	423 (388)	1 (17)	
200	197 (128)	266 (233)	3 (27)	
300	97 (55)	113 (82)	8 (29)	
400	68 (42)	68 (44)	0 (3)	

Table 5: Mechanical properties of metal matrix composites by MA, with the MA matrix material properties shown in parentheses

resulting in composites with low creep rates.[49] In a different direction, a composite Al-4.5Cu-3graphite made by MA has been found to perform well as a bearing material.[50]

Magnesium System

MA has been employed in the development of "supercorroding" alloys.[51] Magnesium, mechanically alloyed with Fe/Cu/Cr/C/Ti, has been used for making links with precisely controllable corrosion rates for releasing deep-sea equipment at specific depths.[52] Such supercorroding alloys have also been of use in other submarine applications, such as a heat source in diver suits and as a hydrogen gas generator. The advantage of MA in generating the supercorroding alloys is the extent of control afforded by the closeness of the anode/cathode metals achieved. By adjusting the alloy composition, the reaction rates (Fig. 7) can be precisely controlled in the device.

The system Mg-M is of great interest as an alloy for hydrogen storage systems for energy applications.[53,54] By alloying with catalyst elements/intermetallic compounds the hydrogenation capability of Mg is substantially enhanced.[55] The additions include elements such as Ni [56,57], Ce [58], La [58] which can be alloyed with Mg conventionally, elements such as Fe [59], Co, Ti, Nb [55] that do not normally alloy with Mg, intermetallics such as $LaNi_5$, Mg_2Cu, $CeMg_{12}$ and even oxides such as TiO_2 [60]. MA is often the only processing technique available at present for such systems that can act as a "hydrogen pump" by absorption and desorption of hydrogen, with the nucleation site for the formation of the hydride provided by the dispersions. Considerable work appears to be in progress on these systems and the development of a hydrogen pump for energy conversion is being conducted jointly between the USSR and France.

Titanium Systems

Although there has been some early work on the application of MA to titanium systems[61,62], experimental constraints appear to have curtailed further research. Currently, several aspects of the application of MA to Ti systems are under investigation. These include extension of solid solubility limits possible by MA, the formation of amorphous phases by MA in favorable systems as a possible route to novel crystalline structures, and the microstructural engineering offered by MA.

Titanium - Magnesium System

Magnesium is an attractive alloying addition to titanium because of its very low density. However, because of its low melting point the solubility of magnesium in titanium obtained by conventional processes has been negligible, even though other metallurgical and physical factors are favorable for the alloying. On a laboratory investigation solid solution of Mg up to 3.1 wt% (6.0 at%) in Ti has been achieved.[63] Lattice parameters of the alloy powder measured by x-ray diffraction show substantial dilation of the hcp lattice. Mg also appears to be a very strong β stabiliser: The beta-transus was 820°C in binary Ti-Mg solid solution, and in the ternary Ti-Mg-Gd system, the beta transus was still further depressed to 799°C.[64]

Scaling up and consolidation of the Ti-Mg based "alloys" is now under development at AFWAL. Neither optimum powder processing for uniform microstructure, nor complete densification has been achieved yet. Considerable work remains to be done in developing the MA processing techniques fully for this system, and in evaluating the system for possible applications.

Amophous Phase Formation

Extensive work has gone into the study of amorphous phase formation by mechanical alloying in the recent past. Studies have shown that MA results in the formation of an amorphous phase in favorable systems, such as Ti-Ni [65-68], Ti-Fe [69-70], Ti-Co [69,70], Ti-Cu [71], Ti-Pd [72], Ti-Cu-Pd [73], and Ti-Ni-Cu [74]. In all reported cases the formation of the amorphous phase has been confirmed by x-ray diffraction, electron diffraction

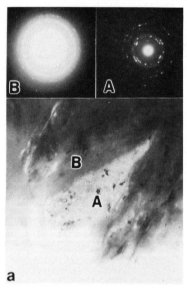
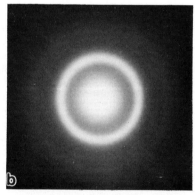

Fig. 8: Ti-Ni-Cu alloy, showing (a) microcrystalline features at an intermediate stage of mechanical alloying, and (b) fully amorphous feature with further MA processing

(Fig. 8) and calorimetry (Fig. 9). Several theories have been suggested for the mechanism of the amorphous phase formation by MA. These include (i) the occurrence of incipient melting under the impact of colliding balls [65], (ii) solid state amorphization of diffusion couples with high negative heat of mixing and anomalously high diffusion coefficients [65,66], and (iii) favoring of the amorphous phase under the conditions of high configurational entropy arising from the high dislocation density [74,75].

Some attempts have been made to utilize the amorphous phase formed in MA titanium alloys. In the Ti-Cu, Ti-Pd-Cu systems, consolidation to 95% density has been achieved while retaining the amorphous nature of the alloy by low temperature hot isostatic pressing.[76] In the Ti-Ni-Cu system, the MA processing has been scaled up and studies are currently in progress on low temperature consolidation of the alloy and its characterization. The microstructure of the Ti-Ni-Cu amorphized alloy powder consolidated at 500°C shows areas recrystallized from the amorphous powder that cannot be resolved optically. The hardness in such regions is very high (770 Knoop) even at higher consolidation temperatures [75], and appears to be a special feature that can be explored further.

Beta Titanium Alloys

Two advanced beta alloys, Ti-1Al-8V-5Fe-1Er and Ti-24V-10Cr-5Er, both with extensive rare earth addition for subsequent oxide dispersions are being developed by a combination of RS and MA.[77] The alloy Ti-1Al-8V-5Fe-1Er consolidated directly from gas atomized powder has coarse beta grains of about 30 μm (Fig. 10a). MA up to 48h did not yield a totally uniform structure, but the majority of the structure on consolidation consisted of sub-micron grains (Fig. 10b) with dispersions 30 to 50 nm in size. In the finer grains, the effect of heat treatments could not be resolved optically though the hardening was measurable (Table 10). The fine-grain structure is seen to be stable, with little grain coarsening and little drop in the hardness on solution treating at 675°C.

The alloy Ti-24V-10Cr-5Er in the as-atomized condition was badly segregated with large chunks of free Er. The structure on consolidation of the MA alloy powder consisted of both very fine grained regions (Fig. 11a), and some coarser grains, with dispersions in the fine grain regions of less than 10 nm in size (Fig. 11b). Hardness is increased in the alloy by MA processing (Table 6) due to the grain refinement and dispersions.

It is seen that fine dispersions and stable grain refinement are possible by MA in both the alloys. However, considerable optimization of the MA processing parameters is necessary to ensure that the structure advantages are uniform.

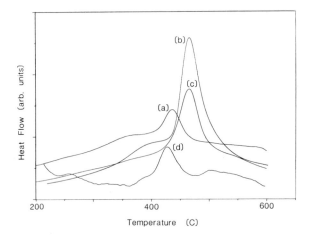

Fig. 9: DSC traces of Ti-Ni-Cu showing recrystallization of the amorphous phase in (a) splat, (b) blended elemental powders Spex milled for MA, (c) prealloyed powder Spex milled and (d) blended elemental powder Attritor milled.

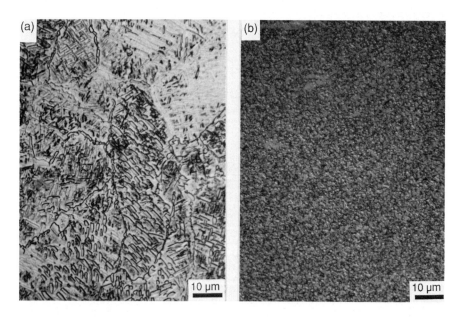

Fig. 10: Optical micrographs of Ti-1Al-8V-5Fe-1Er consolidated (a) from as-atomized powder, and (b) from MA powder.

257

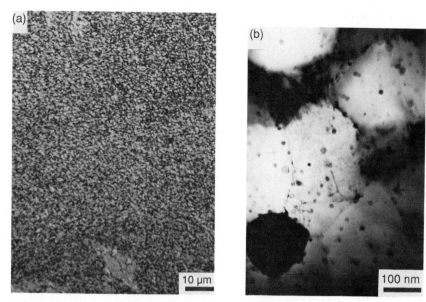

Fig. 11: (a) Optical micrograph of Ti-24V-10Cr-5Er
consolidated from MA powder showing extremely fine grains,
(b) TEM of the consolidated alloy showing fine dispersions.

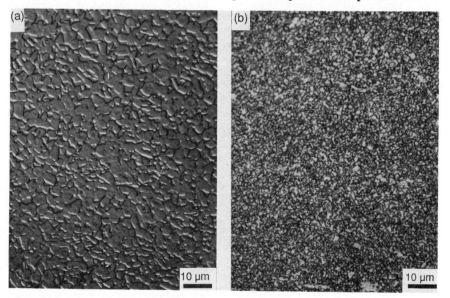

Fig. 12: Optical micrographs of Ti-25Al-10Nb-3V-1Mo (at%)
consolidated from (a) as-atomized powder, and (b) from MA
powder.

Alloy	Consolidated	ST	STA
I. Ti-185			
Atomized	372		
MA	501	487	526
II. Ti-24/10			
Atomized	304	301	292
MA	437	435	440

Table 6: Knoop hardness values on MA titanium alloys

Titanium Intermetallic Alloys

An early study on MA of intermetallic TiAl for dispersion strengthening [62] was not continued. Based on the physical metallurgy of the intermetallic systems, which is now better understood, modification of the microstructure of the alloy Ti-25Al-10Nb-3V-2Mo (at%) by MA was investigated.[78] The microstructure of the alloy consolidated directly from the atomised powder shows a mainly uniform two-phase structure consisting of equiaxed α2 with a grain size of about 4 μm, with intergranular B2 (Fig. 12a) and some occasional patches of lenticular (α2) grains. Consolidation of the MA powders led to a structure consisting mainly of very fine grains, of about 1 μm (Fig. 12b), with some coarser grains of about 3 μm retained. Dispersions in the consolidated alloy containing Gd were coarse, about 0.1 μm in size. The room temperature hardness in the MA

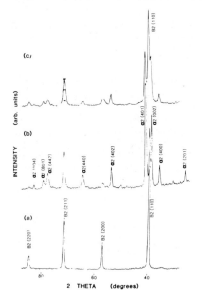

Fig. 13: X-ray diffraction of Ti-25Al-10Nb-3V-1Mo (at%)
(a) atomized powder,
(b) consolidated from as-atomized powder, and
(c) consolidated from MA powder.

material (720 Knoop) is much higher than the material from the gas atomized powder (340 Knoop), a much higher increase than can be expected from grain refinement alone. X-ray diffraction of the consolidated alloy (Fig. 13) shows that the as atomised powder has a completely B2 structure, which turns to a mixture of both α2 and B2 phases on consolidation. The proportion of the more ductile B2 retained in the case of consolidated MA powder is considerably more than in the directly consolidated gas atomized powder.

As in the case of beta titanium alloys, the microstructural features that can be achieved by MA of the atomised powders in the Ti$_3$Al alloy appears to be very promising.[79] In addition to the stable grain refinement possible, greater retention of the more ductile phase in the alloy also seems possible with MA.

Conclusions and Future Thoughts

The current status of the MA of alloys based on the light metals aluminum, magnesium and titanium has been discussed. The aluminum systems are already at the early commercial plant production stage, with newer systems such as Al-Ti looking very attractive in the research laboratory. The magnesium system has been utilized for "supercorroding" and hydrogen storage applications, with no attempts to date to design structural alloys using the MA technique. Work on the titanium system is very much in the research arena but already encouraging results have been obtained on low density Mg containing alloys, and on RS dispersion strengthened terminal and intermetallic alloys.

For the future it is likely that MA of Al base materials will continue to grow with low density alloys and elevated temperature alloys leading the way. It is possible that the difficulties in controlling the microstructure of forgings made from Li containing ingot alloys will result in the MA approach being the processing route of choice for this product form. The excellent stability for MA systems such as Al-Ti (to 510°C) combined with good ductility compares favorably with metal matrix composite concepts which so far suffer from low ductility levels.

The advantages which MA may offer in developing new and improved magnesium base alloys has still to be explored. Incorporation of dispersoids for enhanced elevated temperature capability is one obvious route to be considered. Additionally, the major problem of poor corrosion resistance of magnesium base alloys [1,2] could perhaps be addressed by alloying via MA - using normally insoluble alloying elements to produce a "stainless" magnesium.

The potential for MA of the titanium system has already been demonstrated. However, much further study is needed particularly in the low density alloying addition of Mg, and in dispersion stengthened elevated temperature alloys.

For all three systems the incorporation of a total "engineered material" approach in which MA is one facet of the whole lies in the future. Here, as an example, RS to control initial chemistry, MA to control the microstructure, and addition of ceramic fibers to enhance stiffness and elevated temperature performance will result in a material to compete successfully with ceramics and organic composites.

Acknowledgements

R. Sundaresan acknowledges support form a National Research Council Associateship. We are grateful to Dr. V. Arnhold of Sintermetallwerk Krebsöge, Drs. J.J. de Barbadillo and R.C. Benn of Inco Alloys International, and Dr. G. Staniek of DFVLR for useful discussions and for some of the materials used in the paper.

References

1. F.H. Froes: Materials Edge (May/June 1988) 19.
2. F.H. Froes: "Space Age Metals Technology", Eds. F.H. Froes and R.A. Cull, SAMPE (1988) 1.
3. J. Wadsworth and F.H. Froes: J. Metals 40 (1988), to be published.
4. J.S. Benjamin and M.J. Bomford: Met. Trans. 8A (1977) 1301.
5. J.H. Weber and R.D. Schelleng: "Dispersion Strengthened Aluminum Alloys", Eds. Y.-W. Kim and W.M. Griffith, TMS (1988).
6. R.F. Singer, W.C. Oliver and W.D. Nix: Met. Trans. 11A (1980) 1895.
7. J.A. Hawk and H.G.F. Wilsdorf: Scripta Metall., 22 (1988) 561.
8. J.A. Hawk, P.K. Mirchandani, R.C. Benn and H.G.F. Wilsdorf: "Dispersion Strengthened Aluminum Alloys", TMS (1988).
9. Sumitomo Electric Industries, Jap. Patent No. 6063337 (1985).
10. D.L. Erich: "Development of a Mechanically Alloyed Aluminum Alloy for 450-650 F Service", AFML-TR-79-4210 (1979).
11. J.S. Benjamin and R.D. Schelleng: Met. Trans. 12A (1981) 1827.
12. R.D. Schelleng and S.J. Donachie: Metal Powder Rep. 38 (1983) 10.
13. Y.W. Kim and L.R. Bidwell: Scripta Metall. 16 (1982) 799.
14. A. Renard, A.S. Cheng, R. de la Veaux and C. Laird: Mater. Sci. Eng. 60 (1983) 113.
15. D.L. Erich and S.J. Donachie: Metal Prog. 121 (1982) 22.
16. R.D. Schelleng: Metal Powder Rep. 43 (1988) 239.
17. T.G. Nieh, P.S. Gilman and J. Wadsworth: Scripta Metall. 19 (1985) 1375.
18. T.R. Bieler, T.G. Nieh, J. Wadsworth and A.K. Mukherjee: Scripta Metall. 22 (1988) 81.
19. R.D. Schelleng, P.S. Gilman and S.J. Donachie: "Overcoming Material Boundaries", SAMPE (1985) 106.
20. R.D. Schelleng, A.I. Kemppinen and J.H. Weber: "Space Age Metals Technology", SAMPE (1988) 177.
21. R.T. Chen and E.A. Starke, Jr.: Mater. Sci. Eng. 67 (1984) 229.
22. P.S. Gilman, J.W. Brooks and P.J. Bridges: "Aluminum-Lithium Alloys III", Eds. C. Baker, P.J. Gregson, S.J. Harris and C.J. Peel, The Institute of Metals (1986) 112.
23. W. Ruch and E.A. Starke, Jr.: "Aluminum-Lithium Alloys III", The Institute of Metals (1986) 121.

24. W.E. Quist, G.H. Narayanan, A.L. Wingert and T.M.F. Ronald: "Aluminum-Lithium Alloys III", The Institute of Metals (1986) 625.
25. G. Jangg, F. Kutner and G. Korb: Powder Metall. Int. 9 (1977) 24.
26. V. Arnhold and J. Baumgarten: Powder Metall. Int. 17 (1985) 168.
27. V. Arnhold and K. Hummert: "Dispersion Strengthened Aluminum Alloys", TMS (1988).
28. G. Jangg, J. Vasgyura and K. Schröder: "Horizons of Powder Metallurgy", Eds. W.A. Kaysser and W.J. Huppmann, Part II, Verlag Schmid GmbH (1986) 989.
29. G.J. Brockmann, J. Baumgarten and V. Arnhold: "Horizons of Powder Metallurgy", Part I, Verlag Schmid GmbH (1986) 313.
30. E. Arzt and J. Rösler: "Dispersion Strengthened Aluminum Alloys", TMS (1988).
31. J.A. Hawk, W. Ruch and H.G.F. Wilsdorf: "Dispersion Strengthened Aluminum Alloys", TMS (1988).
32. M.L. Övecoglu and W.D. Nix: Int. J. Powder Metall. 22 (1986) 17.
33. S. Ezz, M.J. Koczak, A. Lawley and M.K. Premkumar: "High Strength Powder Metallurgy Aluminum Alloys II", Eds. G.J. Hildeman and M.J. Koczak, TMS (1986) 287.
34. D.L. Yaney, M.L. Övecoglu and W.D. Nix: "Dispersion Strengthened Aluminum Alloys", TMS (1988).
35. S.S. Ezz, A. Lawley and M.J. Koczak: "Dispersion Strengthened Aluminum Alloys", TMS (1988).
36. G. Staniek, Private communication (1988).
37. P.S. Gilman and K.K. Sankaran: "Dispersion Strengthened Aluminum Alloys", TMS (1988).
38. J.A. Hawk, P.K. Mirchandani, R.C. Benn and H.G.F. Wilsdorf: "Dispersion Strengthened Aluminum Alloys", TMS (1988).
39. P.K. Mirchandani and R.C. Benn: "Space Age Metals Technology", SAMPE (1988) 188.
40. W.E. Frazier and M.J. Koczak: Scripta Metall. 21 (1987) 129.
41. G.S. Murty, M.J. Koczak and W.E. Frazier: Scripta Metall. 12 (1987) 141.
42. W.E. Frazier amd M.J. Koczak: "Dispersion Strengthened Aluminum Alloys", TMS (1988).
43. A.D. Jatkar, R.D. Schelleng and S.J. Donachie: "Metal-Matrix, Carbon and Ceramic-Matrix Composites", Ed. J.D. Buckley, NASA, CP 2406, (1985) 119.
44. J.D. Blucher, C. Fujiwara and J.A. Cornie: 116th TMS Annual Meeting, Denver, 1987.
45. T.G. Nieh, C.M. McNally, J. Wadsworth, D.L. Yaney and P.S. Gilman: "Dispersion Strengthened Aluminum Alloys", TMS (1988).
46. D.L. Davidson: Metall. Trans. 18A (1987) 2115.
47. J. Wadsworth, C.A. Henshal, T.G. Nieh, A.R. Pelton and P.S. Gilman: "High Strength Powder Metallurgy Aluminum Alloys II", Eds. G.J. Hildeman and M.J. Koczak, TMS (1986) 137.
48. T.G. Nieh, C.A. Henshal and J. Wadsworth: Scripta Metall. 18 (1984) 1405.
49. T. Creasy, J.R. Weertman and M.E. Fine:"Dispersion Strengthened Aluminum Alloys", TMS (1988).
50. A. Sharma, P.R. Soni and T.V. Rajan: Metal Pow. Rep. 43 (1988) 37.
51. S.S. Sergev, S.A. Black and J.F. Jenkins: U.S.Pat. 4,264,362 Apr. 28, 1981.
52. S.A. Black: "Development of Supercorroding Alloys for Use as Timed Releases", CEL-TN-1550, (1979).
53. E.Yu. Ivanov, I.G. Konstanchuk, A.A. Stepanov and V.V. Boldier: Dokl. Akad. Nauk. SSSR 286 (1986) 385.

54. M.Y. Song, E.I. Ivanov, B. Darriet, M. Pezat and P. Hagenmüller: Int. J. Hydrogen Energy 10 (1985) 169.
55. E.I. Ivanov, I. Konstanchuk, A. Stepanov and V. Boldyrev: J. Less Common Metals 131 (1987) 25.
56. A. Stepanov, E.I. Ivanov, I. Konstanchuk and V. Boldyrev: J. Less Common Metals 131 (1987) 89.
57. M.Y. Song, E.I. Ivanov, B. Darriet, M. Pezat and P. Hagenmüller: J. Less Common Metals 131 (1987) 71.
58. E.Yu. Ivanov, B. Darriet, A.A. Stepanov, K.B. Gerasimov and I.G. Kostanchuk: Izv. Sibie. Otd. Akad. Nauk, SSSR, Khim. 15 (Sep. 1984) 30.
59. I.G. Kostanchuk, E.Y. Ivanov, M. Pezat, B. Darriet, V.V. Boldyrev and P. Hagenmüller: J. Less Common Metals 131 (1987) 181.
60. M. Khrussanova, M. Terzieva, P. Peshev and E.Yu. Ivanov: Mater. Res. Bull. 22 (1987) 405.
61. T.K. Wassel and L. Himmel: "A Study of Mechanical Alloying of Metal Powders", TACOM-TR-12571 (1981).
62. I.G. Wright and A.H. Clauer: "Study of Intermetallic Compounds: Dispersion Hardened TiAl", AFML-TR-76-107 (1976).
63. R. Sundaresan and F.H. Froes: Int. Conf. PM and Related High Temperature Materials, Bombay, India (1987).
64. R. Sundaresan and F.H. Froes: 6th World Conf. on Titanium, Cannes, France (1988).
65. R.B. Schwarz, R.R. Petrich and C.K. Saw: J. Non-Crystalline Solids 76 (1985) 281.
66. R.B. Schwarz and C.C. Koch: Appl. Phys. Lett. 49 (1986) 146.
67. G. Cocco, S. Enzo, L. Schiffini and L. Battezzati: Mater. Sci. and Engg. 97 (1988) 43.
68. L. Battezzati, G. Cocco, S. Enzo and L. Schiffini: Mater. Sci. and Engg. 97 (1988) 121.
69. B.P. Dolgin, M.A. Vanek, T. McGory and D.J. Ham: J. Non-Cryst. Solids 87 (1986) 281.
70. B.P. Dolgin: "Science and Technology of Rapidly Quenched Alloys", MRS Symposium Proc. Vol. 80 (1987) 69.
71. C. Politis and W.L. Johnson: J. Appl. Phys. 60 (1986) 1147.
72. C. Politis and J.R. Thompson: "Science and Technology of Rapidly Quenched Alloys", MRS Vol. 80 (1987) 91.
73. C. Politis and J.R. Thompson: "Horizons of Powder Metallurgy", Part I, Verlag Schmid GmbH (1986) 141.
74. R. Sundaresan, A.G. Jackson, S. Krishnamurthy and F.H. Froes: Mater. Sci. and Engg. 97 (1988) 115.
75. R. Sundaresan, A.G. Jackson and F.H. Froes: 6th World Conference on Titanium, Cannes, France (1988).
76. W. Krauss, C. Politis and P. Weimar: Metal Powder Report 43 (1988) 231.
77. R. Sundaresan and F.H. Froes: 6th World Conference on Titanium, Cannes, France (1988).
78. R. Sundaresan, A.G. Jackson and F.H. Froes: 6th World Conference on Titanium, Cannes, France (1988).
79. R. Sundaresan and F.H. Froes: "Modern Developments in Powder Metallurgy", Vol. 18-21, MPIF (1988).

Properties and Applications of Dispersion Strengthened Aluminum Alloys

Volker Arnhold and Klaus Hummert, Sintermetallwerk Krebsöge GmbH, Division of Research and Development, Radevormwald, FRG

Abstract

Dispersion strengthening with nonmetallic dispersoids offers special properties without using large amounts of other metallic alloying elements. Optimum properties are achieved for high temperature strength and fatigue strength at elevated temperatures combined with excellent creep resistance. The very uniform and fine microstructures lead to isotropic behaviour and high stability at very low and very high temperatures. Some selected properties are discussed and applications in different fields of technology are explained.

1. Introduction

Several approaches have been made for the production of aluminum alloys containing nonmetallic dispersoids. First attempts were made with SAP in 1948 (1,2) based on ball milling of aluminum powders in oxidizing atmospheres. Derived from the mechanical alloying of superalloys (3) materials containing carbides and oxides are available today (4). The mechanical alloying process is conducted with a process control agent (PCA), e.g. stearic acid, which leads to a content of 1.1 wt.% carbon and 0.8 wt.% oxygen.

Reaction milling (RM) has been developed by Jangg at the Technical University of Vienna (Austria) (5,6,7). This process, attritor milling without any PCA, is now used as an industrial process by Eckart-Werke (FRG), who are one of the leading PM-aluminum powder suppliers in Europe. The materials with the trade name DISPAL (Dispersion strengthened aluminum) have been jointly developed by Krebsöge, Erbslöh-Aluminium and Eckart-Werke and are officially promoted since 1985.

The RM-process generally leads to materials with excellent high temperature properties (8). Consequently applications should be tackled with special demands at temperatures beyond $300^{\circ}C$, as shown in Fig. 1, giving a general comparison of DISPAL with other materials in terms of high temperature strength. Table 1 gives a survey of the most important demands for alloys for combustion engines, compressors or similar high temperature components.

To fulfill these demands and taking into account the specific processing capabilities of the RM-process, a simple and straight forward alloy concept has been chosen:

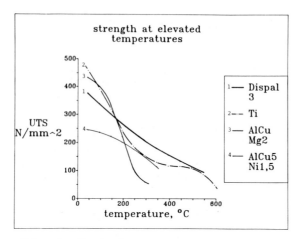

Fig. 1: Typical range of application of DISPAL

Table 1: Performance profile for high temperature applications

x Temperature range between - 50°C and 450°C

x Yield strength (RT)   : >300 MPa
  Yield strength (400°C): >100 MPa

x Max. loss in UTS after annealing (500 h, 450°C): < 10 %

x Elongation:>5 % (all temperatures)

x Thermal shock resistance at least as good as standard I/M
  alloy (hard anodized)

x Fatigue limit (RT)   : > 100 MPa
  Fatigue limit (300°C): > 60 MPa

x Thermal conductivity, thermal expansion similar to standard
  I/M alloy

x Low price

I.   Aluminum + Carbon + Oxygen

     $Al + Al_4C_3 + Al_2O_3$ (DISPAL)

II. Aluminum Silicon 12 + Carbon + Oxygen

     $AlSi12 + Al_4C_3 + Al_2O_3$ (DISPAL Si12)

2. Processing route

Fig. 2 gives the six processing steps normally used for the production of DISPAL materials, schematically:

> Air atomization
>
> Reaction milling
>
> Cold isostatic pressing
>
> (Vacuum-) Heat treatment
>
> Extrusion
>
> Machining, Forging, Finishing

Fig. 2: Consolidation route of DISPAL, schematically

The atomization step is standard air atomization. The proprietary RM-process of Eckart-Werke is defined by adding the carbon via lamp-black or graphite and adjusting the oxygen content by close control of the milling atmosphere (O_2, Ar, N_2, air, etc.). Fig. 3 shows the product of the RM-operation, coarse granules with a size between 0.1 mm and 1 mm in diameter.

.102kx   15kv 700

Fig. 3: SEM picture of reaction milled granules

The micrographs of sections of these granules (Fig. 4) reveal some inhomogeneities, which are still reason for further optimization of the process. The next consolidation steps are CIP of billets (e.g. dia. 245 mm, length 350 mm), (vacuum-) heat treatment – the use of vacuum depends on the specific product demand – and hot extrusion. The extrusion ratio is at least 20:1, if there is no additional forging operation.

The extrusion temperature varies between 450°C (DISPAL Si12) and 500°C (DISPAL). Encapsulation and vacuum degassing has been closely analyzed for these materials. Obviously hydrogen content can be reduced (30 - 50 ppm CIP route, 5 - 15 ppm canning or vacuum hot pressing route), but the higher hydrogen content does not show any significant deterioration on the tested material properties. The explanation could be the higher oxide and carbide content (9), which may lead to trapping of hydrogen at these dispersoid sites, preventing a blistering effect. This is also known for SAP and special steels for nuclear applications (10,11).

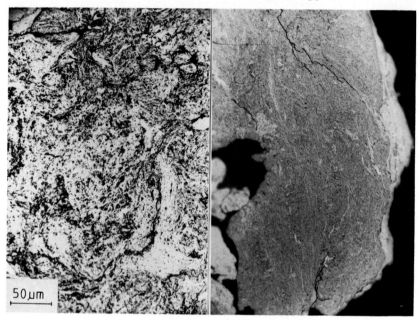

Fig. 4: Micrographs of reaction milled granules, left: DISPAL 2; right: DISPAL 1 Si12

Fig. 5 shows the microstructures of the two alloys taken from transverse and longitudinal sections of an extruded bar (dia. 26 mm, extrusion ratio: 33:1).

DISPAL 2 shows very fine inclusions, which are not fully reacted free carbon, and some cloudy structures. DISPAL 1 Si12 shows very fine, uniformly distributed silicon precipitations. There are no carbon residuals visible.

The basic concept of "custom tailoring" the alloy by adjusting the dispersoid content is expressed by Fig. 6, showing the tensile strength vs. dispersoid content and the corresponding elongation. Up to 11 vol.% of combined dispersoid content ($Al_4C_3 + Al_2O_3$) the strength increases almost linearly with increasing dispersoid content, starting for DISPAL Si12 at a higher level. The corresponding elongations decrease.

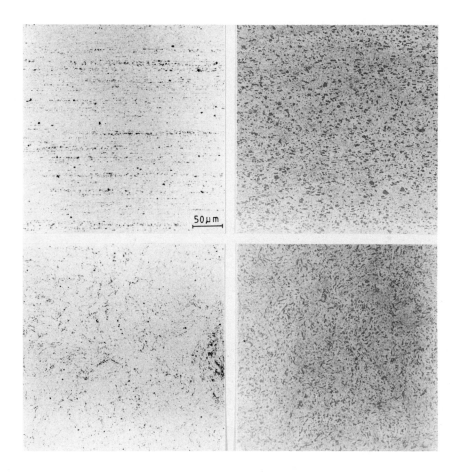

Fig. 5: Microstructures of extruded DISPAL bars
top left: DISPAL 2 (long.) top right: DISPAL 1 Si12
(long.) bottom left: DISPAL 2 (trans.) bottom right:
DISPAL 1 Si12 (trans.)

3. Material properties of DISPAL 2 and DISPAL 1 Si12

The alloys DISPAL 2 and DISPAL 1 Si12 have turned out to be
the best suited compromise for many demands (see Table 1),
containing 2 wt.% carbon or 1 wt.% carbon resp. The oxygen content
is adjustable in both cases.

Fig. 7 "top" shows UTS versus temperature for both P/M alloys
in addition with the data for the classical standard piston alloy
I/M AlSi12CuMgNi in the "as forged" condition (this alloy will be
called "reference" in the following context) (12, 13). The P/M
grades can endure even 400°C, whereas the I/M material has
reached its limits at 350°C.

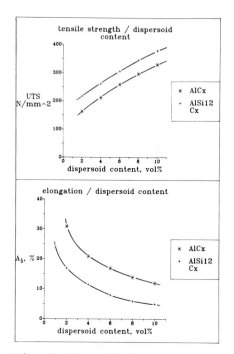

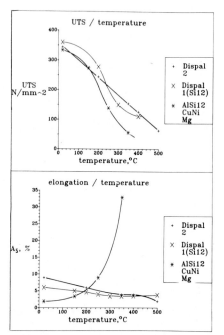

Fig. 6: Strength (top),
elongation (bottom) as function
of dispersoid content

Fig. 7: Comparison of the
ultimate tensile strength and
elongation as function of
temperature
top: UTS, bottom: elongation

The elongation behaviour (Fig. 7, bottom) of DISPAL material
is very typical for materials containing nonmetallic dispersoids
(2, 8), slowly declining with increasing temperature. Here the I/M
alloy shows a superiority. The strength retention (strength
measured at room temperature after annealing) is shown in Fig. 8,
revealing nearly no loss in strength at all, which is caused by
the absolutely stable microstructure.

Hardness as function of test temperature (see Fig. 9) shows a
similar dependance as shown for UTS (compare Fig. 7). The highest
improvements have been achieved for the fatigue strength at
elevated temperatures. Fig. 10 gives a comparison of fatigue
strength at room temperature and at 300°C, measured by rotating
bending test. At room temperature the results are nearly similar
for all alloys, whereas at 300°C the DISPAL materials have
twice the strength of the I/M alloy.

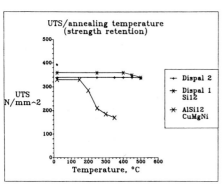

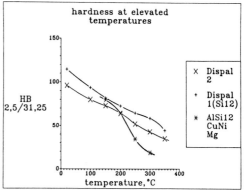

Fig. 8: Strength retention after annealing for 250 h at different temperatures.

Fig.9: Comparison of hot hardness of DISPAL with reference alloy

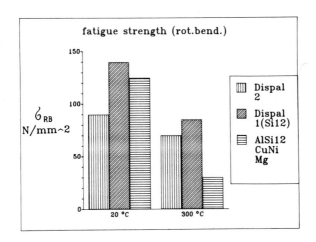

Fig. 10: Comparison of fatigue strength (rotating bending test) for room temperature and 300°C

The next three figures show some special features of the DISPAL materials: good creep behaviour, excellent thermal shock resistance and high sound damping capacity. The creep curve of DISPAL 3 (Fig. 11) proceeds nearly horizontal at the level of pure titanium at 500°C (8) (not normalized to density). The slopes of the titanium alloys are steeper, so that at long times a cross over will occur.

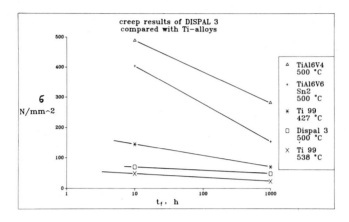

Fig. 11: Creep curve of DISPAL 3 (at 500°C)
compared with several Ti-alloys

Fig. 12 shows the principles of a convenient test rig for
piston manufacturers to compare thermal shock resistance of
different alloys (14). A specimen, shaped like a piston plate is
induction heated to 400°C and cooled by compressed air down to
150°C in fixed cycles. The number of cracks (Fig. 13, top) and
total crack length (Fig. 13, bottom) are examined metallographically after every 1000 cycles. The lowest crack initiation and
crack propagation have been analyzed in DISPAL 1 Si12.

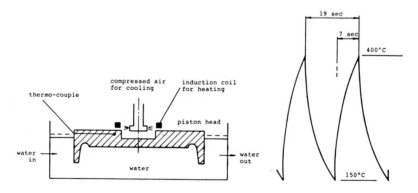

Fig. 12: Test rig and test procedure at IZUMI Automotive Ind.
schematically (14)

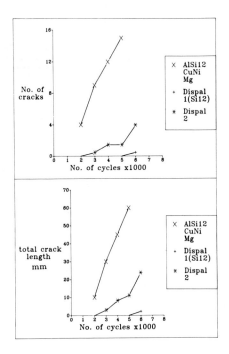

Fig. 13: Comparison of thermal
shock test results of DISPAL and
I/M alloy (14)
top: number of cracks
bottom: total crack length

The extremely fine microstructure of DISPAL materials (8) -
grain size between 0.6 µm and 1.5 µm and dispersoid size between
20 nm and 100 nm - and the very high number of dislocations have
a significant influence on the sound absorption properties (15),
which is shown in Fig. 14. The damping coefficient describes the
decay of an oscillation in the solid according to:

$$\delta = \ln\left(\frac{A_n}{A_{n+N}}\right)$$

giving the loss of energy per cycle when the specimen is allowed
to decay freely.

A_n = amplitude; A_{n+N} = A_n/e; e: = 2.718...Euler´s number

272

The damping coefficient is much higher for the DISPAL grades than for the reference alloy (Fig. 14 top). As has been shown (15) the damping capacity of RS-P/M material of similar composition is at the same level as for I/M grades. δ increases linearly with increasing volume content of dispersoids (Fig. 14, bottom).

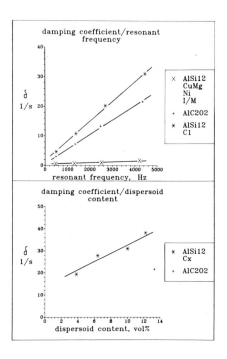

Fig. 14: Comparison of damping behaviour of aluminum alloys
top: damping coefficient vs. resonant frequency
bottom: damping coefficient vs. dispersoid content of DISPAL

Similar damping behaviour has been measured at very low temperatures (16). Fig. 15 shows damping data of 2 conventional alloys compared to DISPAL 2. The goal of application is an interferometer in an European satellite (see Fig. 19).

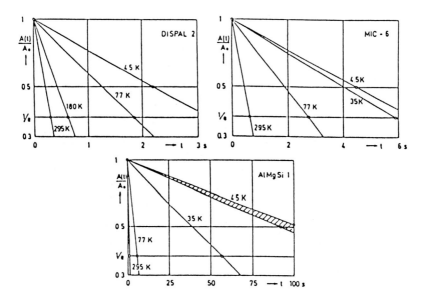

Fig. 15: Temperature dependence of the amplitude decay of oscillating bars made of three different aluminum alloys (16).

4. Discussion

Under the aspect of the performance profile described in Table I the DISPAL materials show a good potential to satisfy most of the demands. What has to be analyzed more closely are notched tensile and notched fatigue properties in comparison with the I/M grades. Improvements have to be made in the milling operation to diminish the number of inclusions and to achieve a better homogeneity.

The question of vacuum degassing is important when DISPAL materials should be welded or joined. To improve these techniques intensive vacuum degassing might decrease hydrogen content to a level which will allow the application of these operations. Friction welding is a convenient method to join DISPAL to aluminum alloys or steels (17).

Some problems arise from the economics of DISPAL materials. Based on attritor milling the today´s price is in the range of 50 - 80 DM/kg. For very specific applications this value is acceptable, but for a wider success, the price should be much lower. Therefore the attritor milling has to be replaced by low energy ball milling. These developments are in progress and have led to first very promising results.

5. Applications

   Two examples are shown for the use in combustion engines:
Fig. 16 demonstrates the possibility of forging DISPAL bars to
piston preforms (here: a motor cycle piston), whereas Fig. 17
gives a new concept of a composite piston for a high performance
racing motorcycle. It is a composite piston made of DISPAL and
special plastic. Also a forged piston of DISPAL 1 Si12 is shown.

Fig. 16: Piston preform (forged) for a motorcycle
(by courtesy of IZUMI Automotive Ind., Japan)

Fig. 17: Composite piston made of DISPAL and special
plastic (by courtesy of GVM, Leverkusen, FRG) and a piston
for the same engine, forged and machined.

Fig. 18 explains the application of DISPAL 2 (thin sheet,
punched pieces) as an electrical conductor used in the newly
developed NaS-battery of ASEA-BBC; the operating temperature is
above 300°C, the demand is good creep resistance.

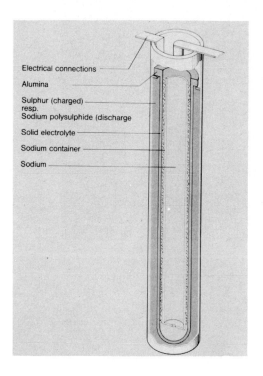

Electrical connections

Alumina

Sulphur (charged)
resp.
Sodium polysulphide (discharge

Solid electrolyte

Sodium container

Sodium

Fig. 18: DISPAL 2 as high temperature
electrical conductor (by courtesy of
ASEA-BBC, Heidelberg, FRG)

The specific, fine and uniform microstructure of DISPAL and
the simple alloy concept lead to a very small distortion during
heating or cooling over a wide temperature range. Together with
the excellent damping properties this especially results in
improvements in the construction of an interferometer for
aerospace applications, which has to be cooled below 10 K (see
Fig. 19).

Fig. 19: Interferometer application for
DISPAL 2 (by courtesy of MPI for Physics
and Astrophysics)

6. Outlook

Besides the improvements in quality and economy of the
described two alloy systems, the future work aims at new alloy
systems to gain other properties (e.g. for lower thermal
expansion, high wear resistance, higher strength, etc.). A
promising approach has been started with alloys of the type AlSi20
and AlSi12Fe5. First results are shown in Fig. 20, giving strength
and corresponding elongation for these materials.

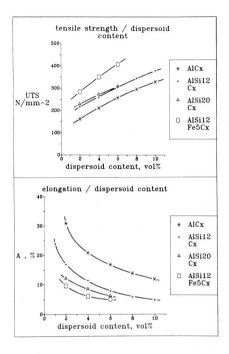

Fig. 20: Strength (top) and elongation (bottom) of some new alloys
under development

Acknowledgements

The authors wish to thank Dr.J.Baumgarten and Dr.G.Brockmann
(Erbslöh-Aluminium) for the cooperation and help in the mutual
development. They thank Mr.H.C.Neubing (Eckart-Werke) for his
cooperation and discussion in powder manufacturing. The cooperation of Alcan Deutschland GmbH, ASEA-BBC, Izumi Automotive Ind.,
Kolbenschmidt AG, Mahle GmbH, MPI for Physics and Astrophysics in
testing DISPAL materials is pointed out. We thank the research
engineers of Sumitomo Electric Industries for their cooperation
and many fruitful discussions.

References

1. E. A. Bloch, Met. Rev. 22 (6), (1961), 193

2. N. Hansen, Powder Met. 20 (10), (1967), 94

3. J. S. Benjamin, Met. Trans. 1, (1970), 2943

4. J. S. Benjamin, R. D. Schelleng, Met. Trans. 12 A, (1981),
   1827

5. G. Jangg, F. Kutner, G. Korb, Aluminium 51, (1975), 641

6. G. Jangg, F. Kutner, G. Korb, Powder Metall. Int. 9 (1),
   (1977), 24

7. G. Jangg, F. Kutner, G. Korb, Wire 28 (6), (1978), 3

8. V. Arnhold, J. Baumgarten, "Dispersion strenghtened aluminium
   extrusions", Powder Metallurgy International, 17 (4), (1985)
   168 - 172

9. J. Meunier, Symposium on "Rapidly Solidified PM-Al-Alloys",,
   Philadelphia, PA, April 4 - 5, 1984

10. P. J. Maziasz, K. Farrell, "The tensile properties of high
    oxide SAP containing helium and tritium" (Paper presented at
    the conference "Fusion Reactor Materials", Miami, Florida,
    January 29 - 31, 1979)

11. W. Lohmann, K. H. Graf, J. Nucl. Mat., 122 - 123, (1984) 1033

12. Technical reference manual "Kolbenkunde", Mahle GmbH,
    P.O. Box 500 769, 7000 Stuttgart 50, FRG

13. Technical reference manual "Kolben", Kolbenschmidt AG,
    P.O. Box 13 51, 7107 Neckarsulm, FRG

14. Y. Suzuki, private communication with the authors, IZUMI
    Automotive Industry Co., Ltd., May 25, 1987

15. J. Töpler, V. Arnhold, "Study of damping properties of
    powder metallurgical aluminium alloys", to be published

16. R. O. Katterloher, "Low Temperatur Decay of Vibrational
    Resonance for Bars made of DISPAL 2 Sintered Al-Powder,
    Alloy MIC-6 and AlMg Sil". (Paper presented at Twelfth
    International Cryogenic Engineering Conference, University
    of Southampton, July 12-18, 1988)

17. H. Kreye et. al., " On the weldability of mechanically
    alloyed aluminium", Proceedings of the 1986 International
    Powder Metallurgy Conference and Exhibition , Düsseldorf,
    July 7 - 11, 1986, 707 - 711, published by Verlag Schmid
    GmbH, Freiburg, FRG

Creep in Dispersion Strengthened Aluminium Alloys at High Temperatures - A Model Based Approach

Joachim Rösler and Eduard Arzt, Max-Planck-Institut für Metallforschung, Stuttgart

Abstract

The mechanism of dispersion strengthening has been studied in dispersion-strengthened aluminium alloys produced by reaction milling. The beneficial effect of second phase particles is usually attributed to the existence of a "threshold stress", below which particles cannot be surmounted by climbing dislocations. But, as will be shown below, predictions of this "threshold stress" concept are widely inconsistent with the experimentally observed creep behaviour. A new constitutive equation is proposed which is based on TEM observations suggesting thermally activated dislocation detachment from the particles to be rate controlling. Creep in several dispersion-strengthened alloys can be consistently described with this concept, which also points to ways of optimizing these alloys.

1. Introduction

Dispersion-strengthened aluminium alloys exhibit excellent creep properties at temperatures as high as 0.8 T_m. Thus they are promising candidates for applications above about 300 °C where precipitation hardened alloys rapidly lose their strength because of microstructural degradation. Although the beneficial effect of a finely distributed, thermally stable dispersion of particles has been well known since the pioneering work of Irmann (1) and Zeerleder (2) it is still debatable why dispersoids can retard the motion of dislocations so effectively. Up to now it is widely believed that "local climb" as proposed by Brown and Ham (3) and Shewfelt and Brown (4) is the rate controlling process causing a threshold stress below which creep should be negligible. In the past, serious objections have been raised against this model. Lagneborg (5) has argued that "local climb" should be an energetically unstable process. A recent analysis of the climb kinetics by Rösler and Arzt (6) does indeed show that climb should never be truly local and that dispersoids should be only weak obstacles to the motion of dislocations at high temperatures if the climb process itself were rate controlling.

An alternative mechanism for the decisive dislocation process was suggested by Srolovitz et al. (7,8). They showed that dislocations can be attracted to the dispersoid-particles at high temperatures because of diffusional relaxation of the stress field at the particle-matrix interface. Arzt and Wilkinson (9) subsequently showed that a significant "threshold stress" σ_d must be exceeded in order to detach dislocations from attractive particles. The kinetics of the detachment process was calculated by Rösler (10) and Rösler and Arzt (11), who propose a new constitutive equation for creep in dispersion strengthened alloys.

The question whether dislocation climb or dislocation detachment is rate limiting is of great practical significance, because the respective models lead to totally different predictions regarding the optimum design of dispersion-strengthened high-temperature alloys. It is the aim of the present paper to analyse the creep data of dispersion-strengthened aluminium alloys in the light of the new constitutive equation. Consequences for the optimum design of these alloys will then be drawn from these results. The application of the concept to other alloys is described in (11-13).

2. The Creep Models

As illustrated in fig. 1 the current dislocation models propose different rate-controlling deformation mechanisms. In the case of the more "classical" concept, "local climb" at the front

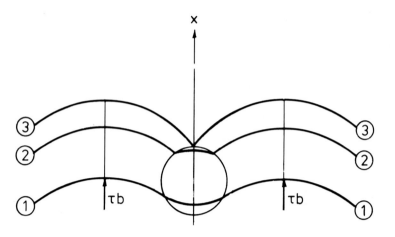

<u>Fig. 1</u>: Schematic illustration of the proposed rate controlling deformation mechanisms for the different creep models; 1: local climb of the dislocation over the dispersoid; 2+3: dislocation detachment from the disperoid.

side of the particle is believed to be rate-limiting. This means that only the dislocation segment at the particle is allowed to climb, whereas the dislocation arms stay in the glide plane. From this geometrical assumption a threshold stress σ_{th} results, which is about 70 % of the Orowan stress for cubes (3) and 40% for spheres (4). Thus the semi-empirical power-law equation for creep can be corrected by introducing an effective stress $\sigma - \sigma_{th}$:

$$\dot{\epsilon}/D = A' \cdot \frac{Gb}{k_B T} \left[\frac{\sigma - \sigma_{th}}{G} \right]^n \tag{1}$$

with creep rate $\dot{\epsilon}$, diffusivity D, shear modulus G, Burgers vector b, dimensionless constant A' and thermal energy $k_B T$. The following predictions regarding the creep behaviour of dispersion-strengthened materials arise from eq. (1):

- The activation energy for creep should be that for volume diffusion. Thus the creep data should be temperature-independent when the normalized strain rate $\dot{\epsilon}/D$ is plotted versus the normalized stress σ/G.
- The influence of the second phase particles should be only of geometrical nature, with the value of σ_{th} depending on the interparticle spacing but not on temperature (except through G). The creep strength is predicted to be independent of the crystal structure and interfacial properties of non-shearable particles.

Because it is difficult to rationalize from a theoretical standpoint why dislocations should climb locally (5,6), attention has focussed on the detachment process as also illustrated in fig. 1. In fact it has been shown by Arzt and Wilkinson (9) that only a slight dislocation line energy relaxation of about 6 % at the particle is necessary to shift the rate controlling mechanism from local climb to the detachment process. This point of turn over would be shifted towards even smaller relaxations if more realistic climb geometries were assumed. The kinetics of the detachment process, which is thermally activated, was recently quantified by Rösler (10) and Rösler and Arzt (11), who derive a new constitutive equation for creep in dispersion-strengthened materials:

$$\dot{\epsilon}/\dot{\epsilon}_0 = \exp\left[-\frac{Gb^2 \cdot r \cdot [(1-k)\cdot(1-\sigma/\sigma_d)]^{3/2}}{k_B T}\right] \qquad (2)$$

with $\dot{\epsilon}_0 = 6\rho\lambda D/b$. Here ρ is the dislocation density, 2λ the interparticle spacing, and r the particle radius. The numerator in the exponential term is the activation energy E_d necessary to detach the dislocation from the back of the particle.

Two important parameters appear in eq. (2). First, the parameter k defined by

$$k = \frac{E_p}{E_m} \qquad (3)$$

where E_m and E_p are the dislocation line energies in the matrix and at the particle respectively. Maximum creep strength is obtained for $k \to 0$ i.e. for maximum relaxation at the particle, whereas the detachment barrier vanishes for $k \to 1$ (= no relaxation). That the creep strength is a strong function of k is illustrated in fig. 2, where the calculated creep rate $\dot{\epsilon}/\dot{\epsilon}_0$ is plotted versus σ/σ_d. A remarkable transition in the creep behaviour occurs at $k \approx 0.9$. For $k < 0.9$ high stress exponents up to insignificant small creep rates are predicted whereas for $k > 0.9$ the creep strength is predicted to degrade significantly at low strain rates and/or high temperatures, resulting in low values of the stress exponent n.

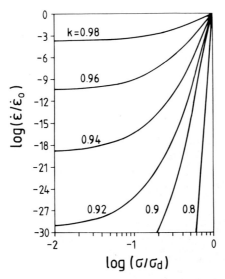

Fig. 2: Calculated stress dependence of $(\dot{\epsilon}/\dot{\epsilon}_0)$ as a function of the interaction parameter k with $Gb^2r/k_B T = 3000$.

The second parameter is the athermal detachment stress σ_d. For stresses $\sigma \geq \sigma_d$ the dislocation can be detached spontaneulsy from the particle back. As shown by Arzt and Wilkinson (9) the athermal detachment stress is related to the Orowan stress σ_0 by

$$\sigma_d = \sigma_0 \cdot \sqrt{1-k^2} \qquad (4)$$

From the concept of thermally activated detachment the following predictions regarding the creep strength of dispersion hardened materials can be made:

- The apparent activation energy for creep should be $Q_{app} = Q_V + E_d$ (Q_V: activation energy for volume diffusion; E_d: activation energy for the detachment process). Thus the creep strength is expected to be temperature dependent even when plotting $\dot{\varepsilon}/D$ versus σ/E.

- The creep strength should be a strong function of the interaction parameter k (fig. 2) and thus of the particle-matrix interfacial properties.

- Because, at a given volume fraction f_V, the athermal detachment stress σ_d is a function of the particle radius r it follows that an optimum particle radius should exist (10,11):

$$r_{opt} = 0.3 \frac{b \cdot \sqrt{1-k^2}}{\sigma/E} \cdot \sqrt{f_V} \qquad (5)$$

These predictions differ substantially from those of the local climb concept in that the interfacial properties, i.e. the particle/matrix combination, are now expected to become important and that an intermediate rather then a minimum diameter should be optimum for non-shearable particles. In the following section an attempt will be made to distinguish between the two models by analyzing experimental creep data in the light of eqs. 1 and 2.

3. Analysis of Creep Data

The alloys investigated in detail were Al-2.2%C-0.8%O (subsequently referred to as "AlC2") and Al-4%Mg-1.4%C-2.6%O ("AlMgC1") (all percentages by weight). Both materials were produced by reaction milling (14,15). The dispersoids in AlC2 are about 8 vol%-Al_4C_3 and 1.5 vol% Al_2O_3, with an average particle size of 43 nm (see also fig. 3). The alloy AlMgC1 is hardened by about 5.2 vol% Al_4C_3 and 4.9 vol% MgO, with a typical particle size of 34 nm (Ref. 10); the magnesium is completely transformed into MgO and is no longer in solid solution. The creep properties of the two alloys are displayed in figs. 4 and 5.

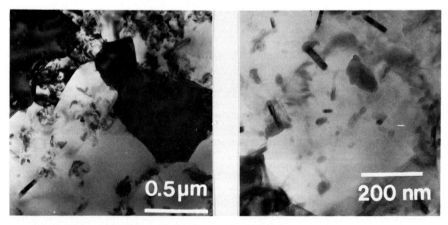

Fig. 3: Micrographs of reaction milled Al-2.2wt%C-0.8 wt%O (alloy "AlC2") showing the fine grained structure (left) and the distribution of carbide and oxide dispersoids (right).

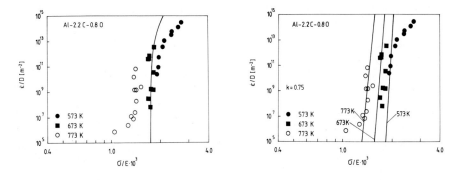

Fig. 4: Creep data of Al-2.2 wt% C-0.8wt% O (alloy AlC2);
left: comparison with the model calculations (solid line) due to the threshold stress concept;
right: comparison with the model calculations (solid lines) due to the detachment concept.

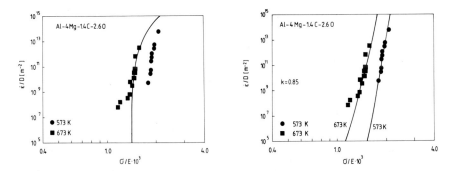

Fig. 5: Creep data of Al4 wt% Mg-1.4 wt% C-2.6 wt% O (alloy AlMgC1);
left: comparison with the model calculations (solid line) due to the threshold stress concept;
right: comparison with the model calculations (solid lines) due to the detachment concept.

A striking feature of the creep data is the extremely high stress sensitivity, especially in the case of AlC2 at low normalized strain rates. In this regime creep is controlled by the inter-action of single dislocations with dispersoids, whereas at high strain rates ($\dot{\varepsilon}/D \geq 10^{12}$ m^{-2}) the formation of dislocation networks is observed. The strong drop in the creep strength at the highest temperature and lowest creep rates was shown to be due to the fine grain size (10,16). Since it is possible to optimize the grain structure by means of secondary recrystalli-zation at least for low particle volume fractions (16), this does not seem to be a principal limitation. Thus we will focus our attention on the direct strengthening contribution of the second phase particles, noting that the models introduced in section 2 were not designed to explain the creep behaviour at the highest and lowest normalized strain rates.

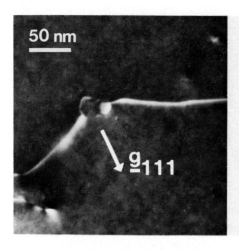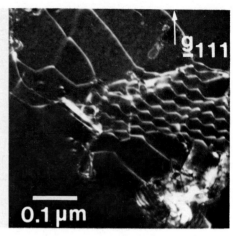

Fig. 6: Typical dislocation configurations after creep deformation (material: AlC2); left: single dislocation-dispersoid interaction is rate controlling at low strain rates; right: formation of dislocation networks at high strain rates $\dot{\varepsilon}/D \geq 10^{12}$ m^{-2}.

It is remarkable in figs. 4 and 5 that the creep data do not lie on single curves despite modulus and diffusivity correction. This shows that the activation energy for creep is higher than that for volume diffusion. Thus the classical threshold stress concept fails to describe the temperature dependence properly.

The theory of thermally activated dislocation detachment gives a natural explanation for the strong temperature dependence because the activation energy Q_{app} is the sum of the activation energy for volume diffusion and that for the detachment process. Knowing Q_{app} and the stress exponent n_{app} the interaction parameter k can be calculated (10,11). It is seen in fig. 4 that the experimental creep behaviour of AlC2 is well described by assuming a line energy relaxation of 25% at the second phase particles (k=0.75). Applying the same analysis to alloy AlMgC1, k=0.85 is found, which corresponds to a line energy relaxation of 15%.

These results suggest that carbide particles should be more effective barriers to the motion of dislocations at high temperatures than magnesium oxide. In fact the 400 °C creep strength of AlC2 is about 20 % higher than that of AlMgC1 although the mean interparticle spacing is about 24 % less in the latter case (10) (77 nm instead of 101 nm). It would be hard to rationalize the inferior creep properties of AlMgC1 with a climb threshold concept according to which σ_{th} should be inversely proportional to the particle spacing.

The detachment model is further supported by the fact that it explains even the creep behaviour of Al-8Fe-4Ce (17), in which dispersoids are formed by rapid solidification. Its creep behaviour differs from a threshold like behaviour in that the stress exponent declines strongly with decreasing normalized strain rate $\dot{\varepsilon}/D$ and a pronounced temperature dependence is observed (fig. 7). One would tend to invoke microstructural instabilities as being responsible for the remarkable loss in strength, but as was shown by Yaney and Nix (17) this contribution is only of minor importance. Instead the concept of thermally activated dislocation detachment offers a natural explanation for this peculiar behaviour. By assuming a very weak particle dislocation attraction (only 5 % line energy relaxation, k=0.95), both the stress and the temperature dependence of the creep rate are satisfactorily described (fig. 7). Although speculative, since detailed TEM investigations are lacking for this type of material, this remarkable result suggests that thermally activated dislocation detachment may be the rate controlling deformation mechanism also in these materials. It is evident from fig. 7 that one would seriously overestimate the creep strength by extrapolating the data points obtained at low temperature by means of the threshold stress concept.

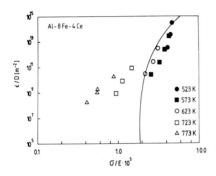 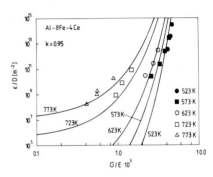

Fig. 7: Creep data of Al-8Fe-4Ce (data from Yaney and Nix (17));
left: comparison with the model calculations (solid line) due to the threshold stress concept;
right: comparison with the model calculations (solid lines) due to the detachment concept.

As is discussed by Rösler (10) and Rösler and Arzt (11) the correlation between model predictions and experimental data is not fully quantitative in that unrealistically high values of the pre-exponential factor in eq. 2 have to be assumed to fit the data. In fact very satisfactory correspondence has been obtained in analyzing single crystal data (11). Thus it seems likely that there is an additional contribution from other deformation mechanisms like diffusional creep to the overall creep behaviour. Because the value of the relaxation factor is insensitive to the magnitude of Q_{app} the conclusions stated above are nevertheless still valid.

4.   Consequences for the Design of Dispersion Strengthened Alloys

From the new constitutive equation for creep (eq. 2) two major requirements for optimum high temperature strength can be deduced. Firstly the interaction parameter k, which appears as a new material parameter, should be as low as possible. Its magnitude must in some way depend on the interfacial properties and should be characteristic for a given particle-matrix combination. Maximum interaction strength was found for the system Al/Al_4C_3 (k=0.75) followed by Al/MgO (k=0.85) and Al/Al_xFe_yCe (k=0.95). Thus Al alloys strengthened by carbide particles seem to have the best potential for applications at very high temperatures. The above analysis suggests that intermetallic particles produced by rapid solidification are less efficient than ceramic particles incorporated by mechanical alloying techniques. Examination of further creep data is required to see whether this observation is part of a general trend.

The second requirement regards particle size. Given a certain particle volume fraction, the detachment model predicts that there should be an optimum particle radius, while the climb model would demand a small spacing. Inserting typical stress levels in eq. (5) one finds that the optimum particle diameter lies generally between 10 nm and 30 nm. Because actual spacings are generally in or above this range, there seems to be some further potential for increasing the creep strength by particle refinement.

5. Conclusions

-   The comparison of experimental creep data with theoretical predictions strongly suggests that dislocation creep in dispersion-strengthened aluminium alloys is controlled by thermally activated dislocation detachment from the dispersoids and not by a climb-related "threshold stress".
-   An optimum dispersoid diameter is predicted which lies typically in the range of 10 - 30 nm. Further refinement of the dispersion would result in inferior creep properties because of enhanced detachment kinetics.
-   Two "quality parameters" (k and σ_d) for the dispersion have been defined. For the purpose of alloy screening, they can be readily obtained from a limited number (about five) of creep tests.

6. Acknowledgements

The authors are grateful to Dr. Arnhold (Sintermetallwerk Krebsöge) and Dr. Brockmann (Erbslöh Aluminium) for providing the alloy material. The work was supported financially by the Bundesministerium für Forschung und Technologie under project number 03 M0010 E4.

7. References

(1) R. Irmann: Techn. Rundsch., 41, (1949) 19.
(2) A. v. Zeerleder: Z. Metallkd., 41, (1950) 228.
(3) L.M. Brown, R.K. Ham: in "Strengthening Methods in Crystals", A. Kelly, R.B. Nicholson, eds., Applied Science Publishers, London, (1971) p. 126.
(4) R.S.W. Shewfelt, L.M. Brown: Phil. Mag., 35, (1977) 945.
(5) R. Lagneborg: Scripta Met., 7 (1973) 605.
(6) J. Rösler, E. Arzt: Acta Met., 36 (1988) 1043.
(7) D.J. Srolovitz, R.A. Petkovic-Luton, M.J. Luton: Acta Met., 31 (1983) 2151.
(8) D.J. Srolovitz, M.J. Luton, R.A. Petkovic-Luton, D.M. Barnett, W.D. Nix: Acta Met, 32 (1984) 1079.
(9) E. Arzt, D.S. Wilkinson: Acta Met., 34 (1986) 1893.
(10) J. Rösler: Ph.D. Thesis, University of Stuttgart (1988).
(11) J. Rösler, E. Arzt: to be published.
(12) E. Arzt: this volume.
(13) E. Arzt: Res Mechanica, in press.
(14) G. Jangg, F. Kutner, G. Korb: Aluminium, 51 (1975) 641.
(15) G. Jangg, F. Kutner, G. Korb: Powder Met. Int., 9 (1977) 24.
(16) R. Joos: Diploma Thesis, University of Stuttgart (1988).
(17) D.L. Yaney, W.D. Nix: Met. Trans. A., 18 (1987) 893.

Elevated Temperature Characterization and Deformation Behaviour of Mechanically Alloyed Rapidly Solidified Al-8.4wt%Fe-3.5wt%Ce Alloy

M. L. Öveçoglu, Developmental Technologies Int. B.V., Heerlen, The Netherlands,

W. D. Nix, Dept. of Materials Science and Engineering, Stanford University, Stanford, U. S. A.

Introduction

High strength aluminum-based alloys produced by the utilization of powder metallurgy techniques are useful for a wide variety of aerospace applications since they have high strength to weight ratio, they are easy to fabricate, relatively inexpensive, and more importantly, they endure elevated service temperatures. Recently, there has been a considerable effort in the development of these alloys for a number of high temperature applications. As a result of U.S. Air Force supported research and development on a new generation aluminum alloys, the ternary Al-Fe-Ce system has been determined to be the most promising alloy from a series of twenty-one binary and ternary alloys(1). Rapid solidification processing (RSP) techniques are utilized in order to introduce a large volume fraction of finely dispersed intermetallic compounds for elevated strength of the Al-Fe-Ce ternary alloy.

Recent investigations on rapid solidification processed (RSP) Al-Fe-Ce alloys have revealed remarkable low temperature mechanical properties but drastic strength losses at high temperatures(2,3,4). These strength losses are attributed to intermetallic coarsening by Angers (2) and Koczak(3) and to intermetallic particle shearing by Yaney(4). In order to circumvent the problem of strength decrease at the required high service temperatures, it is essential that the strengthening mechanisms be based on secondary phases which do not degrade at high temperatures. One of the important ways to achieve this is through the use of particles of very stable refractory phases, such as oxides and carbides.

The primary purpose of this investigation has been to develop a dispersion strengthened (D.S.) RSP Al-Fe-Ce alloy by means of combining the low temperature strength behaviour observed in the case of RSP Al-Fe-Ce alloys and the high temperature properties of dispersion strengthened alloys. Mechanical alloying (MA) was used to create the necessary stable carbide and oxide particles in RSP Al-Fe-Ce powders for the high temperature property enhancement(5). Microstructural characterization and mechanical testing have been performed on MA RSP Al-Fe-Ce at elevated test temperatures and the test results are compared with those for rapid solidification processed (RSP) Al-Fe-Ce and dispersion strengthened (D.S.) Aluminum considering the presence of the refined dispersoid and intermetallic phases and their effect to the mechanical properties of MA RSP Al-Fe-Ce in two temperature regimes.

Experimental Procedure

Mechanical Alloying of rapid solidification processed (RSP) Al-Fe-Ce powders

The material used in this investigation, an Al-8.4 wt%Fe-3.5 wt% Ce alloy, was prepared by ALCOA but supplied by the Air Force Wright Aeronautical Laboratories in the form of rapidly solidified powders. The as-received powder batches weighing 4.9 each were put inside 55 ml. hardened tool steel containers along with 0.1 gram of an organic surfactant (Nopcowax-22) and with 31 6.35 mm diameter steel balls. Mechanical alloying (MA) was accomplished by using a Spex No. 8000 Mixer/Mill at different processing times for each batch. The experimental details of the MA process and the characterization investigations during MA of RSP Al-Fe-Ce are given by the same authors in a previous study which reports the achievement of optimum processing conditions for this alloy(5). Based on this characterization investigation study, two different powder samples are chosen for elevated temperature mechanical testing and microstructural characterization work : i) RSP Al-Fe-Ce MA 25 minutes-which represents the intermediate processing stage, and ii) RSP Al-Fe-Ce MA 180 minutes - which represents the steady-state processing stage.

X-ray Diffractometry

In order to detect the intermetallic phases precipitated during RSP and MA, X-ray Diffractometry scans were conducted on as-received RSP Al-8.4 wt%Fe-3.5 wt% Ce powders and on those mechanically alloyed for 180 minutes, i.e. RSP Al-Fe-Ce MA 180 minutes sample. All X-ray diffraction data were obtained with Cu-Kα radiation in a Rigaku "Geigerflex" D/max-IA X-ray diffractometer which has a theta-theta wide angle goniometer. A microcomputer provided the data acquisition capability for operating, measuring, and data storage and the necessary interfacing with a database which contained JCPDS powder diffraction card data of inorganic compounds. After annealing at 573, 673, and 773 ^oK for 120 minutes, RSP Al-Fe-Ce MA 180 minutes powders were scanned at step sizes of 0.05, 0.02, and 0.01 with a data collection time of 10 seconds. These annealing temperatures are the ones used during mechanical testing of this alloy(6).

Consolidation

Both RSP Al-Fe-Ce MA 25 min. and RSP Al-Fe-Ce MA 180 min. powder samples were vacuum degassed, and hot pressed under the same condition(6). MA 25 minute sample was hot extruded at 723 ^oK at a ratio of 20 to 1. It was impossible to extrude the billet which had been mechanically alloyed for 180 minute at this temperature because of its excessive cold worked structure. So, following an annealing treatment this sample was extruded at a higher temperature of 823 ^oK with a lower extrusion ratio of 10 to 1. Powder consolidation was carried out in Materials Analysis Laboratories of Lockheed Aerospace Corporation in Palo Alto.

Elevated Temperature Compression Testing

Constant total true strain rate compression tests at elevated temperatures were conducted in an Instron electomechanical testing machine which is interfaced with a Hewlett-Packard 3054A Data Acquisition System. A split type Applied Test Systems Model 3320 furnace was used for compression testing at elevated temperatures. The compression testing conditions are in accordance with the experimental technique developed and explained by Yaney(4) and a detailed description of the test apparatus is given elsewhere(6). Compression specimens of 5.08 mm diameter and 7.62 mm length were machined off directly from both extrusions, i.e. as-extruded MA 25 min. and MA 180 min., such that the compression axis was chosen parallel to the extrusion direction.   The compression tests were conducted at temperatures of 523, 623, 673, 723 and 773 ^oK and three different true strain rates of 10^{-5}, 10^{-4}, and 10^{-3} sec^{-1} for each temperature. Samples were deformed 15 to 20 percent and then water quenched in order to preserve the structure. Figure 1 is a representative illustration of the elevated temperature tests. The deformation at all temperatures and strain rates was characterized by flow softening.

Transmission Electron Microscopy

Two different samples were used for further TEM investigations : i) As-extruded RSP Al-Fe-Ce MA 180 minutes and and ii) As-extruded RSP Al-Fe-Ce MA 180 minutes, which was compression tested at a temperature of 773 ^oK and a true strain rate of 10^{-4} sec^{-1}. Disks of about 0.4 mm in thickness extracted from the transverse sections of both samples were dimpled and electropolished until a center perforation was achieved. Electropolishing was performed in a 20% HNO_3 - 80% CH_3OH solution at 243 ^oK using a Fishione 110 Twin Jet Polisher. TEM samples were examined using a Philips EM 400 at an operating voltage of 120 kV. The microscope was operated in the selected area diffraction (SAD) mode with an aperture size of 60 mm for the detection of the dispersoid phases.

Results and Discussions

Microstructural Observations

Figure 2 is an optical micrograph of the as-extruded Al-Fe-Ce MA 180 minute sample taken parallel to its extrusion direction. The grains in the as-extruded alloy are fairly large and are elongated along the extrusion direction. We believe that severe cold work introduced by high

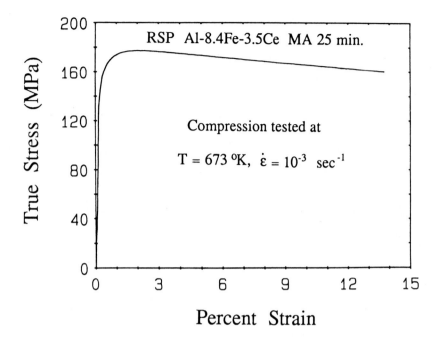

Figure 1: True stress versus true strain for as-extruded RSP Al-8.4%Fe-3.5%Ce MA 25 tested at 673 $^\circ$K and a strain rate of 10^{-3} sec^{-1}.

Figure 2: Optical micrograph of as-extruded RSP Al-8.4%Fe-3.5%Ce MA 180 minutes parallel to the extrusion direction. Etched with Keller's reagent.

energy ball milling is to such an extent that the alloy recystallizes during consolidation.

X-ray diffractometer scan data of RSP Al-Fe-Ce powders mechanically alloyed at various processing times have revealed only the major peaks of α-Al. However, after annealing the RSP Al-8.4wt%Fe-3.4wt%Ce MA 180 min. powder sample, intermetallic phases such as Al-Fe-Ce, Al_6Fe, $Al_{10}Fe_2Ce$ and $Al_{13}Fe_4$ were detected(6). These phases have also been reported by Angers(2) and Yaney(4) in RSP Al-Fe-Ce alloys.

Figure 3 is a Bright Field (BF) TEM micrograph with the accompanying Selected Area Diffraction (SAD) pattern of the as-extruded RSP Al-8.4wt%Fe-3.4wt%Ce MA 180 minutes. The presence of Al_4C_3 can be seen from the (110) and (012) Al_4C_3 rings in addition to the major rings of the α-Al matrix phase. The observation is in agreement with similar findings of Oliver(7) in the case of ODS Al and of Gilman(8) for MA Al-Al_2O_3 alloys. The smooth and continuous nature of the carbide rings implies that the dispersoid phases are fine, numerous and unifomly dispersed throughout the matrix. Figure 4 is a BF TEM micrograph and the respective SAD diffraction pattern of the as-extruded RSP Al-Fe-Ce MA 180 minute sample compression tested at a temperature of 773 °K and a strain rate of 10^{-4} sec^{-1}. From the attached spot pattern, the grain shown in the BF is identified as the Al_6Fe intermetallic phase. The diffraction spots only arise from the Al_6Fe phase. In other words, the white regions shown in the BF micrograph do not give rise to any diffraction spots in the SAD pattern and tilting the crystal did not have any effect on the contrast level of these regions.

Elevated Temperature Deformation Behaviour

Figures 5 and 6 are the double logarithmic representation of diffusion compensated strain rate versus modulus compensated stress for RSP Al-Fe-Ce MA 25 minute and MA 180 minute materials respectively plotted along with RSP Al-Fe-Ce(4) and dispersion strengthened Aluminum (7) all tested in compression. The data points shown by different symbols for different temperatures belong to this investigation, whereas black solid spots to D. S. Al and solid lines with adjacent temperatures to RSP Al-Fe-Ce alloy. As can be seen in both figures, at lower temperatures, represented by 523 °K and 623 °K, the flow stress values of both MA and non-MA Al-Fe-Ce are close to each other. This observation implies that the deformation mechanisms at low temperatures are similar for both cases and further refinement of dispersed second phases for fully-MA material is not manifested in an increased contribution of strength at low temperatures. If the deformation at low temperatures is due to the Orowan bowing mechanism, then one would expect that due to its more closely spaced oxides and carbides MA Al-Fe-Ce is much stronger than the non-MA in which the strengthening might be due to the presence of high volume fraction intermetallic particles. These intermetallic phases are also present in MA RSP Al-Fe-Ce alloy in the same proportion as those in RSP Al-Fe-Ce since both materials are from the same rapidly solidified batch. However, as stated by Yaney and the authors of this article in a recent work(9), unlike MA Al-Fe-Ce, non-MA Al-Fe-Ce does not recrystallize during consolidation. Therefore, the low temperature strength of RSP Al-Fe-Ce is contributed both by high volume fraction intermetallic phases and by heavily cold worked structure which is fairly stable at low temperatures.

At higher temperatures, namely at 673, 723, and 773 °K, the effect of mechanical alloying becomes increasingly important with increasing temperature. The elevated temperature strength of MA RSP Al-Fe-Ce is still due to well-dispersed oxides and carbides which are inherent of mechanical alloying process. The higher flow stresses achieved for RSP Al-Fe-Ce MA 180 minute than those for MA 25 min. test samples at elevated temperatures are due the uni form dispersion of stable dispersoids. We believe that the steady-state processing conditions achieved at the completion of mechanical alloying process(5,6) lead to homogeneous dispersoid and intermetallic particle distribution which is not achieved at the intermediate mechanical alloying stage. In addition, large elongated grains recrystallized during consolidation process reduce the elevated temperature weakening effect due to grain boundary sliding. The strength differences between non-MA and MA RSP Al-Fe-Ce at the temperatures of 723 and 773 °K are due to the presence of these dispersoid phases when the contribution to strengthening from the

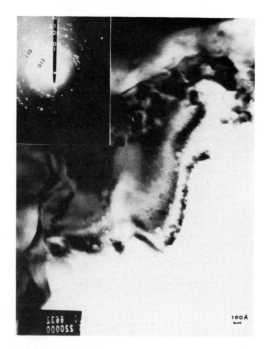

Figure 3: TEM micrograph and diffraction pattern showing the matrix region α-Al taken from the RSP Al-Fe-Ce MA 180 minutes and as-extruded sample.

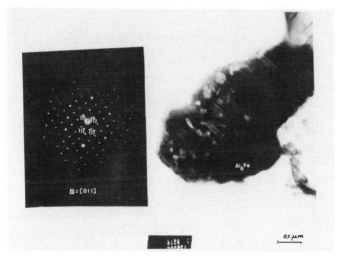

Figure 4: TEM micrograph and diffraction pattern taken from a MA Al-Fe-Ce deformed at 773 °K and a strain rate of 10^{-4} sec^{-1}.

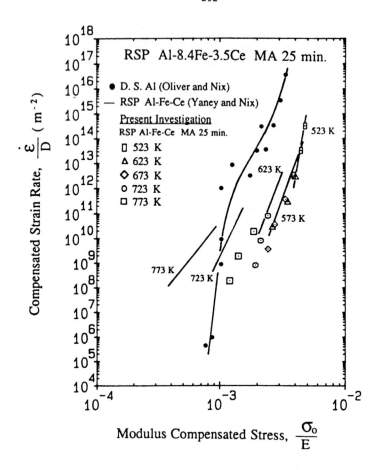

Figure 5: Double logarithmic plot of diffusion compensated strain rate versus modulus compensated stress for RSP Al-Fe-Ce MA 25 minutes compared with the RSP Al-Fe-Ce of the same composition(4) and with D.S. Aluminum(7) tested in compression.

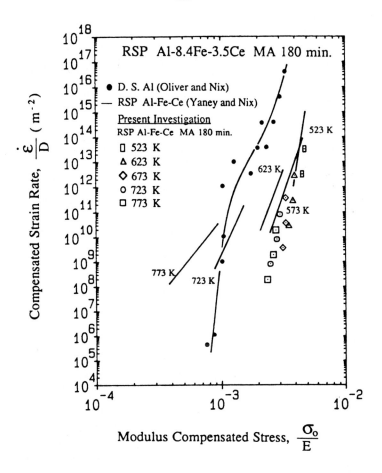

Figure 6: Double logarithmic plot of diffusion compensated strain rate versus modulus compensated stress for RSP Al-Fe-Ce MA 180 minutes compared with the RSP Al-Fe-Ce of the same composition(4) and with D.S. Aluminum(7) tested in compression.

intermetallic phases decreases since they coarsen and/or begin to deform. The addition of these refractory phases in order to supplement the strengthening of the intermetallics which has been the essence of this research work manifests itself in an average strength increase of about six times for MA RSP Al-Fe-Ce compared to non-MA RSP Al-Fe-Ce alloy at the highest testing temperature of 773 °K and the lowest true strain rate of 10^{-5} per second. Thus, the oxide and carbide particles present in MA RSP Al-Fe-Ce resist the plastic deformation and act as the primary sources of strengthening at high temperatures.

As can be seen in Figure 6, MA RSP Al-Fe-Ce is stronger than D.S. Al at all temperatures and strain rates. Also, RSP Al-Fe-Ce is stronger at low temperatures but weaker at high temperatures than D.S. Al. The data for fully-MA RSP Al-Fe-Ce shown in Figure 6 follow a similar trend as those for D.S. Al. In other words, the strength difference of a factor of 2.5 between MA RSP Al-Fe-Ce and D.S. Al is maintained at all temperatures and strain rates. This suggests that the presence of intermetallic phases and uniformly spaced dispersoids produce a combined strengthening effect which is superior to either the effect of dispersion or prepitation strengthening considered alone. This observation confirms the primary goal of producing an alloy which exhibits high strength at both intermediate and high temperatures. That the dispersoid phases produce a combined strengthening effect along with the intermetallic phases can also seen in Figure 4 where the diffraction spots only arise from the Al_6Fe phase and not from the white regions, as mentioned earlier in the text. Since these regions have amorphous nature, they may be the amorphous oxide dispersoid phases which were also detected by Gilman(8) in the case of MA $Al-Al_2O_3$ alloys. The parallel fringes crisscrossing the grain indicate the deformation nature of the intermetallic phase at elevated temperatures. The exact nature of the deformation is not known, but as indicated by Yaney(4) twinning is thought to be the dominant deformation mechanism.

Taking into account of the existence of a threshold stress, a true activation energy for creep deformation of 162 kJ/mole and a stress exponent of 8.5 were estimated for the fully-MA Al-Fe-Ce alloy by extrapolating the strain rate versus flow stress curve at highest test temperature to the lowest strain rate(6). If the dispersoid phases do not readily deform and the deformation mechanism primarily occurs by Orowan mechanism in this alloy, then one would expect that the true activation energy for deformation should be approximately equal to the activation energy for deformation of self-diffusion of aluminum. Although not close, this value is comparable to the activation energy of aluminum self diffusion which is $Q_{S.D.}$ = 142 kJ/mole. We believe that the slight discrepancy is due to the presence of uniformly dispersed aluminum oxides and carbides not only throughout the matrix α-Al phase but also inside some intermetallic phases. Due to their presence in intermetallic grains, the intermetallic phases are also dispersion strengthened as well as the matrix phase.

Summary

Mechanical alloying(MA) process was used to introduce stable aluminum carbide and oxide dispersoids in order to supplement the strengthening by intermetallic phases present in a rapidly solidification processed(RSP) Al-Fe-Ce alloy. MA RSP Al-Fe-Ce powders were compacted by using thermomechanical steps such as vacuum degassing, vacuum hot pressing and hot extrusion. Due to the severe cold working conditions during mechanical alloying, the MA RSP Al-Fe-Ce was recrystallized upon consolidation exhibiting large and elongated grains aligned along the extrusion direction.

By means of X-ray diffractometry investigations, the intermetallic phases such as Al-Fe-Ce, Al_6Fe, $Al_{10}Fe_2Ce$ and $Al_{13}Fe_4$ were detected. Transmission Electron Microscopy investigations confirmed the presence of the Al_4C_3 stable dispersoid phase and revealed the deformation nature of an intermetallic phase.

Constant true strain rate compression tests were conducted on compacted specimens at elevated temperatures. Comparison of the test data for fully-MA RSP Al-Fe-Ce with those for the one mechanically alloyed at an intermediate processing time and for the non-MA RSP Al-Fe-Ce has indicated that the former is stronger than the other two due to the presence of its more refined nature of dispersoids. Also, MA RSP Al-Fe-Ce is stronger than dispersion strengthened Al at a temperature regime of 523 - 773 °K by a factor of about 2.5. MA RSP Al-Fe-Ce have similar strength values as RSP Al-Fe-Ce at low temperatures of 523 and 623 °K because of the heavily deformed microstructure of the RSP Al-Fe-Ce alloy but it is six times stronger than RSP Al-Fe-Ce at high temperatures of 673, 723 °K, and 773 °K. Stable dispersoids such as Al_2O_3 and Al_3C_4 are believed to provide the necessary high temperature strength. The slight discrepancy of the creep activation energy for this material from that for aluminum self diffusion is due to the strengthening effect of dispersoids on the intermetallic phases.

Acknowlegements

The authors would like to thank Dr. T. G. Nieh of Lockheed Aerospace Corporation in Palo Alto for his help in extruding the mechanically alloyed and hot pressed RSP Al-Fe-Ce billets. The initial part of this research was supported by Air Force Office of Scientific Research under the contract AFSOR No. 81-0022C.

References

(1) W. M. Griffith, R. E. Sanders, Jr., and G. J. Hildeman: "Elevated Temperature Aluminum Alloys for Aerospace Applications - High   Strength Powder Metallurgy Aluminum Alloys" - Proceedings of 111th Annual AIME Meeting, Dallas Feb. 1982, Editors: M. J. Koczak and G. Hildeman, AIME Publication (1982), pp. 209.

(2) L. Angers: "Particle Coarsening in Rapidly Solidified Al-Fe-Ce   Alloys", Ph. D. Dissertation, Northwestern University, Evanston, Illinois (August 1985).

(3) A. Lawley and M. J. Koczak: "A Fundamental Study of P/M Processed Elevated Temperature Aluminum Alloys", AFOSR Grant # 82-0010, Department of Materials Engineering, Drexel University, Philadelphia, PA 19104.

(4) D. L. Yaney: "Mechanisms of Elevated Temperature Deformation in Selected Aluminum Alloys and Related Intermetallic Compounds", Ph.D. Dissertation, Stanford University, April 1986.

(5) M. L. Öveçoglu and W. D. Nix: Int. Jr. of Powder Metallurgy 22(1) 17(1986).

(6) M. L. Öveçoglu: "Development of a Dispersion Strengthened Al-Fe-Ce Alloy by Mechanical Alloying and Related Theoretical Modeling of Dislocations in Composite Materials", Ph. D. Dissertation, Stanford University, Stanford, California (June 1987).

(7) W. C. Oliver: "Strengthening Phases and Deformation Mechanisms   in Dispersion Strengthened Solid Solutions", Ph. D. Dissertation, Stanford University, Stanford, California (June 1981).

(8) P. S. Gilman: "The Development of Aluminum-Aluminum Oxide   Alloys by Mechanical Alloying", Ph. D. Dissertation, Stanford   University, Stanford, California(March 1979).

(9) D. L. Yaney, M. L. Öveçoglu and W. D. Nix: The effect of Mechanical Alloying of the Deformation Behaviour of a rapidly solidified Al-Fe-Ce alloy", to be published as AIME Proceedings of 1988 Annual Meeting in Phoenix, Arizona, USA.

C.3:
PERMANENT MAGNETS

Mechanically Alloyed Nd-Fe-B Permanent Magnets

L. Schultz, K. Schnitzke and J. Wecker, Siemens AG,
Research Laboratories, Erlangen, West Germany

Abstract

Microcrystalline Nd-Fe-B powder with high coercivity can be prepared from elemental powders by mechanical alloying and a solid-
state reaction. It can be compacted to dense isotropic magnets by
hot uniaxial pressing. Hot die-upsetting forms magnetically anisotropic samples. With regard to microstructure, magnetic properties, and processing, the mechanically alloyed material is comparable with rapidly quenched Nd-Fe-B. Also, hard-magnetic materials
with the $ThMn_{12}$ structure have been formed by mechanical alloying.

Introduction

The announcement of high-performance permanent magnets based on
Nd-Fe-B in 1983 [1] generated a huge scientific and technological
interest. Since the involved elements are available in large quan-
tities and relatively cheap - at least in comparison with Co-Sm,
so far, the best-performing permanent magnets - this development
enables large-scale applications of rare-earth magnets. The origin
of the excellent magnetic properties is the structure of the
$Nd_2Fe_{14}B$ phase which is tetragonal and has a complex unit cell
containing four formula units (68 atoms). The Curie temperature T_C
of this phase is 314 ^{O}C, which can be improved by a small Co
substitution. The anisotropy field H_A of about 75 kOe enables high
coercivities (10 to 25 kOe) and the high saturation magnetization
$4\pi M$ of 16 kG gives an upper theoretical limit of the maximum
energy product $(BH)_{max}$, which is the technologically most important property, of 64 MGOe. In fact, a record-high $(BH)_{max}$ value
of 45 MGOe has been achieved, which is by far higher than the
values for the Co-Sm magnets (15-28 MGOe) (for a more detailed
review see [2]).

These Nd-Fe-B magnets are usually produced either by the powder
metallurgical [1] or by the rapid quenching processes [3]. Recently we demonstrated that Nd-Fe-B permanent magnets can also be
produced by mechanical alloying and a subsequent solid-state reaction [4]. In this paper, the preparation process and properties of
hard-magnetic powder and of compacted, magnetically isotropic and
anisotropic Nd-Fe-B magnets are described. Also, first results on
the formation of $ThMn_{12}$ structure-type magnets with moderate coercivities are reported.

Preparation process and magnetic properties of Nd-Fe-B

Mechanical alloying of Nd-Fe-B powder is performed in a planetary
ball mill (Fritsch Pulverisette 5) under an argon atmosphere. The
elemental powders (Fe < 325 mesh; amorphous B < 1 μm; Nd < 0.5 mm)

are mixed and poured into a cylindrical milling container together with 10 mm diameter steel balls. The ball milling first produces powder with a layered microstructure of Fe and Nd. The submicron boron powder remains undeformed. It is caught by the colliding Fe and Nd particles which are cold-welded and is, therefore, embedded in the Fe/Nd interfaces (or within the Fe or Nd layers if Fe-Fe or Nd-Nd collisions occur). Further milling leads to a refinement of the layered microstructure until the Fe and Nd layers are no longer resolved by light microscopy. The X-ray diffraction pattern (Fig. 1, upper diffraction diagram) shows broadened intensity peaks of pure Fe. The Nd peaks are smeared out and, therefore, no longer detectable. The crystallite size of the Fe lies in the range of 200 to 400 Å. An atomic mixing by deformation during milling is not possible (see paper 4 in these proceedings [5]). The powder particles consist of very fine Fe and Nd layers with the embedded boron particles. There are no hints that either a crystalline or an amorphous Fe-Nd phase is formed during milling. Thermodynamic calculations show that for Fe-Nd the difference ΔG of the free enthalpies between the amorphous phase and the layered composite is positive over the whole composition range [6], thus preventing an interdiffusional reaction. Therefore, by mechanical alloying, a layered Fe-Nd composite is energetically favored. The formation of the crystalline Nd_2Fe_{17} phase seems to be prevented by nucleation problems.

The hardmagnetic $Nd_2Fe_{14}B$ phase is formed by a successive heat treatment (Fig. 1, lower diffraction diagram). Because of the extremely fine microstructure of the milled powders, the reaction can take place at relatively low temperatures or with short reaction times. Optimum coercivities are obtained after an annealing at 700 °C for 15 to 30 minutes. The grain size of the $Nd_2Fe_{14}B$ phase is then about 50 nm.

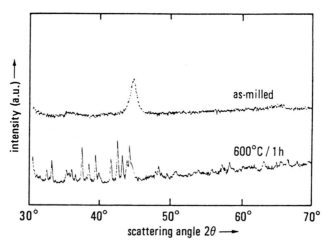

Fig. 1: X-ray diffraction patterns of Nd-Fe-B powder as-milled and after an additional heat treatment at 600 °C for 1 h.

Fig. 2 shows the further processing of the milled powder and the resulting magnet types schematically. Annealing of the as-milled material results in magnetically isotropic powders which can be used to produce resin-bonded permanent magnets (MM1; MM: "Magnemill"). Isotropic compacted magnets (MM2) can be obtained either from as-milled or from pre-reacted powders by uniaxial hot pressing [7]. A pressure of 1 kbar at a temperature above 600 ^{o}C is sufficient to get fully dense material. Magnetically anisotropic

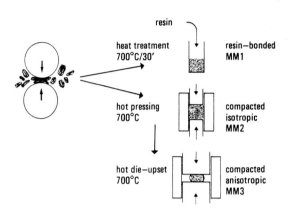

Fig. 2: Preparation of Nd-Fe-B magnets from mechanically alloyed powder.

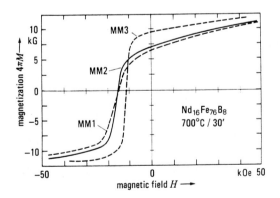

Fig. 3: Demagnetization curves of comparably prepared resin-bonded (MM1), compacted isotropic (MM2), and anisotropic (MM3) $Nd_{16}Fe_{76}B_8$ magnets.

samples (MM3) can be prepared from compacted isotropic magnets by texturing via hot deformation [7,8] in a similar way as has been shown before for rapidly quenched materials [9]. In our experiments, 5 mm diameter samples are uniaxially compressed to half their height within a 7 mm diameter die at 700 °C. Crushing of these samples leads to magnetically anisotropic powder. Fig. 3 shows the demagnetization curves of all three types of magnets produced from mechanically alloyed $Nd_{16}Fe_{76}B_8$ powder. The resin-bonded sample (MM1) exhibits a coercivity of 15.3 kOe (the magnetization values relate only to the magnetic powder). The compacted sample (MM2) shows a similar coercivity (15.4 kOe). The hot deformation to form the anisotropic sample (MM3) reduces the coercivity to 11.5 kOe, but it improves the remanence, the squareness of the magnetization loop and the energy product. The demagnetization curves parallel and perpendicular to the press direction of an anisotropic sample are shown in Fig. 4. The difference in the curve

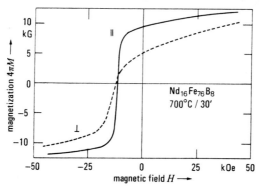

Fig. 4: Demagnetization curves measured parallel and perpendicular to the press direction of mechanically alloyed and die-upset $Nd_{16}Fe_{76}B_8$ magnets.

shape clearly demonstrates that the sample is now magnetically anisotropic. The ratio of the remanences measured parallel and perpendicular to the press direction, which is a measure of the degree of alignment, is 1.8. Applying a further optimized processing, coercivities of 17.2 kOe for the compacted isotropic and of 15.3 kOe for the anisotropic sample have been obtained. Also the degree of alignment of the anisotropic magnets can be considerably improved by increasing the deformation ratio [9,10]. The mechanically alloyed Nd-Fe-B and also (Nd,Dy)-Fe-B samples (a Dy substitution increases the coercivity considerably [11]) show a good temperature dependence of the coercivity which is similar to rapidly quenched material (Fig. 5). Thus, a $Nd_{13}Dy_2Fe_{77}B_8$ sample with a coercivity of over 20 kOe at room temperature still has a coercivity of about 10 kOe at 150 °C.

Magnets with $ThMn_{12}$ structure

After the success of the Nd-Fe-B magnets a search for further intermetallic phases which might be suitable as permanent magnets has started. So far, the most interesting phases are the iron-rich and rare-earth-containing phases with the tetragonal $ThMn_{12}$ cry-

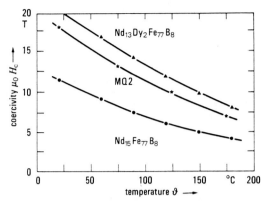

<u>Fig. 5:</u> Temperature dependence of the coercivity of magnetically isotropic, mechanically alloyed Nd-Fe-B and (Nd,Dy)-Fe-B samples in comparison with a rapidly quenched Nd-Fe-B sample (MQ2, after [8]).

stal structure (1:12 magnets), such as $RE(Fe,V)_{12}$ [12], $RE(Fe,Ti)_{12}$ [13] or $RE(Fe,Mo)_{12}$ [14]. As a general trend for these 1:12 intermetallics, the alloys with RE = Sm show by far the highest anisotropy values. Due to the promising intrinsic magnetic properties (for example $Fe_{80}Mo_{12}Sm_8$: $4\pi M = 10$ kG, $T_c = 200$ °C, H_A = 90 kOe) these phases are candidates for magnets with high coercivities. Although the anisotropy field H_A for $Fe_{80}Mo_{12}Sm_8$ is higher than that for $Nd_2Fe_{14}B$, the coercivity of the samples prepared by standard powder metallurgical techniques did not exceed 1 kOe. One of the problems here is the loss of Sm during melting, since Sm has a much higher vapor pressure than the other rare-earth metals at elevated temperatures which must be applied due to the high melting points of Mo or Ti. We, therefore, applied the mechanical alloying process for these materials [15]. Fig. 6 shows the demagnetization curve of a magnetically isotropic resin-bonded $Fe_{78}Mo_{10}Sm_{12}$ sample with a coercivity of 3.5 kOe. The best coercivity values obtained so far are 4.8 kOe for $Fe_{80}Mo_{10}Sm_{10}$ and 4.6 kOe for $Fe_{80}Ti_{10}Sm_{10}$ samples.

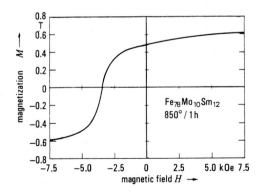

<u>Fig. 6:</u> Demagnetization curve of a mechanically alloyed $Fe_{78}Mo_{10}Sm_{12}$ sample.

Conclusions

These results show that mechanical alloying is a useful technique to prepare Nd-Fe-B magnets with magnetic properties that are comparable to rapidly quenched material. Also, 1:12 type magnets have been produced with considerable coercivities, although these values must be further improved to make the 1:12 magnets technologically interesting. Therefore, the mechanical alloying process has a good chance to get its share on the growing rare earth permanent magnet market.

The authors acknowledge technical assistance by F. Gaube and P. Kummeth, magnetic measurements by F. Friedrich and helpful discussions with K. Wohlleben. This work has been supported by the German Ministry for Research and Technology.

References

[1] M. Sagawa, S. Fujimura, N. Togawa, H. Yamamoto, and Y. Matsuura, J. Appl. Phys. 55 (1984) 2083.
[2] J.D. Livingston, Proc. 8th Int. Workshop on Rare-Earth Magnets and their Applications, Dayton, OH, 1985, p. 423.
[3] J.J. Croat, J.F. Herbst, R.W. Lee, and F.E. Pinkerton, Appl. Phys. Lett. 44 (1984) 148.
[4] L. Schultz, J. Wecker, and E. Hellstern, J. Appl. Phys. 61 (1987) 3583.
[5] L. Schultz, in E. Arzt and L. Schultz (eds.), Proc. DGM Conference on New Materials by Mechanical Alloying Techniques, Calw/Hirsau, October 1988, DGM Informationsgesellschaft, Oberursel, these proceedings.
[6] L. Schultz and J. Wecker, Mater. Sci. Eng. 99 (1988) 127.
[7] L. Schultz, K. Schnitzke and J. Wecker, J. Appl. Phys. 64 (1988) 5302.
[8] R.W. Lee, Appl. Phys. Lett. 46 (1985) 790.
[9] S. Heisz and L. Schultz, Appl. Phys. Lett. 53 (1988) 342.
[10] Y. Nozawa, K. Iwasaki, S. Tanigawa, M. Tokunaga and H. Harada, J. Appl. Phys. 64 (1988) 5285.
[11] L. Schultz and J. Wecker, Proc. 9th Int. Workshop on Rare-Earth Magnets and their Applications, Bad Soden, September 1987, Deutsche Physikalische Gesellschaft, Bad Honnef, p. 301.
[12] K.H.J. Buschow, Proc. 9th Int. Workshop on Rare-Earth Magnets and their Applications, Bad Soden, September 1987, Deutsche Physikalische Gesellschaft, Bad Honnef, p. 453; K.H.J. Buschow, in "High Performance Permanent Magnetic Materials", S.G. Sankar, J.F. Herbst, and N.C.Koon (eds.), Materials Research Society Symp. Proc., Vol. 96, p. 1.
[13] K. Ohashi, T. Yokoyama, R. Osugi, and Y. Tawara, IEEE Trans. Mag. MAG-23 (1987) 3101.
[14] A. Müller, J. Appl. Phys. 64 (1988) 249.
[15] L. Schultz and J. Wecker, J. Appl. Phys. 64 (1988) 5711.

C.4:
NONEQUILIBRIUM PHASES

Development of Nanocrystalline Structures by Mechanical Alloying

W. Schlump, H. Grewe, Krupp Forschungsinstitut, Essen

Introduction

Nanocrystalline structures define a new solid state. This was theoretically predicted and experimentally confirmed by Gleiter and co-workers (1-3). In conventional polycrystals having a grain size of 10 µm the volume-fraction of atoms located in incoherent grain boundaries is about 10^{-4}. This estimate was based on the assumption that the thickness of a grain boundary is on average two layers of atoms.

If the grain size is reduced to 10 nm or less the density of the grain boundaries increases to $10^{18} - 10^{21} \cdot cm^{-3}$. This means that about fifty percent of the atoms are located in grain boundaries. The individual grain boundary is not dis-ordered. It has a two-dimensionally ordered structure determin-ed by the adjacent crystals. If the crystals forming the mate-rial are orientated at random, every grain boundary has a sepa-rate structure with different atomic arrangements. The super-position of these 10^{19} uncorrelated grain boundaries leads to widely distributed distances between atoms. As a consequence, the interfacial component has an atomic arrangement with no preferred nearest-neighbour configuration. This structure has little short- or long-range order. It can be described as gas like.

This means that nanocrystalline materials can be seen as a structural compound one half of which consisting of a crystal-line phase and the other half of a gas-like phase (grain bound-ary phase).

This new solid state entails considerable changes in the physi-cal properties. One special aspect will be discussed here. In nanocrystalline materials self diffusion is enhanced by the factor 10^{19} compared with conventional polycrystalline materi-als. The reason for this is the considerable density of the grain boundaries (grain-boundary diffusion) and the high clean-ness of the boundaries (3), which is in contrast to normal ma-terials where the grain boundaries are to some extent filled up by segregation. It is the high self-diffusion that causes even ceramics to become ductile at room temperature as a result of creep processes governed by diffusion, such as superplasticity.

The common way to produce nanocrystalline materials is to first evaporate a melt in an inert gas atmosphere at a pressure of about 1 kPa. Condensation of the vapour gives a very fine pow-der with a size of less than 10 nm. This powder is then compacted by cold pressing under vacuum in the same apparatus. Another method is the sol-gel process, which is especially used

when nanocrystalline ceramics are to be produced. This paper
shows that nanocrystalline materials can also be produced by
mechanical alloying.

Experimental materials

Powders displaying an internal nanocrystalline structure were
produced by high energy or reaction milling either in a planetary ball mill or in an attritor having a horizontal geometry.
The milling forces, expressed in terms of acceleration, were up
to 8 g in the attritor and 15 g in the planetary ball mill.

Processing was done with elementary powders having particle
sizes between 1 and 100 μm. Milling times varied between 48 h
(planetary ball mill) and 90 h (attritor) with a powder-to-ball
mass ratio of about 1:10. Normally milling was done under vacuum conditions or in a defined atmosphere of inert gas, nitrogen
or oxygen. The powders were then examined by metallography,
x-ray diffraction and electron microscopy in transmission, for
which purpose the powders were cold-pressed, ground and thinned
by ion milling.

Metal-metal systems

Mechanical alloying is normally used to produce oxide-dispersed
alloys (ODS-alloys) (4,5) and amorphous powders (6-9). In these
and all other cases the internal microstructure of mechanically
alloyed powders is determined by the thermodynamics and kinetics of the system in combination with the parameters of the
milling process.

Fig.1: Fe-Ta (70:30, mass-%) milled for 48 h
in a planetary ball mill.

309

Powders with nanocrystalline structures cannot be produced in
alloy systems in a composition falling in the range of solid
solutions. This applies to ODS-alloys on the basis of Ni and Al
and even to systems like Fe-Ta, as far as these were investigated. Characteristic of these systems is the formation of homogeneous solid solution having a lamellar structure (Fig. 1).
The single grains vary in size between 50 and 500 nm. As a result of the milling process they are heavily disturbed and exhibit numerous dislocations and glide bands. The lamellar
structure is frequently accompanied by a weak texture within
the particles.

The range considered for the formation of solid solutions does
not apply to the thermodynamic equilibrium expressed in the
phase diagram. This range can be widely extended by mechanical
alloying. In this investigation it was found that up to 40% Ta
could be dissolved in α-Fe (Fe-Ta is a system with similar
crystal structures). Maximum solubility in the equilibrium
range is 1.9%.

The powders undergo a change in microstructure, if W is added
in an amount of 5% to the Fe-Ta system during milling (Fig. 2).

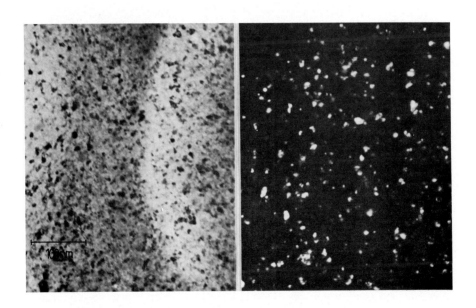

Fig. 2: Fe-Ta-W (67:28:5, mass-%) milled for 48 h in a planetary ball mill. TEM, left: bright field, right: dark
field.

This causes the characteristic lamellar structure to disintegrate and disappear. An equiaxed crystalline structure is formed with a grain size of less than 10 nm. This polycrystal consists of α-Fe and W. While W as a hard phase assists the milling process, it forms no (or less pronounced) alloy with the Fe. If the amount of Ta, exceeds 40% Ta crystals appear, as was shown by electron diffraction.

This structure derived from an extremely high degree of plastic deformation. Nevertheless, no deformation structures such as dislocations or glide bands can be seen in the grains. It is assumed that dislocations were moved to the grain boundaries by image forces. This process begins at grain sizes of about 20 nm. As will be seen in the following, this seems to be characteristic of nanocrystalline structures.

The extreme broadening of the range in which solid solutions form could not be found in systems with dissimilar crystal structures. As in the system Cu-Ta, milling produces a structure consisting of Cu and Ta grains in the nanocrystalline range (Fig. 3). Here, too, no dislocations could be detected in the structure.

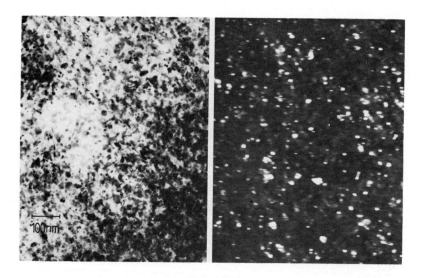

100nm

Fig. 3: Cu-Ta (55:45, mass-%) attritor milled for 90 h.

In alloy systems such as Ti-Ni, which tend to form amorphous phases, a different structure is formed during mechanical alloying. Selecting a composition in the metastable two-phase region amorphous-crystalline gives a mixture of nanocrystalline α-Ti and Ni crystals embedded in an amorphous matrix (Fig. 4). The amount of the amorphous phase can be varied by changing the composition.

311

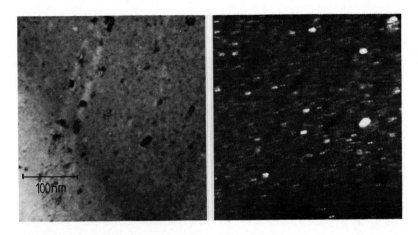

Fig. 4: Ti-Ni (70:30, mass-%) attritor milled for 90 h.

Such mixed microstructures normally appear to be completely
amorphous when examined by x-ray diffraction. The same result
is obtained if powder grain size is less than 3 to 4 nm. Dependence of grain size on milling time can be determined by the
broadening of the x-ray lines (Fig. 5). As can be seen from the
figure, the reduction in grain size occurs in the early stage
of milling. At the same time the lattice constants increase.
This is particularly pronounced in the case of Ti. For Ni this
increase occurs only in the early stage of milling.

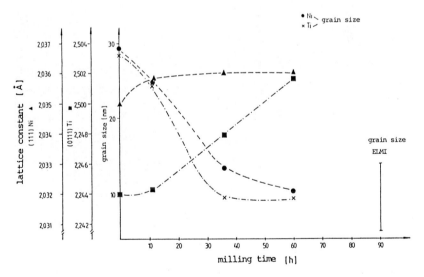

Fig. 5: Lattice constant and grain size as a
function of milling time.

Appearance of the nanocrystalline structures can be seen by
electron microscopy in an early umcomplete stage of milling
(Fig. 6). While the milling is in progress lamellar structures
are formed with a thickness that can go down to a few nm. This
structure also displays dislocations and deformation bands in a
dense pattern.

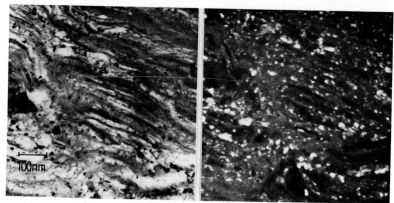

Fig. 6: Ti-Ni (70:30, mass-%) attritor milled for 36 h.

These lamellar structures disappear if their size drops below
a critical value. The nanocrystalline structure is then formed.
The formation of an amorphous phase can also be seen in the
illustration. This occurs in regions parallel to the lamellar
structures.

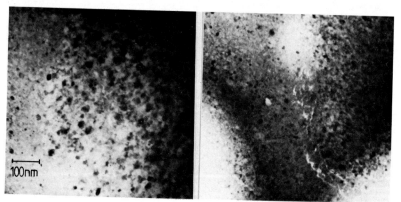

Fig. 7: Ti-Ni (70:30, mass-%) milled for 24 h in a planetary
        ball mill.

The appearance of an amorphous phase in these systems depends
on the milling parameters. If milling takes place with a high
energy input thermodynamically stable intermetallic compounds
such as Ti_2Ni are formed instead of the amorphous phase
(Fig. 7). The size of the crystals is about 10 nm.

313

It was found that this result can be applied to other systems
forming intermetallic compounds. Milling of Al and Ni powders
(composition 70:30) imparts a mixed structure to the powder
particles which consists of Al_3Ni crystals in an Al-matrix. The
size of the components is also in the range of nanocrystalline
structures (Fig. 8).

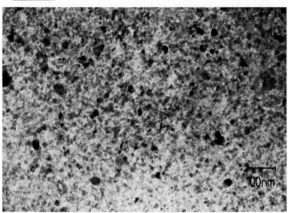

Fig. 8: Al-Ni (70:30, mass-%) attritor
milled for 90 h.

Metal-metalloid-systems

The development of powders with nanocrystalline structures is
not restricted to metallic systems. Even cermets (ceramic-metal
compounds) can be produced by mechanical alloying. This can be
done by milling metal powder together with ceramic powder or by
reaction milling. Reaction milling is defined as a process in
which chemical reactions are activated by communition. (Even
the formation of intermetallic compounds can thus be seen as a
reactive process).

These cermet powders can be produced by milling a mixture of
metallic powders with additions of metalloid-forming elements
such as oxygen, nitrogen or carbon. The metal powder should be
chosen to ensure that one part reacts with the metalloid-forming elements, while the other is not reactive. If, for example,
the system Ti-Ni is milled with the additions of air or nitrogen, the metalloids TiO or TiN are formed by reaction embedded
in a metallic Ni-matrix (Fig. 9). At the end of the milling
process both the metallic and the metalloid components have a
grain size of some nm in the powder particles. Depending on
milling time all Ti can be transformed into the ceramic phase
by this process.

The formation of these compounds and their structures depends
on the enthalpy of the chemical reaction. This can be demonstrated using cermets with a carbide phase as an example. The
milling of Ti and Ni powders with an addition of C (stoichio-
metrically adjusted to bring about the formation of TiC) results in a chemical reaction whose intensity varies with time.

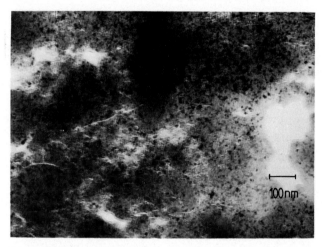

Fig. 9: Ti-Ni (70:30, mass-%) attritor milled
for 90 h in nitrogen.

At the beginning of the process elementary Ti reacts with C to
form coarse carbides with a size of some μm in a metallic matrix (Fig. 10). In a later stage of the milling process even Ti
and Ni react to form, to some extent, an amorphous phase. This
leads to a decrease in the enthalpy of the chemical reaction.
As a result smaller carbides are formed. At the end of the process the Ti of the amorphous matrix reacts with the C forming
carbides of nanometer size.

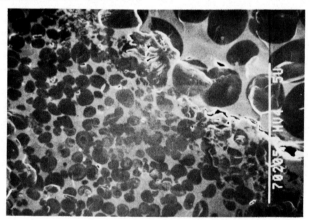

Fig. 10: Ti-Ni-C (62:27:11, mass-%) attritor
milled for 90 h.

Coarser carbides can be prevented from being formed by preal-
loying the Ti and Ni powders mechanically and then adding C.
This gives a cermet powder consisting of TiC in a Ni-matrix
with component grain sizes of less than 10 nm (Fig. 11).

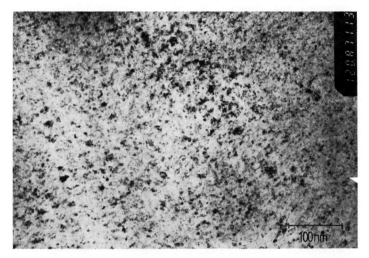

Fig. 11: Ti-Ni-C (62:27:11, mass-%)
        metallic powders premilled in an attritor.

If the free enthalpy for the chemical reactions is smaller than
that for the TiC formation, as in the systems V-Ni-C and
Si-Ni-C, cermet powders with nanocrystalline structures can be
produced without premilling of the metallic powders (Fig. 12).

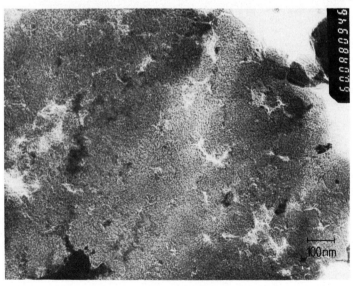

Fig. 12: Si-Ni-C (60:15:25, mass-%)
        milled for 48 h in a planetary ball mill.

In other systems the enthalpies are too small to produce the
desired phases even if activated by mechanical alloying. The
milling of W, Ni and C powders leads to the formation of W_2C,
instead of WC in a Ni-matrix with excess C. By hot isostatic
pressing at temperatures of 1000°C this carbide changes into
WC. At full density this material has an hardmetal-like structure with nanocrystalline grain sizes.

Thermodynamic aspects are also to be considered in the production of cermet powders with an oxide phase. The milling of Al
and Ni in air results in the formation of intermetallic phases
such as Al_3Ni. These phases are resistant to oxidation and the
amount of oxygen does not exceed 2%. Al_2O_3 (β-phase) is only
formed when milling takes place in oxygen.
This cermet powder possesses a nanocrystalline structure
(Fig. 13).

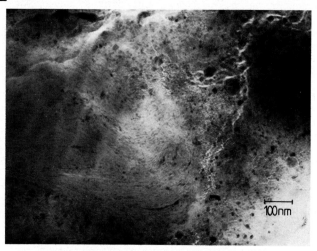

Fig. 13: Al-Ni (70:30, mass-%) attritor milled
for 90 h in oxygen.

Mechanical aspects

Powders and compacted materials with nanocrystalline structures
have a very high hardness compared with normal polycrystalline
materials. This is assumed to be due to the fact that deformation in such materials cannot occur by dislocation gliding.
Excessive energy is required to generate dislocations, if the
grain size is less than about 20 nm. Some initial results showing the hardness of nanocrystalline materials are listed in
Table 1:

system (mass %)	treatment	hardness HV 0.1	HV 30
Cu–Ta 85:15	cold pressed	360	
	annealed 1 h 500 °C	390	
	" 1 h 600 °C	370	
	" 1 h 700 °C	340	
Cu–Ta 70:30	cold pressed	440	
	annealed 1 h 500 °C	470	
	" 1 h 600 °C	440	
	" 1 h 700 °C	400	
Cu–Ta 55:45	cold pressed	500	
	annealed 1 h 500 °C	510	
	" 1 h 600 °C	520	
	" 1 h 700 °C	430	
TiNi–C 62:27:11	cold pressed	1075	
	HIP 1 h 900 °C	1511	952
	HIP 1 h 1000 °C	1624	1178
W–Ni–C 79:16:5	HIP 1 h 900 °C	1885	1165
	HIP 1 h 1000 °C	1985	1270
TiAl6V4 WC–30Co	} conventional	330	220
			850

Table 1: Hardness of nanocrystalline materials

Summary

Nanocrystalline materials can be regarded as a structural compound consisting of a crystalline and a gas-like (grain boundary) phase.

Powders displaying such structures can be produced by mechanical alloying. In metallic systems this requires a composition outside of the region in which metastable solid solutions form (determined by milling), or in the two-phase region amorphous-crystalline.

Reaction milling is another way to produce nanocrystalline powders. These are similar to cermets and have a structure composed of nanocrystalline size ceramics embedded in a metallic matrix. The microstructure of such powders is determined by the enthalpy of the chemical reactions. The powders, especially the cermet powders, are very hard and exhibit a high thermal stability.

References

(1) H. Gleiter, P. Marquardt: Z. Metallkd. 75 (1984), 263.
(2) R. Birringer, H. Gleiter: Adv. in Mater. Sci. and Eng., Pergamon Press (to be published).
(3) H.W. Schäfer, R. Würschum: J. Less. Com. Met. 140 (1988), 161.
(4) J.S. Benjamin: Metall. Trans. 1 (1970), 2943.
(5) M.F. Fleetwood: Mater. Sci. and Techn. 2 (1986), 1176.
(6) R.B. Schwarz, R.R. Petrich, C.K. Saw: J. Non-Crystal. Sol. 76 (1985), 281.
(7) P.Y. Lee, C.C. Koch: Appl. Phys. Lett 50 (1987), 1578.
(8) E. Hellstern, L. Schultz: Mat. Sci. and Eng. 93 (1987), 213.
(9) J.R. Thompson, C. Politis: Europhys. Lett. 3 (1987), 199.

(The development work described in this paper is funded by the West German Defence Ministry.)

Amorphous and Nano-meter Order Grained Structures of Al-Fe and Ag-Fe Alloys Formed by Mechanical Alloying

P.H.Shingu[a], B.Huang[a], J.Kuyama[a], K.N.Ishihara[a] and S.Nasu[b]

a) Department of Metal Science and Technology, Kyoto University, Yoshida Sakyoku, Kyoto 606, JAPAN
b) Department of Metal Physics, Faculty of Engineering Science, Osaka University, Toyonaka, Osaka 560, JAPAN.

1. Introduction

In the previous paper, nano-meter order grained microstructure in Al-Fe alloys formed by mechanical alloying(MA) was reported(1). Present paper reports the results of further studies on the formation process of microstructures in Al-Fe alloys. It was found that the process of gradual refinement of crystal size and the formation of amorphous phase takes place concurrently. Experimental results on the formation of nano-meter order grained structure formed in Ag-Fe system, which is a typical mutually immissible system in contrast to the compound forming Al-Fe, will also be reported. Results on examinations of structure and mechanical properties in consolidated Al-Fe powders will also be given.

2. Experiment

Conventional ball-milling method was deliberately chosen in order to insure the low energy MA process. The advantage of such low energy MA process exists in the larger possibility of retaining the non-equilibrium **structures** formed by the mechanical energy. The temperature of a vial during MA process was about 340K in the present experiment. The details of experimental procedures of ball-milling and consolidation were given in the previous papers(1,2).

3. Result and Discussion

3.1 Amorphous phase formation in Al-24.4at%Fe sample.
    Examination of X-ray powder diffraction patterns of MA treated Al-Fe samples of various composition, the tendency of occurrence of halo-like broad peak appeared most strongly in the composition range near Al-20at%Fe. Fig.1 shows the change in diffraction peaks of Al-24.4at%Fe sample for various MA time. After 180hrs of MA the diffraction peak from the fcc Al became almost negligible. The corresponding calorimetric measurements by DSC are given in Fig.2. The exothermal peak for the 454hrs MA treated sample is sharper in shape and lower in temperature than other samples MA treated for shorter time.
    Fig.3 shows that by the MA treatment of over 180hrs the size of dispersed Fe particles becomes too small to be identified by the scanning electron microscope(SEM). At this stage, the transmission electron microscope(TEM) images show the refinement of grain size in the range of several tens of nano-meter. The TEM image of the sample which was MA treated for 454hrs revealed almost no images of crystal grains except for scattered grains of roughly ten nano-meters in size as shown in Fig.4. The electron

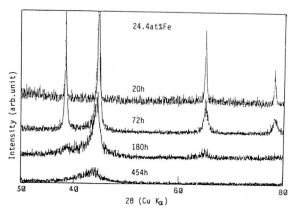

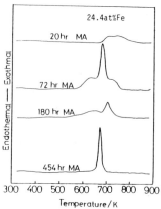

Fig.1 X-ray powder diffraction patterns of Al-24.4at%Fe samples ball-milled for various length of time.

Fig.2 DSC curves of Al-24.4 at%Fe samples ball-milled for various length of time.

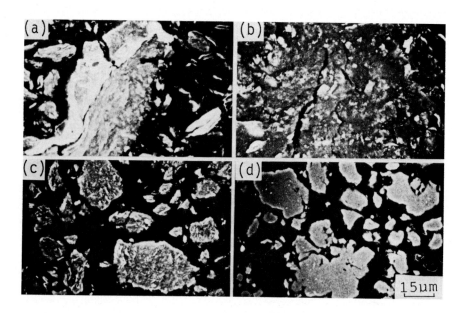

Fig.3 SEM images of the Al-24.4at%Fe powders ball-milled for (a) 20hrs,(b) 36hrs,(c) 72hrs and (d) 180hrs.

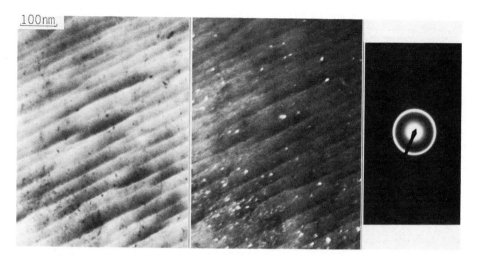

<u>Fig.4</u> Al-24.4at%Fe, ball-milled for 454hrs. Bright and dark field TEM images and electron diffraction pattern. The line patterns running left to right upward are the wrinkles of the sample formed by the knife edge of the thin slicing apparatus.

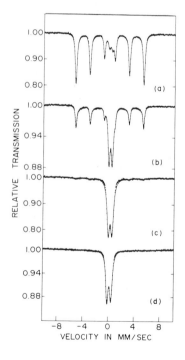

<u>Fig.5</u> Mössbauer spectra of Al-24.4at%Fe samples ball-milled for (a) 72hrs, (b)180hrs and (c)454hrs, and (d) Al-40at%Fe-4at%Ti sample ball-milled for 800hrs.

diffraction patterns shown in Fig.4 show that the matrix activates only the halo-ring pattern the position of which matches with that of the broad halo-pattern of X-ray diffraction as shown in Fig.1. Several bright spots around the halo-ring in Fig.4 are diffraction spots from the Fe crystal grains.

The fraction of Fe atoms incorporated in the amorphous phase can readily be visualized by the Mössbauer spectra as shown in Fig.5. In Fig.5, the spectrum (c) is obtained from the same sample as that shown in Fig.4. Consequently, the paramagnetic doublet absorption peak in Fig.5-(c) can be interpreted as due to the amorphous phase. This doublet absorption, which was stated as due to an unknown phase in the previous paper(1), also appeared in the MA treated Al-40at%Fe-4at%Ti system as shown in Fig.5-(d). When small amount of Ti is added to Al-Fe system, the amorphization tendency becomes stronger, as revealed by the TEM observations. The spectrum (d) shows almost complete disappearance of ferromagnetic component suggesting that Fe atoms are incorporated almost entirely in the amorphous phase.

It should be noted that in Fig.5, the doublet absorption peak can be identified already in the sample which was MA treated for 72hrs. This result indicates that the amorphous phase starts to form at an early stage of MA treatment. However, by the TEM observation, the identification of amorphous phase was difficult even in the 180hrs MA treated sample where major part(56%) of Fe atoms were in the amorphous phase as calculated from the absorption peak area of the spectrum in Fig.5-(c).

Summarizing these results, it may be deduced that the formation of amorphous phase in Al rich Al-Fe alloys by MA treatment starts at the interface of Fe particles and Al matrix. Minimum size of Fe particles before the occurrence of complete amorphization seems to be about ten nano-meter. However further experimental confirmation is needed to clarify the final stage of complete amorphization. Finally, it should be noted that the large amorphous phase forming tendency around Al-20at%Fe composition was also reported by the sputtering method(3).

3.2 Nano-meter order crystalline structures in Ag-56.3at%Fe alloys

In contrast to the compound forming tendency of Al-Fe system, the Ag-Fe system is a typical mutually immissible system. The heat of solution of Ag and Fe in liquid state calculated by Miedema's formula is 28kJ/mol and even in liquid state Ag and Fe are immissible and separate out like water and oil. Prolonged ball-milling, however, revealed the disintegration of Fe particles into Ag matrix as shown by SEM images shown in Fig.6. X-ray diffraction patterns in Fig.7 show the noted broadening of diffraction peaks from fcc Ag and bcc Fe phases by the MA treatment of 600hrs. The TEM images of corresponding sample show fine grains of several tens nano-meter as shown in Fig.8. The Mössbauer spectrum of 600hrs MA treated sample indicates the appearance of Fe atoms with internal magnetic field which is quite different from that of pure bcc Fe. These results indicate the occurrence of colloidal dispersion of Fe particles and Ag particles. The crystal grain size is similar to that of the MA treated Al-Fe alloys, however, at the interface of Fe and Ag particles solid solution of varied concentration of two elements appears to be formed but not the amorphous phase. Such solid solution formation was reported by sputtering experiments(5).

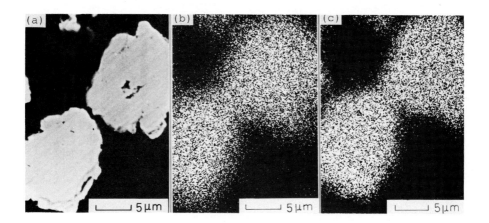

Fig.6 Ag-56.3at%Fe, ball-milled for 370hrs. (a) SEM image,
(b) Fe Kα image, (c) Ag Kα image.

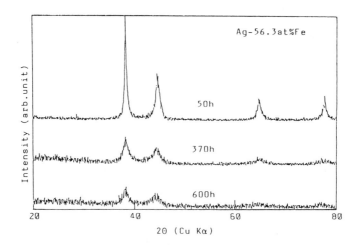

Fig.7 X-ray diffraction patterns of Ag-56.3at%Fe samples
ball-milled for various length of time.

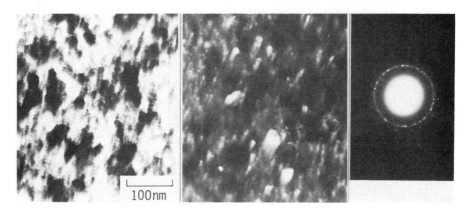

Fig.8 Ag-56.3at%Fe, ball-milled for 600hrs. Bright and
dark field TEM images and diffraction pattern.

3.3 Consolidation of mechanically alloyed powder
      In order to keep the metastable or non-equilibrium state of
MA treated powder, the strained powder rolling method(2,6) was
used for consolidation. SEM images of as MA treated powder and the
consolidated sample are given in Fig.9. TEM images of a sample
consolidated and heat treated up to 400°C are given in Fig.10.
These images show that after the heating up to 400°C the Al matrix
grain size remains sub-micron in size and, moreover, the
diffraction contrast inside these grains indicates recovery did
not take place in appreciable extent. Such a temperature resistant
and highly strained state is possible due to the presence of
finely dispersed Fe particles as shown in Fig.9 and also possibly
due to the amorphous phase formed at these particles. Some results

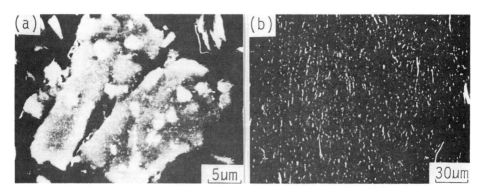

Fig.9 SEM images of Al-3at%Fe powders ball-milled for
36hrs.(a) as ball-milled, (b) after consolidation.

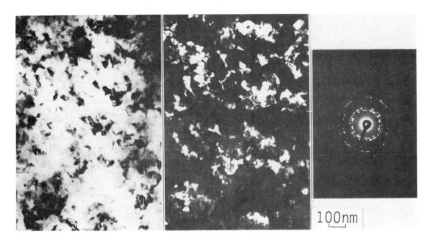

Fig.10 Al-3at%Fe, ball-milled for 36hrs, consolidated and heated up to 637K. Bright and dark field TEM images and electron diffraction pattern.

of mechanical testing of MA treated and consolidated samples are shown in Figs.11 and 12. Powders MA treated for longer time became harder but more brittle after consolidation so that tensile test was not performed. Further improvement in the consolidation procedure is necessary in order to extract the still higher quality of MA treated powders.

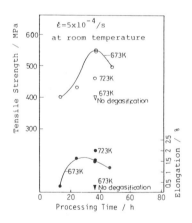

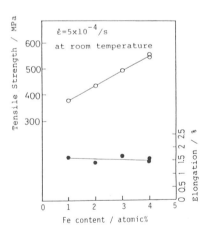

Fig.11 Effect of processing time on tensile properties of Al-4at%Fe. Temperature of consolidation are indicated in the figure.

Fig.12 Effect of Fe content on tensile properties. Samples are ball-milled for 36hrs and consolidated at 637K.

326

4. Conclusion

Mechanical alloying of Al-Fe and Ag-Fe alloys were performed by the conventional ball-milling method. Both Al and Ag disintegrated Fe particles to disperse in colloid like state in the matrix. Both the particle size and the matrix crystal grain size ultimately became about several tens nano-meter.

In the case of Al-Fe, which is a compound forming system, the formation of amorphous phase begins at early stage of MA treatment at the interface of Fe particles and Al matrix. For the concentration range near 20at%Fe, entire sample can be made amorphous by ball-milling. In the case of Ag-Fe, which is a mutually immissible system, amorphous phase does not appear but partial solid solution formation takes place.

Quick heating and fast rolling method, the strained powder rolling process, was applied for the consolidation of MA treated Al-Fe powders. Mechanical property measurements revealed some good results, however, for the powders MA treated for prolonged time further improvement is needed in the consolidation condition.

Reference
(1) P.H.Shingu, B.Huang, S.R.Nishitani and S.Nasu: Suppl. to Trans. JIM 29 (1988) 3.
(2) P.H.Shingu: Materials Science and Eng., 97 (1988) 137.
(3) M.Shiga, T.Kikawa, K.Sumiyama and Y.Nakamura: Japan J. Appl. Magnetism, 9 (1985) 187.
(4) A.K.Niessen, F.R.de Boer, R.Boom, P.F.de Chatel, W.C.M.Mattens and A.R.Miedema: CALPHAD, 7 (1983) 51.
(5) K.Sumiyama, Y.Kawawake and Y.Nakamura: Trans. JIM 29 (1988) 191.
(6) B.Huang, K.F.Kobayashi and P.H.Shingu: J. Japan Inst. Light Metals, 38 (1988) 165.

Mechanically Alloyed Metals with Non-equilibrium Phases

Akihisa Inoue and Tsuyoshi Masumoto
Institute for Materials Research, Tohoku University
Sendai 980, Japan

I. Introduction

Recently, mechanical alloying (MA) technique[1] has attracted great attention as a new type of solid state reaction method of producing non-equilibrium phase materials including amorphous phase[2,3]. The reason for the great attention is probably because of unique homogeneization and/or alloying processes of elemental powders as well as expectation of new materials with functional characteristics which have not been produced by other processing techniques. Further development of the MA technique is thought to be attributed to the possibility that engineeringly useful characteristics are found in MA materials, even though there is scientific interest on the mechanism of amorphization of elemental powders by MA. A number of papers on the amorphization process have previously been presented and will also be presented in the present conference. Accordingly, this paper attempts to review superconducting, magnetic and mechanical properties of MA materials with non-equilibrium phase which cannot be synthesized by other processing techniques.

II. Experimental Procedures

Powders of pure metals and/or pre-alloyed ingots were mixed to give desired average compositions and sealed in a cylindrical metallic (copper, steel, stainless steel or WC-Co alloy) vial (75 mm x 70 mm diameter) under an Ar atmosphere in a glove box. Their powder sizes were usually below 44 μm. The MA was carried out in a conventional laboratory mill with a ball to a powder weight ratio of 20:1. Balls (10 mm diameter) made from the same metals as those for the vials were used in the MA. The vial was opened after each several hours processing in the glove box and small samples of the powders were removed for structural analysis. The MA powders were pressed to a disk shape of about 0.5 to 5 mm high and 5 to 10 mm in diameter for 60 to 600 s in the temperature range of room temperature (R.T.) to 1423 K under a uniaxial compressive pressure of 60 to 300 MPa by using a conventional hot-pressing machine. Subsequently, the pressed compacts were subjected to different annealing treatments to obtain the best characteristics resulting from optimum structural states. The structure of the MA powders and the annealed compacts was examined by X-ray diffractometry and metallographic techniques. In addition, changes in the structure and weight of the MA samples upon annealing in various atmospheres were examined by differential thermal and thermogravimetric analyses (DTA and TGA). Measurements of superconducting properties T_c, $H_{c2}(T)$ and $J_c(H)$ were made by the DC method using the four electrical probes in the temperature range 4.2 K to R.T. under an applied field up to 7.5 T. Magnetic properties were measured for the bonded Nd-Fe-B products made from the MA powders and epoxy resin with a vibrating sample magnetometer

(VSM) at R.T. under an external field of 1.592 MA/m (20 kOe). The Vickers hardness of hot-pressed products was measured between R.T. and 1073 K and under a load of 1 kg.

III. Superconductors Prepared by MA of Immiscible Si-Pb and Si-Pb-Bi Alloys[4]

1. Object

It has recently been clarified[5] that rapidly quenched Ge-rich alloys have duplex structures consisting of immiscible M metal and/or alloy particles finely and homogeneously embedded in Ge matrix and the duplex phase alloys exhibit anomalous electrical resistivity and superconducting properties which are not obtained for Ge-M alloys produced by a conventional solidification method. However, even by the rapid quenching technique, a homogeneously mixed duplex structure for Si-M alloys could not be produced because of large differences in specific weight and boiling temperature as well as the existence of large liquidus miscibility gap resulting from a large difference of melting temperature. A combined method of MA and warm-pressing is expected to produce a homogeneously mixed multiphase structure even in the Si-M systems. This section aims to present the microstructure and electrical properties of Si-M compacts produced by warm-pressing of MA powders and the possibility of an artificial synthesis of new materials exhibiting hybrid-type properties.

2. MA Structure

Figure 1 shows X-ray diffraction patterns of $Si_{80}Pb_{20}$ and $Si_{80}Pb_{12}Bi_8$ compacts produced by warm-pressing of MA powders at 573 K. Diffraction peaks correspond to cubic Si and fcc Pb for the Si-Pb alloy and Si, hcp ε(Pb-Bi), Pb and hexagonal Bi for the Si-Pb-Bi alloy, indicating formation of co-existent phases of Si and Pb or ε(Pb-Bi). The ε phase is a supersaturated solid solution of the equilibrium hcp phase which exists in the range of 24 to 33 at% Bi at R.T.[6]. The lattice parameters determined from the X-ray diffraction patterns shown in Fig. 1 were 0.5430 nm for Si, 0.4953 nm for Pb, a=0.3510 nm and c=0.5716 nm for ε and a=0.4547 nm and c=1.1861 nm for Bi. These are nearly the same as those[7] of pure elements and $Pb_{60}Bi_{40}$ alloy. Furthermore, these lattice parameters remain almost independent of alloy concentration for the Si-base compacts containing 10 to 30 at% Pb and/or Bi. This indicates that no detectable solution of Si into the second phases has taken place, in conformity with the equilibrium phase diagrams for the Si-Pb and Si-Bi systems. Optical micrographs of warm-pressed $Si_{100-x}Pb_x$ (x=10 and 20 at%) compacts presented in Fig. 2 reveal the distribution of Pb in Si matrices. It can be seen that Pb distributes homogeneously and lies discontinuously and in isolation with the Si matrix. The grain size of the Si matrices is about 0.1 to 1 μm and the average particle size and interparticle spacing are about 2 μm and 1 μm, respectively, for Pb and about 2 μm and 1 μm, respectively, for ε(Pb-Bi).

3. Superconducting Properties

Figure 3 shows the normalized electrical resistance (R/R_{10}) curves in the vicinity of T_c for warm-pressed $Si_{100-x}Pb_x$ and $Si_{100-x}(Pb_{.6}Bi_{.4})_x$ (x=10, 20 and

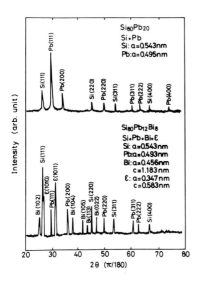

Figure 1: X-ray diffraction patterns of $Si_{80}Pb_{20}$ and $Si_{80}Pb_{12}Bi_8$ compacts produced by warm-pressing of MA powder.

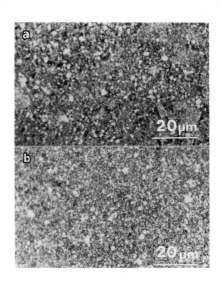

Figure 2: Optical micrographs of $Si_{90}Pb_{10}$ (a) and $Si_{80}Pb_{20}$ (b) compacts produced by warm-pressing of MA powders.

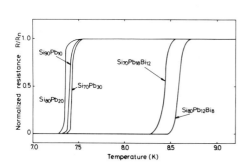

Figure 3: Normalized resistance ratio R/R_{10} as a function of temperature for warm-pressed Si-Pb and Si-Pb-Bi compacts.

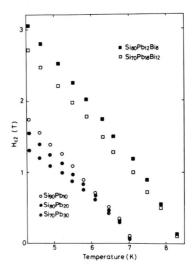

Figure 4: H_{c2} as a function of temperature for warm-pressed Si-Pb and Si-Pb-Bi compacts.

30 at%) compacts. The transition occurs rather sharply with a temperature width of less than 0.10 K. The T_c, which was taken as the temperature at which $R/R_{10}=0.5$, is 7.35 to 7.42 K for $Si_{70-90}Pb_{10-30}$ and 8.45 to 8.60 K for $Si_{70-90}(Pb_{.6}Bi_{.4})_{10-30}$. The change in resistivity by the transition is as large as 2.94 $\mu\Omega m$ for $Si_{80}Pb_{20}$ and 11.65 $\mu\Omega m$ for $Si_{80}Pb_{12}Bi_8$ and their normal resistivities increase significantly with increasing temperature. The resistivity change in the range 10 to 273 K reaches 11.8 $\mu\Omega m$ for $Si_{80}Pb_{20}$ and 11.9 $\mu\Omega m$ for $Si_{80}Pb_{12}Bi_8$. It is thus characterized that the warm-pressed Si-Pb and Si-Pb-Bi compacts made from the MA powders are a superconductor with high normal electrical resistivities.

The upper critical magnetic field H_{c2} was measured at various temperatures ranging from 4.2 K to T_c. Figure 4 shows the temperature dependence of H_{c2} for $Si_{100-x}Pb_x$ and $Si_{100-x}(Pb_{.6}Bi_{.4})_x$ (x=10, 20 and 30 %) compacts. Here we define H_{c2} to be the applied field at which the resistance of the sample reaches 0.6 R_n. The H_{c2} increases with decreasing temperature and the temperature dependence has a positive curvature with the steeper slope of the curves at higher fields. The H_{c2} at 4.2 K is as high as 1.7 to 1.8 T for the Si-Pb alloys and 2.7 to 3.2 T for the Si-Pb-Bi alloys. It is particularly notable that the H_{c2} values are much larger than those[8] of pure Pb and $Pb_{60}Bi_{40}$ and the enhancement factor reaches about 24 to 36 for the Si-Pb compacts and 1.8 to 2.0 for the Si-Pb-Bi compacts.

The critical current density J_c was measured at 4.2 K under an external applied field for $Si_{80}Pb_{20}$ and $Si_{80}Pb_{12}Bi_8$ compacts. Here, the overall cross-sectional area of the samples was used in evaluating J_c. The J_c in the absence of an applied field was about 1.4×10^3 A/m^2 for $Si_{80}Pb_{20}$ and 4.1×10^4 A/m^2 for $Si_{80}Pb_{12}Bi_8$ and decreased to 17.6 A/m^2 at 1 T for the Si-Pb and 2510 A/m^2 at 2.5 T for the Si-Pb-Bi compact. The decrease of J_c with increasing applied field was much sluggish for the Si-Pb-Bi compact, indicating the existence of a number of fluxoid pinning sites.

4. Reason for the High-Field Superconductivity

Cohen and Douglass[9] have predicted that the appearance of superconductivity for the sandwich material consisting of metal-insulator-metal phases is possible in principle for all the superconducting metals by an electron tunneling effect from the metal into the insulator. Furthermore, the prediction has been confirmed for a number of thin film materials consisting of metal and oxide[10]. The microstructure of the Si base compacts prepared by warm pressing of MA powders is not always the same as the duplex structure consisting of metal and oxide insulator prepared by evaporation technique. However, the appearance of superconductivity for the Si base compacts is presumed to be similar to the mechanism proposed for two superposed metal films separated by an insulating barrier such as oxide. That is, the conducting electrons in Pb and ε(Pb-Bi) can tunnel so that a superconducting circuit is formed in the Si-metal duplex materials.

The reason for the significant enhancement of $H_{c2}(T)$ for warm-pressed Si-Pb and Si-Pb-Bi compacts has been investigated on the basis of equations (1)[11,

12] and (2)[13] applicable for the present compacts belonging to a "dirty"
type-II superconductor.

$$H_{c2}(T) = \kappa\sqrt{2}\,H_c = \phi_0/2\pi\xi^2(T)$$
$$= \phi_0/2\cdot[0.85\{\xi_0\,\ell_{eff}T_c/(T_c - T)\}^{1/2}]^2 \qquad (1)$$
$$H_{c2}(T) = (3/2\pi^2)\phi_0/\ell_{eff}\cdot\{\xi_0(n/n_1)^{1/3}T_{c1}/T_c\}^{-1} \qquad (2)$$

Here κ is the Ginzburg-Landau parameter, $H_c(T)$ the thermodynamic critical
field, ϕ_0 the flux quantum, $\xi(T)$ the temperature dependent coherence length,
ℓ_{eff} an effective mean free path, ξ_0 the BCS coherence length, and n_1 and n the
electron concentrations of the pure material and the alloy. Using the data of
King et al.[14], ξ_0/ξ_{0p} is expected to be about 0.8. The ℓ_{eff} values evaluated
from equations (1) and (2) are as small as 2 to 3 nm and 1 to 10 nm,
respectively, for the Si-Pb compacts and 7 to 8 nm and 5 to 6 nm, respectively,
for the Si-Pb-Bi compacts. Accordingly, the anomalously high H_{c2} values for
the Si base compacts are presumably due to an extremely short distance of the
effective mean free path of electrons. Furthermore, the short mean free path
is thought to result from a unique structural modification caused by warm-
pressing of MA powders, i.e., the superconducting phase of Pb or ε(Pb-Bi)
disperses finely and isolately in semiconducting Si matrix and contains a high
density of MA-induced internal defects.

IV. High-T_c Superconductors Prepared by MA of Immiscible Ba-(Y, Gd, Ho or Er)-
Cu and Bi-Sr-Ca-Cu Alloys and oxidation process

1. Object

Since the discovery of a novel $Ba_2YCu_3O_y$ oxide superconductor with a
transition temperature at about 90 K[17], a great effort has been made toward
the synthesis of similar high-T_c oxide superconductors as well as the
clarification of their fundamental superconducting properties. The present
authors have reported that oxidation of rapidly quenched Ba-Ln-Cu (Ln=Yb or Eu)
alloys gives oxide superconductors with zero resistance at about 90 K in the
sample form of tape[18,19] or thin foils[20]. This method has attracted strong
attention as a process of producing a high-T_c oxide superconductor because of
the good ductility of amorphous Ba-Ln-Cu alloy tapes produced by liquid
quenching. However, the liquid quenching method of alloys has the disadvantage
that the process is limited to Ba-Yb-Cu[18,21] and Ba-Eu-Cu[19] alloys where
the constituent elements are soluble in an equilibrium state. Application of
MA to the immiscible Ba-Ln-Cu alloys is expected to result in formation of a
homogeneous non-equilibrium phase which cannot be obtained by any processes
through liquid. In this section, we attempt to clarify (1) the possibility of
the formation of a homogeneous phase in immiscible Ba-Ln-Cu (Ln=Y, Gd, Ho or
Er) alloys by MA, (2) the structural change of the MA phase by heating in air
and oxygen, and (3) the superconductivity of the resultant oxides.

2. MA Structure

X-ray diffraction patterns of the mixed Ba, Cu and $Ln_{80}Cu_{20}$ (Ln=Y, Gd, Ho
or Er) powders with a nominal composition of Ba_2LnCu_3 after 10 h milling are
shown in Fig. 5. After a 10 h milling, the peaks of Ba and Cu metals and

Ln$_{80}$Cu$_{20}$ alloys disappear and only two diffraction peaks which are attributed to an fcc Cu solid solution are seen. The lattice parameters of the non-equilibrium fcc phase are about 0.366 nm, being considerably larger than that (0.3615 nm) of pure Cu. It is thus concluded that MA is useful for the formation of the homogeneous phase even in immiscible alloy systems.

3. Oxidation Behavior of MA Phase and Oxidation-induced Phases

Figure 6 shows the DTA and TGA curves of the MA Ba$_2$YCu$_3$ powder heated at 10 K/min in air. Two large exothermic peaks are seen in the temperature ranges of 425 to 470 K and 475 to 590 K and the low-temperature peak is extremely sharp. The weight of the powders increases in two steps and these steps appear to correspond to the two exothermic reactions. The increase in the powder weight is larger in the second step ranging from 470 to 590 K. In addition, we confirm that the MA Ba$_2$YCu$_3$ powder heated in an Ar atmosphere shows only an exothermic peak with high intensity in the range of 425 to 475 K. Accordingly, the first exothermic reaction shown in Fig. 6 is thought to be due to the transformation of the fcc solid solution to a more stable metallic phase. The second exothermic reaction is probably due to the oxidation of the stable metallic phase. The increase in the powder weight upon oxidation is complete at about 550 K and the oxygen content decreases with further increasing temperature up to 1223 K. As will be described later, the oxide prepared by heating to 600 K does not exhibit superconductivity at temperatures above 77 K and the appearance of high-T$_c$ superconductivity with T$_c$ above 77 K is limited to the oxide heated at temperatures above 1123 K. Accordingly, the high-temperature annealing leading to the gradual decrease of oxygen content seems essential for formation of the Ba$_2$LnCu$_3$O$_y$ with high-T$_c$ superconductivity.

Most of the peaks in the X-ray diffraction patterns of Ba$_2$LnCu$_3$O$_y$ oxides prepared by annealing the MA Ba$_2$LnCu$_3$ powders at 1193 K for 24 h in oxygen can be identified as an orthorhombic phase as indexed in Fig. 7. Thus, the orthorhombic Ba$_2$LnCu$_3$O$_y$ phases with the same structure as that[22] prepared by the conventional sintering method are obtained as the main phase by oxidizing the MA Ba$_2$LnCu$_3$ powders in oxygen. Optical microscopy showed that the orthorhombic Ba$_2$LnCu$_3$O$_y$ phases prepared by annealing at 1193 K for 24 h in oxygen have grain sizes ranging from 1 to 5 μm.

4 Superconducting Properties

As shown in Fig. 8 for the Ba$_2$LnCu$_3$O$_y$ (Ln=Y, Gd, Ho or Er) oxides produced by oxidation of the MA Ba$_2$LnCu$_3$ powders for 24 h at 1193 K in oxygen, the resistance first decreases gradually at temperatures above about 92 K, and then rapidly in the range 91 to 89 K, and becomes zero in the range 88 to 86 K. The T$_c$ values at midpoint and zero resistance for the Ba$_2$LnCu$_3$O$_y$ oxides are 91 and 88 K for Y, 90 and 88 K for Gd and Ho, and 89 and 86 K for Er.

The J$_c$ at 77 K in the absence of applied fields was examined as a function of annealing time at 1193 K in oxygen and air for Ba$_2$YCu$_3$O$_y$. The J$_c$ value increases rapidly in the range 5 to 15 h and becomes constant after 24 h. The highest J$_c$ is 138 A/cm^2 in oxygen and 18 A/cm^2 in air, indicating a significant

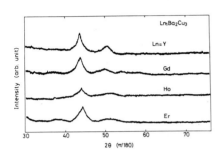

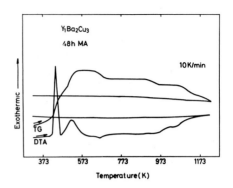

Figure 5: X-ray diffraction patterns of MA Ba_2LnCu_3 (Ln=Y, Gd, Ho or Er) alloys.

Figure 6: Differential thermal (DTA) and thermogravimetric (TGA) curves of a MA Ba_2YCu_3 alloy heated at 10 K/min in air.

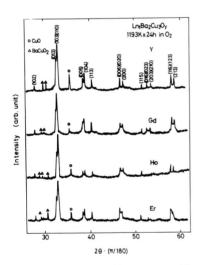

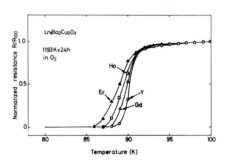

Figure 7: X-ray diffraction patterns of $Ba_2LnCu_3O_y$ prepared by annealing the cold-pressed compacts of the MA powders for 24 h at 1193 K in oxygen.

Figure 8: Normalized electrical resistance R/R_{100} of $Ba_2LnCu_3O_y$ prepared by annealing the cold-pressed compacts of the MA powders for 24 h at 1193 K in oxygen.

difference in J_c for different annealing atmospheres. This is probably due to the difference in oxygen concentration of the resultant oxides and it may therefore be concluded that the high oxygen concentration caused by annealing in oxygen is necessary for a high-J_c superconductor. The highest J_c values at 77 K in the absence of field for the other $Ba_2LnCu_3O_y$ oxides prepared by oxygen annealing at 1193 K for 24 h are 105 A/cm^2 for Gd, 110 A/cm^2 for Ho and 90 A/cm^2 for Er. From the comparison with the data shown in Fig. 8, one can notice the tendency for the J_c values to increase with increasing T_c value. The J_c values are of the same level as those for the oxides prepared by the conventional sintering technique and no significant increase in J_c by MA is seen, even though the method is expected to lead to refinement of grain size and a homogeneous distribution of the constituent elements.

A similar high-T_c oxide has been obtained even in an immiscible $BiSrCaCu_2$ system by the same process consisting of MA of Bi metal and Sr-Ca-Cu alloy powders and oxidation in air. The normalized resistance of the $BiSrCaCu_2O_y$ oxide decreases by 10 % at about 110 K and then disappears at 79 K. Although zero resistance at 110 K is not achieved for the present oxide, the superconductivity with zero resistance at 110 K is nearly the same as that for the $BiSrCaCu_2O_y$ oxide prepared by the conventional sintering process. The J_c value at 77 K is about 20 A/cm^2 in the absence of applied field. No appreciable enhancement of J_c was obtained for the MA powders probably due to the necessity of oxidation treatment at high temperatures above 1000 K.

V. Hard Magnetic Alloys Prepared by MA of Glass-forming Fe-Nd-B Alloys[23]

In a series of researches on the synthesis of materials with functional properties by MA, Fe-Nd-B powders with a non-equilibrium phase produced by MA of conventionally solidified Fe-Nd-B ingots were found to exhibit a hard magnetic property, in spite of the absence of the property for the as-cast Fe-Nd-B ingots.

1. MA Structure

The as-cast ingot with a composition of $Fe_{77}Nd_{15}B_8$ consists mainly of a tetragonal $Fe_{14}Nd_2B$, though a small amount of hexagonal Fe_7Nd phase is coexistent. After MA for 6 h, the crystalline peaks of $Fe_{14}Nd_2B$ and Fe_7Nd almost disappear and the increase in MA time to 24 h causes further progress of the disappearance of diffraction peaks which are attributed to the crystalline phases, suggesting the formation of a mostly amorphous phase. The MA powder size decreases from about 2 to 10 μm for 5 h to about 0.1 to 0.5 μm for 48 h. On the DTA curve of the $Fe_{77}Nd_{15}B_8$ powder subjected to MA for 24 h, a broad exothermic peak due to crystallization is seen in the range 670 to 720 K, indicating that the MA Fe-Nd-B powder consists of non-equilibrium phases including an amorphous phase.

2. Magnetic Properties

Figure 9 shows the hysteresis curves of the as-cast ingot and the MA powder for $Fe_{77}Nd_{15}B_8$ alloy. The intrinsic coercivity (iH_c) is as small as 6 kA/m for the as-cast ingot, but the MA for 24 h causes a significant increase of iH_c to 462 kA/m. This clearly indicates that the hard magnetic property can be improved by MA. The change in the iH_c values of $Fe_{72}Nd_{20}B_8$, $Fe_{77}Nd_{15}B_8$ and $Fe_{82}Nd_{12}B_6$ powders by MA is shown in Fig. 10. Although the iH_c value is as small as 6 to 10 kA/m in as-cast state, it increases rapidly to 178 to 219 kA/m after 9.5 h MA and gradually to 334 to 509 kA/m with increasing MA time to 48 h. The iH_c value of $Fe_{77}Nd_{15}B_8$ powder tends to increase relatively largely even after 48 h MA, suggesting that a high coercivity above 550 kA/m might be obtained by longer time MA. Although the reason for the difference in the MA-induced change of iH_c among the three alloys remains unknown, the difference indicates the necessity that an appropriate alloy composition must be chosen for the achievement of high iH_c by MA. Here, it is important to point out that no significant increase of iH_c is seen in the case of MA of elemental Nd, Fe and B powders, being consistent with the previous data[24]. Consequently, it may be concluded that the high-energy ball-milling treatment to induce a hard magnetic property is effective only against the pre-alloyed Fe-Nd-B ingots. The significant difference in iH_c values between the elemental powder and the alloyed powder is presumably due to a lower degree of oxidation of the pre-alloyed powder as compared with that for the MA elemental powder, even though similar great attention was payed to suppress the oxidation.

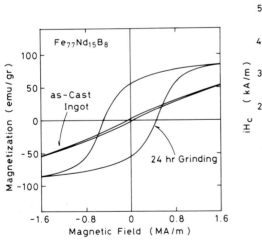

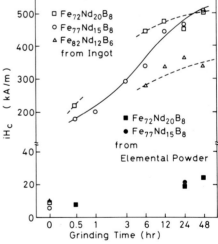

Figure 9: Hysteresis curves of $Fe_{77}Nd_{15}B_8$ alloys in as-cast state and on MA state for 24 h.

Figure 10: Change in iH_c of the MA Fe-Nd-B powders with MA time.

A relatively high coercivity has been obtained for the melt-spun ribbon and for the powders produced by pulverizing the ribbon[25]. By controlling a cooling rate in melt spinning, one can produce the ribbon having a mixed structure of amorphous matrix and tetragonal $Fe_{14}Nd_2B$ precipitates with a size of submicron which is comparable to a theoretical single domain size (0.3 µm for $Fe_{14}Nd_2B$)[26] without domain wall in the grain. Furthermore, it has been thought[27,28] that the formation of the duplex structure, in which the tetragonal $Fe_{14}Nd_2B$ precipitates are surrounded by an amorphous thin layer, is necessary to obtain a high iH_c value and the pinning effect due to the amorphous phase is a dominant factor in enhancing the intrinsic coercivity. The structural similarity between the melt-spun ribbon and the MA powder allows us to infer that the increase in iH_c by MA is due to the formation of coexistent amorphous and tetragonal $Fe_{14}Nd_2B$ phases within a fine scale comparable to a powder size (0.1 to 0.5 µm). Although observation of the microstructure in the MA powder is rather difficult because of its high oxidation tendency, further detailed clarification of the MA-induced micro-structure is in progress.

VI. Heat-Resistant Alloys Prepared by MA of Glass-forming Fe-Cr-Mo-C Alloy and Al_2O_3[29]

1. Object

Hot-pressed Fe-Cr-Mo-C products made from atomized amorphous powders have been reported[30] to have fine and homogeneous structures consisting of martensite and carbides and to exhibit high elevated temperature strength and good wear resistance. It can be expected that a homogeneous dispersion of additional oxide or nitride particles in the mixed structure will give a further significant improvement of the elevated temperature strength. MA is useful for a homogeneous mixing of different materials, in addition to the formation of non-equilibrium phases. It can thus be expected that MA of Fe-Cr-Mo-C and Al_2O_3 powders will produce multiphase powders consisting of Al_2O_3 particles embedded in an amorphous Fe-Cr-Mo-C matrix. Furthermore, hot consolidation of these MA powders into a bulk material is relativey easy because of the MA-induced metastable state. This section deals with the MA structure of $Fe_{58}Cr_{16}Mo_8C_{18}$ plus Al_2O_3 powders and the structure and elevated temperature strength of bulk material made from the MA powders.

2. MA structure

The X-ray diffraction patterns shown in Fig. 11 represent the effect of Al_2O_3 particles on the MA structure. The X-ray diffraction pattern of Fe-Cr-Mo-C powder reveals the existence of crystalline phases even after 72 h milling, while the crystalline peaks of the Al_2O_3-containing alloys decrease significantly, and a mainly amorphous phase is formed for the Fe-Cr-Mo-C powders containing 20 vol% Al_2O_3 after 72 h milling. This clearly indicates that the amorphization by MA is accelerated by the addition of Al_2O_3. There was no significant difference in the X-ray diffraction pattern from the Al_2O_3 particle size, probably because the Al_2O_3 particles are also fragmented by MA.

Furthermore, it was confirmed that the average powder size decreases significantly by the coexistence of Al_2O_3 particles. For instance, the size after MA for 72 h is 17.2 μm for the Fe-Cr-Mo-C powder and between 7.7 to 8.8 μm for the powders containing 10 vol% Al_2O_3 with diameters ranging from 0.02 to 20 μm. There is no appreciable difference in the final MA powder particle size with Al_2O_3 particle size. Thus, the acceleration of the amorphization by the coexistence of Al_2O_3 can be attributed to the acceleration of the pulverization of the alloy.

Figure 12 shows the etched structure of Fe-16Cr-8Mo-18C and Fe-16Cr-8Mo-18C+10 vol% Al_2O_3 compacts produced by hot pressing at 1353 K. No distinct pores are seen in the samples, indicating fully dense hot-pressed products. Furthermore, fine second phase precipitates can be seen homogeneously distributed in the matrices of both samples and a homogeneous distribution of Al_2O_3 particles with a size between 1 and 10 μm is seen in the Al_2O_3-containing alloy. The X-ray diffraction pattern of the Fe-Cr-Mo-C alloy with 10 vol% Al_2O_3 particles indicates that the structure consists of martensite(α'), austenite(γ), $M_{23}C_6$, M_2C and Al_2O_3 and these phases (except Al_2O_3) agree with those of the hot-pressed Fe-Cr-Mo-C alloy without Al_2O_3.

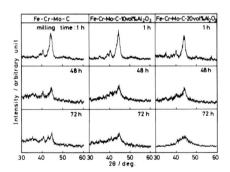

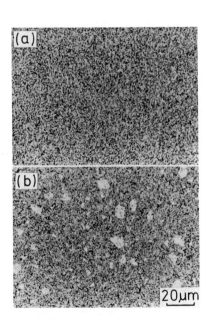

Figure 11: X-ray diffraction patterns showing the change in the MA structure of as-cast Fe-16Cr-8Mo-18C alloy containing 20 vol% of Al_2O_3 particles.

Figure 12: Optical micrographs of Fe-16Cr-8Mo-18C (a) and Fe-16Cr-8Mo-18C+10 vol% Al_2O_3 (b) alloys prepared by pressing the MA powders at 1353 K and 62.4 MPa for 1.8 ks.

3. High-Temperature Hardness

Vickers hardness (H_v) of the hot-pressed Fe-Cr-Mo-C+10 vol% Al_2O_3 alloy has a low value of 310 at T_p=1203 K, increases very rapidly with rising T_p and reaches a maximum value of 1010 at 1353 K, where full density is obtained. A further increase of T_p results in a decrease of H_v to about 690 at T_p=1403 K. The rapid increase of H_v in the range 1203 to 1353 K is due to densification of the material and the significant decrease of H_v at T_p=1403 K is caused by the rapid increase of carbide size. It has furthermore been confirmed that the largest values of H_v and packing factor are obtained in the range 10 to 15 vol% Al_2O_3. This indicates that an appropriate volume fraction to increase both H_v and packing factor is about 10 vol%.

It was thus determined that high density, homogeneous and dense distribution of fine carbides and high hardness are obtained at pressing temperature of 1353 K and a composition of 10 vol% Al_2O_3. Accordingly, the subsequent study on high-temperature hardness was focussed on the hot-pressed material prepared under these conditions. Figure 13 shows the change in H_v as a function of testing temperature of the hot-pressed Fe-Cr-Mo-C and Fe-Cr-Mo-C+10 vol% Al_2O_3 materials. H_v of the Fe-Cr-Mo-C material decreases from 955 at R.T. to 540 at 873 K, while that of the Al_2O_3-containing materials decreases from 1280-1455 at R.T. to 720-760 at 873 K. No significant difference in H_v with the particle size of Al_2O_3 is seen at higher temperatures. The hardness at temperatures below 873 K is significantly improved by the coexistence of Al_2O_3, but this effect tends to decrease rapidly at temperatures above about

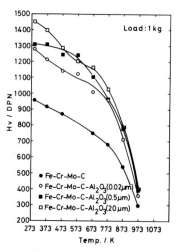

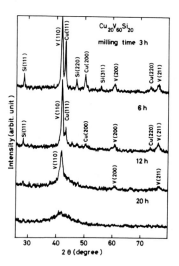

Figure 13: Temperature dependence of H_v for Fe-16Cr-8Mo-18C and Fe-16Cr-8Mo-18C+Al_2O_3 alloys prepared by pressing the MA powders at 1353 K and 62.4 MPa for 1.8 ks.

Figure 14: Change in the X-ray diffraction patterns of MA $Cu_{20}V_{60}Si_{20}$ powders with MA time.

900 K where the softening of the matrix phase will be the dominating factor determining the hardness. In addition, it has been confirmed that the Fe-Cr-Mo-C alloys containing Al_2O_3 with particle size of 0.02 and 1 μm have good wear resistance. Thus, the Fe-Cr-Mo-C+Al_2O_3 materials with a fine and homogeneous composite structure prepared by hot pressing the MA amorphous powders containing Al_2O_3 particles are attractive as a new type of heat-resistant material because of their high hardness at elevated temperatures.

VII. New Amorphous Alloy Systems Produced by MA[31,32]
1. Object
An amorphization by MA of elemental powders has been reported to take place in a number of alloy systems[3,33]. Most of the amorphized alloys reported so far were limited to the alloy systems in which amorphization by rapid solidification has been known. This section was focussed on the clarification of the formation process of the amorphous phase by MA in $Cu_{20}V_{60}$(Si or Ge)$_{20}$ and $Cu_{44}Nb_{42}$(Si, Ge or Sn)$_{14}$ alloys with large liquidus miscibility gap where a homogeneous mixing of the constituent elements by solidification method is very difficult.

2. Amorphization Process of Immiscible-Type Alloys
As an example, Fig. 14 shows the X-ray diffraction patterns of the mixed powders after 3, 6, 12 and 20 h milling for $Cu_{20}V_{60}Si_{20}$. After a 6 h milling, the crystalline peaks of Si except $(111)_{Si}$ have completely disappeared, resulting in the formation of bcc V(Si) and fcc Cu(Si) solid solutions. The further increase of MA time gives the disappearances of the fcc peaks after 12 h milling and the bcc peaks after 20 h milling, resulting in the appearance of a broad diffraction peak (which we attribute to an amorphous phase). The formation of a mostly single amorphous phase was also confirmed for $Cu_{20}V_{60}Ge_{20}$ and $Cu_{44}Nb_{42}$(Si, Ge or Sn)$_{14}$ alloys subjected to 20 h milling.

In order to clarify the microstructure change in the amorphization process by MA, TEM observation was carried out for $Cu_{20}V_{60}Si_{20}$. Figure 15 shows bright-field micrographs and selected area diffraction patterns of the mixed powders after 3 (a and b), 6 (c and d), 12 (e and f) and 20 (g and h) h milling. The powders after a 3 h milling have a multiphase structure with an average grain size of about 200 nm and their phases include a high density of tangled dislocations. The corresponding diffraction pattern consists of many reflection rings which can be indexed as bcc V, fcc Cu and cubic Si. After a 6 h milling, the multiphase structure becomes homogeneous and the main reflection rings can be indexed as bcc V phase. The further increase in the MA time gives the change from the bcc phase including a small amount of fcc Cu phase to amorphous phase including a small amount of bcc V phase (Fig. 15 e and f). The bcc V phase decreases with further increasing MA time and the powders after a 20 h milling consist of a mostly single amorphous phase. Lack of contrast characteristic to a crystalline phase in the bright-field image (g) and broad diffuse haloes in the diffraction pattern (h) taken with an aperture of 150 nm diameter reveal clearly the formation of a homogeneously amorphous phase. The

amorphous nature of the MA powders was also confirmed from the result that
there was no change in the contrast of the bright-field image (g) by tilt of
the sample. Based on the results of X-ray diffractometry and TEM, the progress
of the amorphization was presumed to occur through the following three stages;
(1) the dissolution of Si into V and Cu phases, (2) the dissolution of Cu(Si)
into V(Si) phase, and (3) amorphization of supersaturated V(Cu,Si) solid
solution. The MA amorphous powders had an irregularly polygonal shape and
their average size was as small as about 5 μm.

3. Reason for the Amorphization of the Immiscible Alloys

The Cu-V-Si alloy is an engineeringly important system as a superconducting
material and the structure formed by long-time annealing has been known to
consist of fcc Cu(Si) and A-15 type V_3Si. The equilibrium phase diagrams of
each binary alloy show that Cu-V system has a large liquidus miscibility gap
while the other two alloys include a number of compounds. It has been shown in
La-Au, Ni-Zr, Ni-Hf and Zr-Rh systems that an amorphous phase is formed by a
solid-state diffusion reaction of pure polycrystalline metals[34] at elevated
temperatures. The amorphization has been thought[34-36] to require
simultaneously the following three criterions; (1) fast diffusivity of solute
atom, (2) a negative heat of mixing in the alloys providing the necessary
chemical driving force for the reaction, and (3) the reaction to amorphize must
occur at sufficiently low temperatures to suppress either the nucleation or the
growth of thermodynamically preferred crystalline compounds. It is presumed in

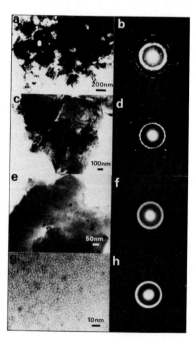

Figure 15:
Changes in the bright-field micrographs and selected area diffraction patterns of MA $Cu_{20}V_{60}Si_{20}$ powders with MA time.

Cu-V-Si and Cu-V-Ge systems that ternary compounds $(V,Cu)_xSi_y$ and $(V,Cu)_xGe_y$ consisting of binary $V_x(Si$ or $Ge)_y$ and $Cu_x(Si$ or $Ge)_y$ compounds with negative heat of mixing exist in a metastable state and the Cu-V-Si and Cu-V-Ge alloys have a chemical driving force enough to form an amorphous phase. Additionally, the high energy caused by MA results in increases in temperature and stress in local region of the powders and permits fast diffusion of Cu, V and Si or Ge elements even at low temperatures where the nucleation and growth of thermodynamically preferred crystalline compounds do not occur, leading to the formation of an amorphous phase. Furthermore, it was pointed out in Fig. 15 that the MA gave the formation of artificial multiphase powders consisting of Cu- and V-base solid solutions. The diffusion distance leading to amorphization is much shorter for the multiphase powders. The short diffusion distance satisfies the condition of fast diffusivity of solute atoms and enables us to amorphize.

VIII. Concluding Remarks

It was clarified in sections 3 to 6 that MA is useful for the formations of homogeneously dispersed structure and homogeneous solid solution including amorphous phase even in immiscible alloy systems where the structural modification by rapid solidification cannot be achieved. As a result, the unique structural modification enabled us to prepare superconducting materials. In addition, application of MA into the glass-forming alloys in Fe-Nd-B and Fe-Cr-Mo-C-Al_2O_3 systems gave rise to the hard magnetic alloy with fine powder morphology and the high heat-resistant materials with homogeneously dispersed Al_2O_3 particles, respectively. These data clearly demonstrate that engineeringly useful materials with functional properties can be prepared when an appropriate alloy design is made by utilizing the features of MA. Accordingly, the importance of the MA technique is expected to increase further in the field of materials science where alloy design on the basis of liquid or vapor quenching cannot be made.

References:
(1) J.S. Benjamin: Met. Trans. 1 (1970) 2943.
(2) A.E. Remakov, E.E. Yurchikov and V.A. Barinov: Phys. Met. Metallography 52 (1981) 1184.
(3) C.C. Koch, O.B. Cavin, C.G. McKamey and J.O. Scarbrough: Appl. Phys. Lett. 43 (1983) 1017.
(4) K. Matsuzaki, A. Inoue and T. Masumoto: "87 Int. Symposium & Exhibition on Science and Technology of Sintering", Tokyo, November, 1987, in press.
(5) A. Inoue, M. Oguchi, K. Matsuzaki and T. Masumoto: Int. J. Rapid Solidification 2 (1986) 175.
(6) H. Hofe and H. Hanemann: Z. Metallkunde 32 (1940) 112.
(7) W.B. Pearson: "Handbook of Lattice Spacings and Structures of Metals and Alloys", Pergamon Press, London, 1958, pp.124, 127, 128 and 156.
(8) R.W. Roberts: "Properties of Selected Superconductive Materials", 1978 Supplement, NBS Technical Note 983, US Department of Commerce, Washington.

(9) M.H. Cohen and D.H. Douglass: Phys. Rev. Lett. 19 (1967) 118.

(10) M. Strongin and O.F. Kammerrer: J. Appl. Phys. 39 (1968) 2509.

(11) R. Koepke and G. Bergmann: Z. Physik 242 (1971) 33.

(12) For example, P.G. Gennes: "Superconductivity in Metals and Alloys", Benjamin Inc., New York, 1966, p.217.

(13) N.K. Hindley and J.H.P. Watson: Phys. Rev. 183 (1969) 52.

(14) H.W. King, C.M. Russell and J.A. Hulbert: Phys. Lett. 20 (1966) 600.

(15) K. Matsuzaki, A. Inoue and T. Masumoto: Jpn. J. Appl. Phys. 27 (1988) L779.

(16) K. Matsuzaki, A. Inoue and T. Masumoto: "Int. Meeting on Advanced Materials", Materials Research Society, Tokyo, May 1988, in press.

(17) M.K. Wu, J.R. Ashburn, C.J. Torng, P.H. Hor, R.L. Meng, L. Gao, Z.J. Huang, Y.Q. Wang and C.W. Chu: Phys. Rev. Lett. 58 (1987) 908.

(18) K. Matsuzaki, A. Inoue, H.M. Kimura, K. Aoki and T. Masumoto: Jpn. J. Appl. Phys. 26 (1987) L1310.

(19) K. Matsuzaki, A. Inoue, H.M. Kimura and T. Masumoto: Jpn. J. Appl. Phys. 26 (1987) L1610.

(20) N. Kataoka, S. Furukawa, K. Matsuzaki, A. Inoue and T. Masumoto: Jpn. J. Appl. Phys. 27 (1988) L70.

(21) K. Matsuzaki, A. Inoue, H.M. Kimura, K. Moroishi and T. Masumoto: Jpn. J. Appl. Phys. 26 (1987) L334.

(22) Y. Syono, M. Kikuchi, K. Oh-ishi, K. Hiraga, H. Arai, Y. Matsui, N. Kobayashi, T. Sasaoka and Y. Muto: Jpn. J. Appl. Phys. 26 (1987) L498.

(23) T. Nakamura, A. Inoue, K. Matsuki and T. Masumoto: J. Mater. Sci. Lett. in press.

(24) L. Schultz, J. Wecker and E. Hellstern: J. Appl. Phys. 61 (1987) 3583.

(25) J.J. Croat, J.F. Herbst, R.W. Lee and F.E. Pinkerton: J. Appl. Phys. 55 (1984) 2078.

(26) J.D. Livingston: J. Appl. Phys. 57 (1985) 4137.

(27) F.E. Pinkerton: IEEE Trans. Mag. MAG-22 (1986) 922.

(28) R.K. Mishra: J. Mag. Magn. Mater. 54-57 (1986) 450.

(29) T. Ogasawara, A. Inoue and T. Masumoto: "Int. Meeting on Advanced Materials", Materials Research Society, Tokyo, May 1988, in press.

(30) A. Inoue, L. Arnberg, M. Oguchi, U. Backmark, N. Backstrom and T. Masumoto: Mater. Sci. Eng. 95 (1987) 101.

(31) A. Inoue, H.M. Kimura, K. Matsuki and T. Masumoto: J. Mater. Sci. Lett. 6 (1987) 979.

(32) K. Matsuki, A. Inoue, H.M. Kimura and T. Masumoto: "87 Int. Symposium & Exhibition on Science and Technology of Sintering", Tokyo, November, 1987, in press.

(33) E. Hellstern and L. Schultz: Appl. Phys. Lett. 48 (1986) 124.

(34) R.B. Schwarz and W.L. Johnson: Phys. Rev. Lett. 51 (1983) 415.

(35) W.L. Johnson, M. Atzmon, M. Rossum, B.P. Dolgin and X.L. Yeh: "Rapidly Quenched Metals", eds. S. Steeb and H. Warlimont, Elsevier Science Pub., Amsterdam, 1985, p.1515.

(36) E. Hellstern and L. Schultz: J. Appl. Phys. 63 (1988) 1408.

Structural and Thermal Analysis of New Systems Prepared by Mechanical Alloying

G. Cocco*, S. Enzo[+], L. Schiffini* and L. Battezzati**

*   Dipartimento di Chimica, via Vienna 2, 07100 SASSARI (ITALY)
[+]   Dipartimento di Chimica Fisica, D. D. 2137, 30123 VENEZIA (ITALY)
** Dipartimento di Chimica Inorganica, Chimica Fisica e Chimica dei Materiali, via P. Giuria 9, 10125 TORINO (ITALY)

Abstract

Crystalline elemental powders of high purity were milled with the aim to explore capabilities and limitations of the mechanical alloying technique for synthesis of new amorphous systems. Amorphous powder formation was observed in the Cu-Ti, Al-Ni and Mo-Ni systems. On the other hand, a crystalline intermetallic formed in Al-Ni and a metastable cubic phase in Cu-Ru. The structural evolution and thermal behaviour of the powders at selected milling times are reported.

Introduction

Early amorphization of metal systems by melt quenching (MQ) was performed at compositions close to deep eutectics according to the relevant equilibrium phase diagrams. Since then, the glass forming composition ranges were widened by means of new techniques exploiting mainly solid state amorphizing reactions. As generally maintained, the parameters controlling the process are a large negative heat of mixing and an anomalously fast diffusion of one component in the host matrix of the other (1), provided that crystalline intermetallic phases cannot be formed solely by the smaller atom diffusing through the crystalline lattice of the larger atom. Moreover, an atomic size ratio rule appears to apply both to systems prepared by MQ and solid state reactions (2).

Specifically, from a technological point of view, mechanical alloying (MA) appears well suited to explore systems of new composition which do not conform to the conventional rules for amorphization. This also offers the opportunity to enlarge the knowledge on the basic parameters leading to metastable alloys. Along these lines we present here attempts to synthesize by MA systems which are not yet amorphized by MQ and/or escaping the atomic size ratio rule, (Al-Ti, Al-Ni, Mo-Ni), and a system showing complete immiscibility both in liquid and solid state, (Cu-Ru). As a reference system for amorphization we have also studied the Cu-Ti system, which is an easy glass former independently of the technique employed (3).

Experimental

Crystalline elemental powders of high purity were milled at various compositions in a hardened tool steel vial under Ar atmosphere, with a SPEX mixer-mill model 8000. The milling was performed at room temperature in a glove box and heat was removed from the vial with a forced convection of Ar flow.

At selected times of milling part of the powder was withdrawn to check the structural situation by X-ray diffraction (XRD). The XRD spectra were collected in the reflection mode with a Philips powder goniometer using a Cobalt anode (CoKα=0.179026 nm).

Differential Scanning Calorimetry (DSC) experiments were carried out with a DUPONT 1090 apparatus. Powders were inserted into an aluminum pan and the cell was evacuated before the runs, during which pure Ar was employed as purging gas. The heating rate was 30 K/min.

Results and Discussion

We show in figure 1 the XRD patterns of the $Ti_{40}Cu_{60}$ system milled up to 28 h. A solid state amorphization reaction starts immmediately and is accomplished almost totally after 11 h. This condition seems to be maintained till 16 h of mechanical milling. A prolonged treatment favours a crystallization process as already apparent in the specimen milled for 28 h. In fact, the first halo at about 28.5 nm^{-1} is sharper than the corresponding previous profiles and also the second maximum at around 50 nm^{-1} emerges more neatly from the noise scatter. Note the weak presence of the (100) reflection of hexagonal close packed (HCP) Ti in all diagrams.

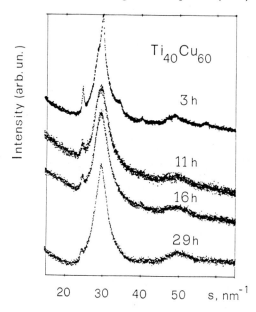

Fig. 1: XRD patterns of $Ti_{60}Cu_{40}$ powders milled for the quoted times as a function of $(4\pi/\lambda)\sin\Theta$.

Figure 2 shows the DSC traces of the samples milled for the quoted times: one or more exothermal events take place in every sample. XRD shows that the major signal is due to an amorphous-to-crystal transformation. The main crystallization product is the Ti_2Cu_3 compound, except for the sample milled 3 h, where a substantial proportion of Ti_3Cu_4 compound is found. This is reflected in the DSC trace: a double stage crystallization is observed, as in MQ ribbons of the same composition, where Ti_3Cu_4 is formed as well. As crystal nuclei are often quenched-in during preparation of ribbons, it is not surprising to observe Ti_3Cu_4 as a crystallization product, because it is the highest melting compound which can be involved in transformations of the $Ti_{40}Cu_{60}$.

The formation of the same phase in ball milled powders seems to indicate that the amorphous phase formed at this stage does not have the average composition of the alloy. Only at a longer milling time the amorphous phase reaches the expected composition and crystallizes as Ti_2Cu_3.

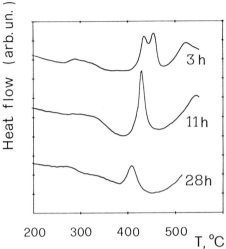

Fig. 2: DSC traces of some $Ti_{60}Cu_{40}$ powders.

In order to check the effect of composition on the amorphization by MA of the Ti-Cu system we have milled crystalline powders in the atomic ratio $Ti_{13}Cu_{87}$. The XRD spectra of this system after short times of milling display a single crystalline cubic phase with complete absence of Ti peak profiles. This structural situation maintains up to 28 h of milling as witnessed in figure 3, where the familiar spectrum of a face centered cubic (FCC) phase is recognizable. We measured the lattice parameter of the cubic phase a=0.363 nm, a figure slightly larger than 0.361 nm quoted for pure copper, suggesting that titanium atoms are incorporated in the Cu frame. This result does not conform to the range of amorphization reported by Politis and Johnson (4) for the same system; hence, it is very likely that the apparatus used for milling plays a significant role for this composition.

The effects of milling crystalline elemental powders of atomic composition $Al_{25}Ti_{75}$ are reported in Figure 4. The lower pattern refers to a mixture of crystalline aluminum and titanium before any mechanical treatment. After 3 h we note a significant intensity loss of titanium peaks. This is quite surprising, because Ti is the major element. Furthermore, the remaining line profiles of Al appear drastically broadened. Finally, after 9 h of MA, an amorphous single phase is observed.

Schultz (5) was the first to observe the behaviour of Ti_xAl_{100-x} system under MA [Sundaresan et al. (6) observed amorphization starting from the Ti_3Al intermetallic] and reported evidence of amorphization at compositions X=40 and X=52, but did not study the composition X=75. Thus, to achieve a more direct comparison with his results we checked also the $Al_{75}Ti_{25}$ composition. From figure 4 (right side) we can infer that at 9 h of MA most of the Ti atoms are incorporated in the Al lattice. After 21 h of MA only a distorted FCC lattice can be found; further milling does not attain an amorphous phase, in agreement with results by Schultz (5). Note that, for this composition, we opened the vial and after 3 h found massive metal in the form of pellets, suggesting that a partial melting of powders occurred in the mill.

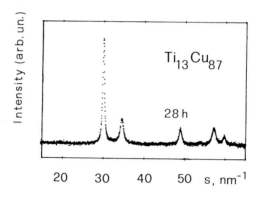

Fig. 3: XRD pattern for the $Ti_{13}Cu_{87}$ system milled for 28 h: the spectrum of a sigle FCC phase is recognizable.

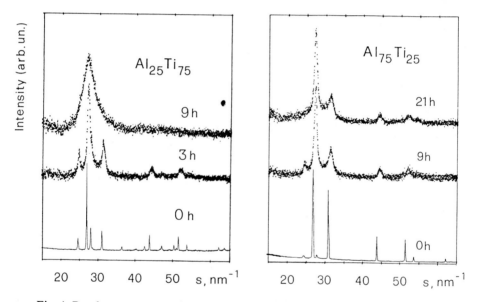

Fig. 4: Powder patterns of $Al_{25}Ti_{75}$ and $Al_{75}Ti_{25}$ systems at quoted milling times.

In the case of a $Cu_{40}Ru_{60}$ composition, it was verified (see figure 5, left side, lower curve) that after 16 h of MA, the two crystalline components are still unalloyed, though largely disordered and fragmented. However, after 92 h of MA a single cubic phase is present (figure 5, left side, upper curve). The lattice parameter obtained using the 4 most intense reflections gives a=0.367 nm, definitely higher than the value for pure Cu.

A somewhat different behaviour holds for the $Ru_{30}Cu_{70}$ composition, as reported in figure 5, right side. Although after 37 h we can note in the XRD pattern a single HCP phase, similar to pure ruthenium but with a different axial ratio c/a, nonetheless after 130 h of milling we ob–

serve formation of a cubic phase with lattice parameter a=0.368 nm, together with a minor secondary phase, not yet indexed. This new phase is metastable as it decomposes into the elements on heating, giving an exothermal reaction in the DSC.

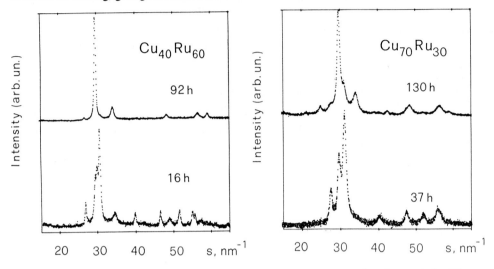

Fig. 5: Structural variations brought about the Cu-Ru system by the quoted milling treatments for the indicated atomic compositions.

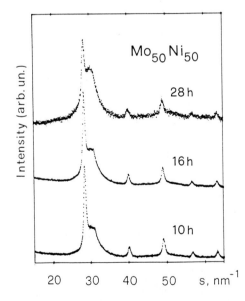

Fig. 6:
XRD patterns for the $Al_{50}Ni_{50}$ compositions testifying occurrence of the AlNi intermetallic.

Finally we present our investigations on two systems which are supposed to be readily glass formers on the basis of their strong negative enthalpy of formation, namely Ni-Al and Ni-Mo. Our results show that the $Al_{50}Ni_{50}$ composition leads soon to the formation of AlNi intermetallic and no amorphous phases are observed proceeding with the milling (see figure 6).

The XRD Patterns reported in figure 7 refer to the $Mo_{50}Ni_{50}$ mixture; they show that an amorphous phase is actually synthesized by MA, although a crystalline fraction of residual Mo is still present after 29 hours of milling. The line positions of the crystalline Mo phase do not change appreciably in our patterns, suggesting that no significant quantity of Ni is dissolved into the Mo habit. Note however, that the crystalline Mo phase is progressively disappearing and it is arguable that, on increasing the milling time, a single amorphous phase can be synthesized. It is obvious that, in order to rationalize this behaviour the knowledge of various thermodynamical and chemical parameters is required such as temperature in the vial, diffusivity and solubility of the elements, role of impurities, and so on. This matter is currently under investigation.

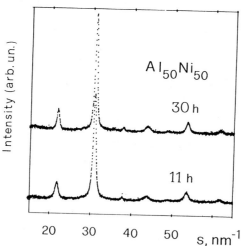

Fig. 7: Powder patterns of the $Mo_{50}Ni_{50}$ system showing clearly a progressive development of an amorphous phase.

Acknowledgement

This work was supported financially by ENEA under contracts # 3965 and # 3405. We thank Dr. S. Daniele for analytical determinations and Mr. D. Cannoletta for XRD measurements.

References

(1) R. B. Schwarz and W. L. Johnson: Phys. Rev. Lett. 51 (1983) 415.
(2) S. H. Liou and C. L. Chien: Phys. Rev. B 35 (1987) 2443; T. Egami and Y. Waseda: J. Non-Cryst. Solids 64 (1984) 113.
(3) A. F. Marshall, Y. S. Lee and D. A. Stevenson: J. Non-Cryst. Solids 64 (1984) 399.
(4) C. Politis and W. L. Johnson: J. Appl. Phys. 60 (1986) 1147.
(5) L. Schultz: Mat. Sci. Eng. 97 (1988) 15.
(6) R. Sundaresan, A. G. Jackson, S. Krishnamurthy and F. H. Froes: Mat. Sci. Eng. 97 (1988) 115.

D:
TOPICAL DISCUSSION

TOPICAL DISCUSSION

On Tuesday, October 4, 1988 a discussion was held on the "Application and Future Perspectives of Mechanical Alloying". It was moderated by L. Schultz and E. Arzt.

The topics were:

1.   Processing Problems
2.   Material Requirements and Applications
3.   Economics
4.   Future Research

1)   PROCESSING PROBLEMS

Q Schultz: What are the problems associated with the powders during and after milling?

A Benjamin: Major problems are the welding of the powder to the balls and the container walls. Some critical parameters are ball to powder ratio, ball size, and in particular additives such as alcohol, stearic acid, and carbon and oxygen. However these additves should be neutral or beneficial to the material. Additionally temperature should be controlled in relation to the brittleness of the material.

Q Schultz: What is the situation with protective atmospheres for commercial production of reactive materials? We know that this is o.k. in the lab, but can this be up-scaled?

A Arnhold: In principle it is possible, but there are economic problems.

A Tank: We would like MA-magnesium alloys for light weight, good corrosion resistance and high-temperature strength.

Q Fischmeister: Would the application you think of allow the price for PM?

A Tank: No.

A Benjamin: MA-Mg with 5% Fe have already been produced for deep sea instruments made from powder and forged components using CO_2 as inert atmosphere.

Q Schultz: Has there been any experience with wet milling?

A Benjamin: GE in the States have milled coarse powders with Heptane as a slurry to refine particles.

Q Arnhold: What are the reasons for the lower high temperature strength of the INCO alloy IN 9052 compared to the reaction milled Jangg processed material?

A Benn: We don't know exactly why but recent work by Wilsdorf, presented at the 1988 TMS Annual Meeting at Phoenix, compared these two alloys. It appears that the carbide distribution formed in the Jangg process is better.

Q Arzt: How does one assess the quality of powder from the mill?

A Benn: This is not easy, the production is very process dependent and therefore difficult to isolate each processing step, there are no definite tests at present.

Q Froes: How do you distinguish good from bad powder?

A Benjamin: Metallographical investigations can be helpful. For example in certain alloys it is easy to observe particles rich in chromium under the optical microscope but we simply can't do this for every particle and statistics play a large role; we can't look at every particle.

Q Arzt: What about stringers and other heterogeneities. Will we have to live with them?

A Benjamin: Certainly we should try to find the reasons e.g. particle sticking to balls and in particular powders being carried over from one charge to the other.

Q Schultz: What about the different compaction techniques?

A Shingu: The costs will be different. We have powder in a metastable condition and we need to preserve this during compaction. Therefore the temperature must not be too high but at the same time we require high density. We therefore should aim for high pressure and low temperature techniques.

Q Schultz: Is dynamic compaction a reality and what about costs?

A Shingu: Yes, this is a very good process to maintain metastability in the powders and this is a well established process in the lab.

2) MATERIAL REQUIREMENTS AND APPLICATIONS

Q Arzt: What are the requirements for MA materials from the user industries? Is that ductility, strength, fatigue resistance, corrosion etc.

A Tank: It depends on application, e.g. we like MA 956 because it is the only MA alloy that can be fusion-welded. We retain 25% of the strength and also brazing is o.k. MA 956 is also nice for us because we can superplastically deform it into shapes in the fine grained condition and sheets have good isotropic properties which is a big advantage. This alloy is a candidate for stationary parts in gas turbine engines. Of course, it must compete with ceramics but MA 956 has advantages, e.g. we can make hollow products unlike from ceramics which crack more easily. We believe this is a good material for use in a future 300 or 400 kW turbine engine for trucks. The competitor is the diesel engine but these are more noisy and produce soot.

Q Arzt: Are there no more wishes left for MA 956?

A Tank: Yes, the oxidation resistance could be a bit better and the argon problem is still a drawback.

Q Fischmeister: What about heat exchangers. Is this not easier using sheet steel and honeycomb structures than using tubes?

A Tank: No, we have an internal pressure which is too high and the thermal fatigue resistance above 1000 °C is better in tube than in plate and fine honeycomb assemblies.

Q Arzt: What about the requirements from aerospace industry?

A Froes: The Al alloys are important in the aerospace industry especially with regard to joining techniques. We would like to move away from fasteners because this increases the weight. There are troubles with joining RS Al alloys. The MA alloys should be joinable using friction welding.

C Benn: We are still low down on the learning curve in MA alloy joining technology.

Q Arzt: Can we hear from the power generation industry, too?

A Jongenburger: Well, price is a major question for us. Specifically transverse properties should be improved. The longitudinal properties are adequate. With respect to thermal fatigue, the MA Ni-base alloys are better than conventionally cast alloys. However, because we are at higher temperatures the thermal stresses are also higher and therefore we believe some improvements in thermal fatigue resistance are required.

Q Schultz: Are MA alloys commercially available and what are the applications today?

A Tank: Yes, we have in production an engine using a pin made from MA 956 in the diesel chamber. But the price is a problem. For the recombuster we use Nimonic 80A but at the moment MA 956 is still too expensive.

C Hack: For their military engines GE uses MA 754 as vanes which are critical parts. For several military aircraft we deliver 170 tons per year of MA 754.

Q Schultz: Are MA alloys presently being used in passenger aircraft?

A Hack: No, not yet.

C Froes: We must remember that it is not unusual for the introduction of new materials to take 25 years. This is the normal industrial resistance.

C Benn: For MA 6000 there was a 16 year maturation period. Of course, other alloys are being developed and it is likely that MA 6000 would be used in pilot studies.

Q Schultz: What about Al-alloys, is there a market?

A Hack: We have two MA Al-alloys. One for room temperature to replace alloy 7075 which has better corrosion resistance and forgeability. The other for high temperature use. This is a promising Al-Ti alloy which has the very high stiffness of Ti but the low density of Al. But this is the future.

Q Schultz: But what about today?

A Hack: Today the US Navy are using an MA Al alloy for torpedo hulls.

C Froes: Really we are waiting for a new series of vehicles and we don't replace old systems with new materials.

Q Schultz: Are the applications today only for the military?

A Froes: First applications in the aerospace industry are normally for the military.

C Arnhold: It is not always military, e.g. the pin in the diesel engine.

C Benjamin: There are possible applications for MA 956 in the nuclear power generating industry.

C Hack: As producers we can't cover all applications such as diesel, aero, nuclear power. All we can do is publicise properties and hope the users come to us.

Q Schultz: What are the applications for amorphous MA alloys?

A Johnson: They have interesting properties, exceptional corrosion resistance and unusual wear characteristics. For example, the W-Mo-Nb - B system has very high hardness, much better than the crystalline materials. We don't necessarily need these materials in bulk form but the question is how can we get amorphous materials onto a surface. There are also good possibilities for amorphous powders to be used as catalysts because of their high surface energy. But how can we coat a tool surface with amorphous powder?

C Schultz: Plasma spraying is a possibility.

Q Schultz: What about hydrogen storage and electrodes?

A Johnson: Yes, there is a good possibility for these materials to be used as electrodes in chlorine reduction plant.

C Froes: Well, there seems to be plenty of applications as Allied Chemical has documented but this is probably just a question of cost.

Q Arzt: What are the advantages of MA in relation to RS?

A Johnson: Well, we can produce very fine structures from any combination of elements by MA. The reason is we are entering the phase diagram not from the top (the liquid state) but from the bottom (solid).

C Arnhold: Yes, but what for?

C Johnson: We will find applications, these fine microstructures are very different from those produced by RS. We are talking about nm size range here!

Q Battezzati: Are there any differences in properties before and after compaction?

A Shingu: They are similar.

Q Schultz: What's happening in Japan and are there any applications?

A Shingu: Of course there are some groups in universities working on MA.

Q Schultz: Yes, but is there a government programme with sponsorship?

A Shingu: Not in Japan.

C Benn: Well, this is also so in Europe, but there is plenty of company work.

3) ECONOMICS

Q Arzt: What are the prices for MA materials and how do they compare to RS?

A Hack: I don't know what the price is for RS materials.

A Benjamin: The costs are comparable but there is an up-scaling problem.

Q Fischmeister: Can prices be reduced if large applications can be found for these materials?

A Hack: Of course and the more complex the product the more chances we have for lowering costs.

C Arnhold: We should remember that the MA and RS routes produce materials for different applications, e.g. MA processing for creep resistant materials.

Q Tank: Are there possibilities for up-scaling ball milling equipment?

A Benn: Yes, and really the dream is to have a continuous production process and this is being looked at now.

C McColvin: Actually the powder and metal costs are not too high. The cost are mostly incurred in the later processing steps.

C Hack: Costs are high because of the slow processes involved.

C Benjamin: That's right, the challenge is really to speed up the process.

Q Schultz: What is the patent situation? How much of a hindrance is this to commercial production and also research? If there is only one producer it is likely that a user will not want to be dependent on only one supplier.

A Benn: This is not always so. For example, Pratt & Whittney had all the patents for single crystal technology but other companies were able to get round these.

C Benjamin: Most patents will have run out in about 5 to 7 years' time. There will always be specific patents for special materials but by the end of the century there should be lots of producers.

4)   FUTURE RESEARCH

C Schlump: Concerning the amorphous materials we still don't really know enough about these materials.

C Shingu: An important point is that the internal energy is very high for these materials, perhaps as high as 50 kJ per mole which is much higher than RS.

C Johnson: In these materials there is a high interfacial density with excess enthalpy and this modifies properties e.g. superplasticity. Science on these materials is not easy, they are so inhomogeneous.

C Arnhold: We want to reach good quality at the lowest possible price. We should try to increase the dimensions of products, e.g. in the Al-ODS alloys we are at present limited to 100 mm diameter section. Research should also look into combining the RS and MA processes to attempt a superposition of properties.

C Benjamin: I suggest the challenge for the future is a machine for MA alloy production which is fast, cold and pure. I have some ideas how this may be achieved........ New discoveries are likely if we expand into multicomponent-systems.

C Tank: We have to improve the materials for gas turbines and diesel engines.

C Arzt: We have to stop here. I would like to thank everybody who contributed to the discussion. I think we have seen that many new materials will yet be produced by MA, its future seems to be starting now!

AUTHOR INDEX

Author Index[*]

[*] The page numbers refer to the 1st page of the relevant article

SUBJECT INDEX
AND
ALLOY INDEX

Subject Index[*]

Subject Index[*]

[*] The page numbers refer to the 1st page of the relevant article

Alloy Index[*]

[*] The page numbers refer to the 1st page of the relevant article

Alloy Index[*]

M

Mg alloys 243
Mn-Ni 85
Mn-Ti 53
Mo-Ni 343

N

Nb-Ge 101
Nb-Ni 53, 73
Nb-Ni-B 53
Nb-Sn 101
Nd-Fe-B 299, 327
Ni-Si-C 307
Ni-Ti 53, 73, 143, 307
Ni-Ti-C 307
Ni-Ti-N 307
Ni-Ti-O 307
Ni-V 85
Ni-V-C 307
Ni-W-C 307
Ni-Zr 53, 73, 95, 111
Nimonic 80 A 3, 129

P

Pb-Si 327
Pb-Si-Bi 327

R

R-400 201

S

SAP 19, 263
Steel powder 129
Stellite 129, 299

T

Th Mn_{12} 299
Ti-Al-Nb-V-Mo 243
Ti-Al-V-Fe-Er 243
Ti-Ni-Cu 243
Ti-Si 79
Ti-V 53
Ti-V-Cr-Er 243
TD Ni 3, 19, 201
TD Ni-Cr 19, 185, 201

V

V-Zr 53

W

W 129
W alloys 129

Y

Y-Ba-Cu-O 327
YD-Ni-Cr-Al 201

[*] The page numbers refer to the 1st page of the relevant article

LIST OF PARTICIPANTS

LIST OF PARTICIPANTS

Dr. Volker Arnhold
Sintermetallwerk KREBSÖGE GmbH
Postfach 5100
D - 5608 RADEVORMWALD
WEST GERMANY

Dr. G.-J. Brockmann
Julius & August Erbslöh GmbH & Co.
Siebeneicker Str. 235

D - 5620 VELBERT 15

Dr. Eduard Arzt
Max-Planck-Institut für Metallforschung
Institut für Werkstoffwissenschaften
Seestraße 92
D - 7000 STUTTGART 1

Dr. Giorgio Cocco
University - Dipartimento di Chimica
Via Vienna 2
I - 07100 SASSARI

Dr. Livio Battezzati
Dipartimento Chimica Materiali
Università die Torino
Via P. Churia 9
I - 10125 TORINO

Dr. Manfred Däubler
MTU - München GmbH
Dachauer Str. 665
D - 8000 MÜNCHEN 50

Dr. John Benjamin
Aluminum Company of America
Alcoa Technical Center
Alcoa Center, PA 15069
U S A

Jürgen Eckert
SIEMENS AG
EFE TPH 12
Postfach 32 40
D - 8520 ERLANGEN

Dr. Raymond Benn
INCO ALLOYS Int., Inc.
P.O. Box 1958
Huntington, West Virginia 25720
U S A

Bård Eftestøl
RAUFOSS A/S

N- 2831 RAUFOSS

Dr. Rüdiger Bormann
Inst. f. Metallphysik
Hospitalstr. 3-5

D - 3400 GÖTTINGEN

Dr. Stefano Enzo
Dept. Physical Chemistry
Calle Larga S. Marta
D.D. 21 37
I - 30123 VENEDIG

Prof. Dr. Hellmut Fischmeister
MPI für Metallforschung
Institut für Werkstoffwissenschaften
Seestraße 92
D – 7000 STUTTGART 1

Dr. Godfrey Hack
INCO ALLOYS Int. Ltd.
Holmer Road
Hereford HR4 9SL
GREAT BRITAIN

Dr. Francis H. Froes
Department of the Air Force (AFCS)
Metals and Ceramics Division
Wright–Patterson Air Force Base
Ohio 45433–6533
U S A

Dr. H.-D. Hedrich
DAIMLER BENZ AG
Abt. FG/TL
Postfach 600 202
D – 7000 STUTTGART 60

Eric Gaffet
CECM/CNRS
15 Rue G. Urbain
F – 94407 VITRY/Seine Cedex
FRANCE

Dr. Burkhard Heine
Max–Planck–Institut für Metallforschung
Institut für Werkstoffwissenschaften
Seestraße 92
D – 7000 STUTTGART 1

Dr. Denis Girardin
Laboratoire Mixle CNRS Sr. Gobain
BP 109

F – 54704 Pont à Mousson Cedex

Dr. Einar Hellum
RAUFOSS A/S

N – 2831 RAUFOSS

Dr. Johannes Grewe
Krupp Forschungsinstitut
Münchnerstr. 100

D – 4300 ESSEN

Dr. Akihisa Inoue
Institute for Materials Research
Tohoku University
Katahira 2-1-1
Sendai 980
JAPAN

Prof. E.Y. Gutmanas
TECHNION – Israel Institute of Technology
Dept. of Materials Engineering
Technion City
Haifa 32 000
ISRAEL

Prof. Dr. Gerhard Jangg
Technische Universität Wien
Inst. f. Chem. Technologie Anorg. Stoffe
Getreidemarkt 9
A – 1060 WIEN

Dr. Christoph Gutsfeld
Kernforschungszentrum Karlsruhe
IMF I, Bau 681
Postfach 3640
D – 7500 KARLSRUHE

Prof. William L. Johnson
present address:
1. Physikalisches Institut
der Universität Göttingen
Bunsenstraße 9
D – 3400 GÖTTINGEN

Peter Jongenburger
ASEA BROWN BOVERI
Corporate Research Center, CRBF 1

CH - 5401 BADEN

Pierre Le Brun
K.U. Leuven
Dept. of Metallurgy and Mat. Sci.
De Croylaan 2
B - 3030 HEVERLEE

Rainer Joos
Max-Planck-Institut für Metallforschung
Institut für Werkstoffwissenschaften
Seestraße 92
D - 7000 STUTTGART 1

Klaus Lempenauer
Max-Planck-Institut für Metallforschung
Institut für Werkstoffwissenschaften
Seestraße 92
D - 7000 STUTTGART 1

Prof. Hiroshi Kimura
National Defense Academy
1-10-20 Hashirimizu
Yokosuka, 239
JAPAN

Petra Loeff
Universiteit van Amsterdam
Natuurkundig Laboratorium
Valckenierstraat 65
NL - 1018 XE AMSTERDAM

Masayoshi Kimura
National Defense Academy
1-10-20 Hashirimizu
Yokosuka, 239
JAPAN

Dr. Michael J. Luton
EXXON Research and Eng. Co.
Route 22 East
Annandale, New Jersey 08801
U S A

Joachim Kinder
Max-Planck-Institut für Metallforschung
Institut für Werkstoffwissenschaften
Seestraße 92
D - 7000 STUTTGART 1

Dr. Mauro Magini
TIB-CHIA
ENEA
Via Anguillarese 301
I - 00060 S. Maria di Galerla (ROMA)

Prof. Carl Koch
North Carolina State University
P.O. Box 7907
Raleigh, NC 22695
U S A

Dr. William Mankins
INCO ALLOYS Int.
P.O. Box 1958
Huntington, West Virginia 25720
U S A

Dr. Georg Korb
Metallwerk Plansee GmbH, R&D

A - 6600 REUTTE

Dr. Gordon McColvin
INCO Alloys Ltd.
Holmer Road
HEREFORD, HR4 9SL
GREAT BRITAIN

Andree Mehrtens
I. Physikalisches Institut
der Universität Göttingen
Bunsenstraße 9
D - 3400 GÖTTINGEN

Prof. Dr. Martin Roth
FH Osnabrück
Fachbereich Werkstofftechnik
Albrechtstraße 30
D - 4500 OSNABRÜCK

Prof. Dr. David G. Morris
University of Neuchatel
Inst. of Structural Metallurgy
Ave. de Bellevaux 51
CH - 2000 NEUCHATEL

Dr. Manfred Rühle
Metallgesellschaft AG
Reuterweg 14
D - 6000 FRANKFURT 1

H.-C. Neubing
Fa. Eckart-Werke
Werk Guentersthal
D - 8560 VELDEN

Dr. Konrad Samwer
1. Physikalisches Institut
der Universität Göttingen
Bunsenstraße 9
D - 3400 GÖTTINGEN

Dr. Inken Nielsen
Batelle Institut e.V.
Am Römerhof 35
D - 6000 FRANKFURT/MAIN 90

Hugo Schlich
DFVLR
Linder Höhe
D - 5000 KÖLN-PORZ

Dr. M. Lütfi Öveçoglu
DEVTECH
Sourethweg 9
NL - 6422 PC HEERLEN

Dr. Wolfgang Schlump
Krupp Forschungsinstitut
Münchenerstr. 100
D - 4300 ESSEN

Manuela Perret
Institut für Metallforschung
Universität Münster
Wilhelm-Klemm-Straße 10
D - 4400 MÜNSTER

Dr. Kurt Schnitzke
SIEMENS AG
ZFE TPH 11
D - 8520 ERLANGEN

Dr. Frank Petzold
Fraunhofer Institut (IFAM)
Lesumer Heerstraße 36
D - 2820 Bremen 77

Dr. Johannes Schröder
Max-Planck-Institut für Metallforschung
Institut für Werkstoffwissenschaften
Seestraße 92
D - 7000 STUTTGART 1

Dr. Ludwig Schultz
SIEMENS AG
ZFE TPH 12
Postfach 3240
D - 8520 ERLANGEN

Dr. Eberhard Seitz
KFA Jülich GmbH
Projektleitung Material- und
Rohstofforschung
Postfach 1913
D - 5170 JÜLICH

Dr. Paul Hideo Shingu
Kyoto University
Dept. of Mat. Sci. and Technology
Yoshida Sakyoku
KYOTO
JAPAN

Dr. H.-R. Sinning
TU Braunschweig
Institut für Werkstoffe
Langer Kamp 8
D - 3300 BRAUNSCHWEIG

Eggert Tank
DAIMLER BENZ AG
Mercedesstraße 136
Postfach 600202
D - 7000 STUTTGART 60

Dr. Manfred Thumann
DFVLR
Linder Höhe
D - 5000 KÖLN-PORZ

Dr. Russell Timmins
Max-Planck-Institut für Metallforschung
Institut für Werkstoffwissenschaften
Seestraße 92
D - 7000 STUTTGART 1

Dr. Georg Veltl
Fraunhofer Institut für
Angew. Materialforschung
Lesumer Heerstraße 36
D - 2820 BREMEN 77

Kersten von Oldenburg
MPI für Eisenforschung
Max-Planck-Straße 1
D - 4000 DÜSSELDORF

Dr. J. Daniel Whittenberger
NASA Lewis Research Center
21000 Brookpark Road
CLEVELAND, OH 44135
U S A

Johann Zbiral
Technische Universität Wien
Inst. f. Chem. Technologie Anorg. Stoffe
Getreidemarkt 9
A - 1060 WIEN